# THE GROWTH AND FORM OF
MODULAR ORGANISMS

# THE GROWTH AND FORM OF MODULAR ORGANISMS

PROCEEDINGS OF
A ROYAL SOCIETY DISCUSSION MEETING
HELD ON 27 AND 28 JUNE 1985

ORGANIZED AND EDITED BY
J. L. HARPER, F.R.S., B. R. ROSEN AND J. WHITE

LONDON
THE ROYAL SOCIETY
1986

Printed in Great Britain for the Royal Society
by the
University Press, Cambridge

ISBN 0 85403 281 9

First published in *Philosophical Transactions of the Royal Society of London*,
series B, volume 313 (no. 1159), pages 1–250

**Copyright**

© 1986 The Royal Society and the authors of individual papers.

It is the policy of the Royal Society not to charge any royalty for the production of a single copy of any one article made for private study or research. Requests for the copying or reprinting of any article for any other purpose should be sent to the Royal Society.

Published by the Royal Society
6 Carlton House Terrace, London SW1Y 5AG

# CONTENTS

[Four plates]

|  | PAGE |
|---|---|
| J. L. Harper, B. R. Rosen and J. White | |
| Preface | 3 |
| J. B. C. Jackson and A. G. Coates | |
| Life cycles and evolution of clonal (modular) animals | 7 |
| D. J. Hughes and R. N. Hughes | |
| Metabolic implications of modularity: studies on the respiration and growth of *Electra pilosa* | 23 |
| A. R. Watkinson and J. White | |
| Some life-history consequences of modular construction in plants | 31 |
| J. S. Ryland and G. F. Warner | |
| Growth and form in modular animals: ideas on the size and arrangement of zooids | 53 |
| F. Hallé | |
| Modular growth in seed plants | 77 |
| D. Barthelemy | |
| Establishment of modular growth in a tropical tree: *Isertia coccinea* Vahl. (Rubiaceae) | 89 |
| A. P. J. Trinci and E. G. Cutter | |
| Growth and form in lower plants and the occurrence of meristems | 95 |
| B. R. Rosen | |
| Modular growth and form of corals: a matter of metamers? | 115 |
| A. D. Bell | |
| The simulation of branching patterns in modular organisms | 143 |
| R. C. Hardwick | |
| Physiological consequences of modular growth in plants | 161 |
| G. O. Mackie | |
| From aggregates to integrates: physiological aspects of modularity in colonial animals | 175 |
| L. D. Gottlieb | |
| The genetic basis of plant form | 197 |
| M. Franco | |
| The influence of neighbours on the growth of modular organisms with an example from trees | 209 |
| P. E. J. Dyrynda | |
| Defensive strategies of modular organisms | 227 |
| G. C. Williams | |
| Retrospect on modular organisms | 245 |

# PREFACE

Contributors to this Discussion Meeting were given only the most general guidance to what is meant by 'modular'. The letter of invitation said "By 'modular' we imply growth of the genetic individual by the repeated iteration of (multicellular) parts – modules. This means that we can consider together at the meeting the growth and form of corals, higher plants, fungi,..., ascidians, ..., bryozoans, etc.".

Modular growth contrasts with 'unitary' growth, in which the zygote develops to a determinate structure that is repeated only when a new life cycle is started from a single-celled stage, usually a zygote. Of course most unitary organisms bear repeated parts, such as legs, wings and body segments, but they are formed in early embryogenesis, not by continued or prolonged somatic iteration, and their numbers are usually very strictly determinate. We might count the numbers of flies in a population of *Drosophila* by counting the numbers of legs and dividing by six, or of birds by counting the wings and dividing by two. There is no such determinate divisor for the genetic individual (genet) of a modular organism – no determinate number of leaves or buds on a tree, polyps on a coral, hyphae in a fungal colony or branches on a root system. Modules themselves, however, may be quite strictly determinate in form, e.g. the number of petals on a flower, the number of leaflets on a leaf.

In many of the papers in this volume authors have attempted their own more precise definition of 'module', usually one especially appropriate for the class of organisms in which they are especially interested. Students of particular groups tend to search for homologies among structures and to define modules as homologues. When comparisons are made across the plant and animal kingdoms, it is analogies rather than homologies of structure that give unity to the concept of modular organisms. In this broader vision the module is the multicellular unit of structure that is iterated in the process of growth. There is no obvious homology between the polyp of a coral and the leaf with its axillary bud on a tree, but the analogy is clear. Both are repeated units of structure.

Most modular organisms are branched; even those palms that fail to branch above ground bear branched root systems and inflorescences. It is the pattern of branching or budding that gives most modular organisms their overall form. This is at its most dramatic in the stark winter outlines of deciduous trees, which are often quite distinctive of various species, though fearfully difficult to describe in words. There is the same difficulty in describing the form of corals; indeed, the most effective description may involve likening them to particular plants, e.g. *Pectinia lactuca*, *Pectinia paeonia*, *Acropora hyacinthus*, *Pavona cactus*; mythical monsters, e.g. *Hydra*, *Medusa*, *Gorgonia*; or to the organs of unitary organisms, e.g. *Diploria cerebriformis* (a 'brain' coral), *Echinopora mammiformis*, *Psammocora digitata*, *Cycloseris patelliformis*, *Heteropsammia cochlea*.

Growth forms that develop by branching or budding may iterate multicellular structures at a variety of levels. Most trees are composed of branches, branchlets and twigs, which in turn bear leaves with axillary buds. Repeated units of structure – modules – may be recognized at each of these levels so that the form of the organism comes from a hierarchy of different orders of repeating units. This is in essence the same sort of hierarchical modularity that an architect recognizes in the repeated form of apartment blocks, apartments within blocks, rooms in the apartments and modular furniture within the rooms.

We can recognize two broad categories of form among modular organisms: that in which the modules remain attached to each other, and that in which modules or groups of modules become separated – organisms that fall to pieces as they grow. In situations in which modules remain attached to each other the form of the whole organism may depend on the accumulation of dead modules from the past. Quite unlike any unitary organisms, a large part of the body of some modular organisms is the accumulated 'necromass' of dead modules. It is this feature that gives trees, corals and mosses their peculiar role in creating structure in a community: forests, coral reefs and peat bogs have a community architecture that is an accumulation of dead modules bearing a thin skin of active feeding modules (polyps or leaves). The fact that most modular organisms occupy fixed positions in space means that they give a continuity to the spatial structure of communities that mobile unitary organisms cannot do.

An organism that is fixed in position in its habitat depletes resources locally. It cannot escape from the resource depletion zones that is creates except by growing away from them. Competition between individuals will commonly occur where such organisms are close enough to enter into each other's resource depletion zones. One effect is that different parts of a single genet may meet quite different environments – a neighbour on one side and none on the other, or a neighbour of one species on one side and a different species on the other. The fitness of a single genet is then simultaneously tested in a variety of environments. The progeny that it leaves may derive almost entirely from a single branch that grows into an open space or into the territory of a weaker neighbour. Modular organisms of this sort allow the interaction between individuals to be studied in a reductionist detail that is virtually impossible in motile (i.e. most unitary) organisms.

The life cycles of modular organisms that fall to pieces as they grow are exemplified by many freshwater plants, e.g. among species of *Lemna, Eichhornia, Salvinia*, and among animals by *Hydra* spp. and *Fungia* corals. Old dead modules are sloughed off and, in marked contrast with trees and most corals, the population of modules remains permanently young. In these cases modules are free to move passively in the water, escape from each other's resource depletion zones, and even disperse from habitat in floods (or, in the case of *Lemna*, on birds' feet). The opportunity for a single genet to colonize extensive areas is maximized in these situations; in Britain, the whole population of a modular organism, such as the aquatic *Stratiotes aloides*, may perhaps be represented by just one single genet. In terrestrial communities there is less opportunity for modules to separate and disperse, though the bulbils of species such as *Allium vineale* are shoot modules that are dispersed in a manner similar to seeds. More usually terrestrial plants that grow to form clones become fragmented by the decay of their interconnecting parts, but the modules become dispersed only to the extent by which they have *grown* into new territory. However even in these cases, clones may be broken by burrowing animals or by ploughing, and some of the most aggressive weeds of agriculture, such as *Elymus repens* and species of *Oxalis* and *Cyperus*, are stimulated to multiply by such damage.

It is easy to see that there are some environments in which the fitness of a genet will be enhanced if the development of its parts is highly integrated and correlated and others in which it is to the benefit of the organism to allow its parts to respond very locally and independently. In *Hydra* and *Lemna* the loss of continuity between modules is part of the programmed course of development; hormonal and nutritional interdependence is wholly sacrificed. At the other extreme the highly organized structure and behaviour of siphonophores is ensured by specialized signalling between the modules. The continuum of forms between these two

# PREFACE

extremes represents the range of compromises between freedom of the individual modules and subservience to the organized genet that has obvious parallels in political systems. How far modular organisms are composed of physiologically independent parts or of correlated integrated wholes may have to be answered differently for different species and may represent compromise responses to conflicting selection pressures.

Overwhelmingly modular animals and plants occupy fixed positions in their habitats or are moved passively by water or other agents. They are therefore denied the opportunity to search for and chase after prey or to run away from predators, seek suitable habitats or escape from competitors. They cannot actively search out mates (there are of course some unitary animals that face the same constraints, such as mussels and barnacles). Almost all the activities that involve mobility in unitary organisms become the responsibility of form in modular organisms and form, in turn, is the consequence of the pattern of growth. If it is not possible to run away from prey, natural selection may well favour the evolution of passive defences such as spines. If it is impossible to search and capture resources, structures may be favoured that attract or intercept them. If it is impossible for individuals to move in search of mates, some other force may be favoured that will move gametes or microspores between potential parents (or outbreeding must be sacrificed).

One of the most important of the properties of unitary organisms, the segregation of the germ plasm, does not hold for modular organisms. Germ plasm is repeatedly regenerated as module gives way to daughter module during the process of growth. This means that, at least in theory, somatic mutation could be a significant force in the evolution of modular organisms. This suggestion may be a red herring but the possibility has to be faced.

A further important quality conferred by modular growth is that the Fisherian reproductive value of a genet may increase with age. Modular growth allows, at least in theory, an exponential increase in the rate at which progeny are produced. Under these conditions the evolution of senescence may be greatly delayed or perhaps not occur at all. Instead it appears that the phenomenon of senescence has become expressed at the level of the module rather than the genet. There is little evidence that the life of the genet in most modular organisms has a natural span; in many species it appears potentially immortal.

The proceedings of this meeting show that there is a great area of biological research, only beginning to be explored, in which botanists and zoologists find much in common. Widely held general models of evolutionary processes, based on the study of unitary organisms, may be seriously upset by increased knowledge of how modular organisms behave. It may be that the distinction between the biology of modular and unitary organisms is more profound than the classic distinction between animals and plants.

*February 1986*

J. L. HARPER
B. R. ROSEN
J. WHITE

# Life cycles and evolution of clonal (modular) animals

By J. B. C. Jackson[1] and A. G. Coates[2]

[1] *Smithsonian Tropical Research Institute, Apartado 2072, Balboa, Republica de Panama*
[2] *Department of Geology, The George Washington University, Washington, D.C. 20052, U.S.A.*

Life cycles of clonal benthic animals are more complicated than those of aclonal species because there are more parameters to vary, and because an individual clone can grow in disconnected bits and pieces and widely different shapes. Consequently, the schedule of life-history events in clonal animals is as closely tied to their size as to their age. The bigger the animal the more likely it is to survive, and there is usually no limit to this advantage. Senescence is typically absent, or unmeasurable. Reproduction and recruitment of new individuals into local populations occurs by both sexually and asexually produced larvae, and by fragmentation. In general, asexual recruitment is more frequent. Larvae of clonal species are strongly philopatric whereas those of aclonal species are dispersed over considerable distances. These differences, and the high incidence of asexual reproduction, mean that parents, siblings, and clonemates may become mixed together within a small area. Under these conditions inbreeding is likely, and there is even evidence of selection for inbreeding in several clonal phyla. Nevertheless, clonal species tend to persist as long in the fossil record as do aclonal species. Thus the relative frequency of sexual reproduction among benthic animals, if not its incidence *per se*, would appear to have little or no macroevolutionary significance.

## Introduction

Most animals are modular at one or more levels of organization (Beklemishev 1969). Cells are universal modules of multicellular animals. Likewise, the segments of tapeworms and polychaetes are modular organ systems, and the polyps of corals and zooids of bryozoan colonies are modular 'individuals'. A fundamental characteristic of such modular individuals, and one that distinguishes them from modules at lower levels of organization, is their usual ability to function, survive, and reproduce sexually alone or in small groups if separated from each other by injury or fission. Common examples are the fragmentation of staghorn corals (Tunnicliffe 1981; Highsmith 1982), budding of hydras, or fission in sea anemones. This *clonal* replication of totipotent individuals has considerable ecological and evolutionary significance, as reflected in the striking differences that exist between the life histories of clonal and aclonal species.

In this paper we review four of the most important of these distinctions in life histories between clonal and aclonal animals, using as examples bottom-dwelling, aquatic animals such as sponges, hydroids, corals, bryozoans, and ascidians. These factors are: (i) complication and multidimensionality of life cycles; (ii) confounding of the relation between individual size and individual age; (iii) variations in the relative contributions of asexual and sexual reproduction of individuals in populations; and (iv) the tendency towards philopatric dispersal of sexual and asexual propagules. We then consider the implications of these factors for the genetic structure of clonal populations, and some of their evolutionary consequences. We have excluded planktonic and terrestrial animals because we know less about them, and because we suspect that their clonal behaviour may have different evolutionary bases (cf. Bell 1982).

Terminology for clonal organisms is not standardized, so we will define ours briefly. We use the term individual to refer to one or more physically and physiologically interconnected modules, regardless of their genetic identity, which is usually unknown and is often unknowable in natural populations of clonal organisms. Thus an individual is an organism that can be counted. These are often called ramets or colonies and usually, but not always, the terms are synonymous. In contrast, a genetic individual, or genet, is the sum of all the individuals, usually but not always genetically identical, that are produced clonally from the same zygote. A genetic individual may be uniquely located in space, or spread out as many individuals over a considerable distance. Thus, operationally, genets are difficult things to study or count, and border on abstractions in our present ignorance of the population structure of most clonal animals. Notwithstanding these difficulties, however, and excepting potentially important somatic mutations within clonal individuals (Buss 1982, 1983), genets are the fundamental units of populations upon which natural selection acts. Compilation of accurate demographies of genets is essential to eventual understanding of the evolutionary bases of life-history variations in clonal animals.

## MULTIDIMENSIONALITY OF LIFE CYCLES

Life cycles of benthic clonal animals are more complex than those of comparable aclonal animals because there are more kinds of features to vary, and thus more possible patterns of life history (Jackson 1977, 1979, 1985; Buss 1979; Hughes 1984; Caswell 1985; Jackson & Hughes 1985).

Numbers of individuals in a population can increase by asexual as well as sexual reproduction, either by injury or fission, and decrease by fusion of previously separated individuals as well as by death. Individual size can decrease markedly over time by fission or injury, stay constant, or increase by fusion or growth. This greatly complicates demographic studies of clonal organisms, although standard techniques, such as the population projection matrices of Leslie (1945), can be generalized to include them (Sarukhán & Gadgil 1974; Hughes 1984; Caswell 1985).

The shape or 'growth form' of a clonal animal depends on the spatial arrangement of its tissues or modular components just as in aclonal animals. The flexibility of these arrangements, however, is much greater in clonal animals. Different species in the same genus, different individuals of the same species, or sometimes even different individuals of the same genet can grow in arborescent, plating, massive, encrusting, or linear forms, depending on a variety of both intrinsic and extrinsic factors (Dustan 1979; Foster 1980; Willis & Ayre 1985).

Life histories of clonal animals vary in relation to their growth form, and the dynamics of their 'mobility' over the substratum due to growth, partial mortality, and passive dispersal of fragments (Buss 1979; Jackson 1979; Coates & Jackson 1985; Jackson & Hughes 1985). Clonal animals may be attached or free-living, the former usually on hard substrate and the latter on sediments or up in the water column. Among attached forms, animals differ in the ways modules (or masses of tissue in organisms like sponges that lack distinct modules) are arranged relative to each other (uniserial or multiserial) and in their primary directions of growth, which can be encrusting, massive, or erect.

Uniserial animals grow as single or loosely branching runners or vines that are highly directional forms adapted primarily for location of spatial refuges (Buss 1979; Jackson 1979). The probability that any module will die is high, but so also is the probability that some other modules will find a refuge and survive. In contrast, multiserial animals grow laterally as well as distally to form more or less continuous surfaces more likely to be able to persist on a given bit of substratum. Their survival is more dependent on maintenance of the genet as fewer, more highly integrated individuals occur than in uniserial forms. The probability that any module will die is relatively low, but so also are the chances of locating different habitats (refuges) by growth.

Encrusting growth on a stable substratum sets no special mechanical constraints for support, and so is potentially infinite, but at a cost of all the potentially harmful processes more likely to occur on the surface of the substratum than above it (Jackson 1979). Erect growth places stringent mechanical demands for support and attachment, but with the potential advantages of greater access to resources in the water, escape from risks on the substratum, and increase in surface area, feeding, and reproductive capacity beyond that possible on the bottom surface (Cheetham 1971; Cheetham & Thomsen 1981). Costs and benefits of massive growth are intermediate between encrusting and erect habits in all respects.

Quantitative relationships between growth forms and growth, maintenance, and reproduction have not been well established, but circumstantial evidence suggests different patterns of investment consistent with the above interpretations. For example, growth rate and larval recruitment both decrease with increasing massiveness of colonies of cheilostome bryozoans on Caribbean reefs. The growing tip of uniserial cheilostomes like *Aetea* sp. advances over the substratum as fast as 72 cm per year compared with about 3–11 cm per year for multiserial encrusters like *Steginoporella* and *Reptadeonella*, and only 1 or 2 cm per year for multilaminate encrusters like *Trematooecia* (Jackson & Hughes 1985; M. Gleason, unpublished data; J. B. C. Jackson and K. Kaufmann, unpublished data). Similarly, in Gleason's recent study of larval recruitment onto fouling panels placed on Venezuelan reefs, 590 *Aetea* settled onto six different sets of fouling panels within six weeks of submergence, as opposed to a total of only five recruits of the three multiserial genera combined.

Growth rate and recruitment also vary within and between growth forms of Caribbean reef corals (Hughes & Jackson 1985; Jackson & Hughes 1985; Hughes 1985). Massive encrusting corals like *Montastrea annularis* and brain corals extend their boundaries at about 1 cm per year and show very little larval recruitment (Bak & Engel 1979; Rylaarsdam 1983). Branching erect species like staghorn and elkhorn *Acropora* grow extremely fast, up to 20 cm per year, but unlike rapidly growing bryozoans they also show little larval recruitment (Adey 1978; Tunnicliffe 1981). Exclusively plating species, which represent a kind of intermediate morphology, have intermediate growth rates, and depending on the species, high or low rates of larval recruitment (Hughes & Jackson 1985). Of course, recruitment rates are not equivalent to reproductive effort, and many corals that show negligible recruitment produce enormous numbers of gametes, including *Montastrea annularis*, brain corals, and other large, massive species (Szmant-Froelich 1985a). What the cost of reproduction of these broadcasting species might be, compared with those with high recruitment, which typically brood large planula larvae, is unknown. In contrast, bryozoans that are common on reefs brood larvae that apparently recruit in numbers roughly proportional to their standing crop and production in the local population (Jackson & Wertheimer 1986; J. B. C. Jackson, unpublished data).

The relative 'mobility' of clones across and above the substratum is another important component of their life histories which varies considerably both within and between growth forms (Jackson & Hughes 1985). This is because mobility depends not only on growth rate, but also the continuous persistence of a clone in a particular place. Uniserial species like *Aetea* commonly fail to regenerate after injuries and grow extremely fast, so that they may appear in and disappear from an area within a month or less. *Steginoporella* is another highly mobile encrusting bryozoan. It grows fast and regenerates only within the younger regions of colonies, whereas older zooids senesce, so that the location of encrusted substratum changes entirely from one year to the next (Palumbi & Jackson 1982, 1983; Jackson & Hughes 1985). In contrast, slower growing encrusting species like *Reptadeonella* and *Trematooecia* regenerate over their entire surfaces at rates at least as fast as they extend their margins. Erect branching staghorn coral *Acropora cervicornis* and plate-like *Leptoseris cucullata* also combine extremely rapid distal growth with failure to regenerate proximal tissues as in *Steginoporella*, whereas massive and plating *Montastrea annularis* regenerate rapidly everywhere on the colony (Hughes 1985).

Despite such striking differences in mobility and persistence, genets of all of the above species (except perhaps uniserial forms like *Aetea*) may live very long, up to decades or centuries for the bryozoans and for centuries to millenia for the corals as calculated from population projection matrices (cf. Hughes 1984). Over time they may come to dominate large areas of their habitats at the apparent expense of other bryozoans or corals. Furthermore, because of their potentially large size, these spatially dominant species may also invest more in reproduction than more ephemeral species, despite relatively late reproductive maturity and poor recruitment. On the other hand, the bryozoan *Membranipora membranacea*, which lives on ephemeral substrata like kelp in the temperate zone, combines the potential for early reproductive maturity and high recruitment rates with exceptionally rapid growth and strong ability to compete for space and ward off predators (Lutaud 1961; Bernstein & Jung 1979; Yoshioka 1982a, b; Harvell 1984). Concepts of r- and K-selected species, tenuous at best for aclonal animals (Wilbur et al. 1974), tend to break down completely for clonal species which typically include in the same animal characteristics attributed to both opportunistic and specialist species.

This extensive variability does not mean, however, that all conceivable combinations of life-history traits exist among clonal animals, any more than they do among aclonal forms. For example, we do not know of any uniserial species that can persist long in any one place through a combination of slow growth and strong resistance to competitors and predators. It is likely that the recognition of theoretical patterns that do not exist in Nature may eventually tell us more about the constraints acting on the evolution of clonal life histories than those that do.

## Size, age and senescence

Decrease in size due to injury, fragmentation, and fission, confound relationships between the size and age of individual clonal animals (Hughes & Jackson 1980, 1985; Highsmith 1982; Hughes 1984; Jackson & Hughes 1985). Chances of these events vary with time, size, and location of the animal. They also vary greatly between species, so that size is probably a good predictor of individual age in some clonal species, individuals of which are unlikely to regress or divide, and an extremely poor predictor for others. This is the basis for the confused relationship between individuals and genets; among species likely to undergo frequent division, the genetic age of newly formed, small individuals in a population may vary by thousands of years.

As a result of such differences in positive and negative growth of individuals, their schedule of life-history events are often more closely tied to their size than to their age (Connell 1973; Hughes & Jackson 1980, 1985; Highsmith 1982; Jackson 1985; Jackson & Hughes 1985). Competitive ability; regenerative capacity; fecundity; and resistance to predators, catastrophic storms, and many diseases; all increase with individual size. The bigger the animal, the more likely it is to survive. Moreover, except for some forms of epidemic disease or geological catastrophes such as changes in sea level, there appears to be no real limit to this benefit of increasing size. Although we live in a period of exceptional environmental instability due to rapid fluctuations in sea level, there are still many examples of scleractinian reef corals 5–10 m in extent and almost certainly thousands of years old (Potts 1984; Potts *et al.* 1985; Hughes & Jackson 1985). In more stable times in the Earth's history, the potential for increase and dominance of local populations by a few genets of extraordinary age must have been far greater.

These observations suggest that reef corals do not senesce, and there is now much evidence to this effect for many other clonal animals (Palumbi & Jackson 1983; Jackson 1985). Individual modular elements of colonies such as hydrozoan polyps and polypides of bryozoans may degenerate permanently or undergo cycles of degeneration and regeneration that are often referred to as modular senescence, but the colonies typically persist. Likewise, the seasonal degeneration of individual sponges, hydrozoans, and bryozoans is usually accompanied by the production of some form of resistant mass that regenerates under more favourable conditions (Crowell 1953; Hartman 1958; Gordon 1977; Ryland 1979; Dyrynda & Ryland 1982; Frost *et al.* 1982). Although there are proven examples of semelparous species, especially among colonial ascidians (Grosberg 1982), few clonal individuals or genets seem to degenerate irreversibly with age, at least not to any measurable degree.

Two sorts of quantitative evidence support this view. First, there is no age-related decrease in physiological capacity among most clonal animals. Fecundity or the standing crop of reproductive products increases proportionally to surface area or numbers of polyps or zooids (Davis 1971; Hayward 1973; Sugimoto & Nakauchi 1974; Babcock 1984; Jackson & Wertheimer 1986; Szmant-Froelich 1985*b*). Similarly, growth rate as measured by the rate of linear extension, does not appear to vary with size or age except among very small individuals. This has been suggested for all clonal groups, and has been measured in the field for foliaceous corals (Hughes & Jackson 1985). In addition, although not specifically tested for, we know of no reports of decreased rates of regeneration of injured tissues as a function of size, except among small 'juvenile' individuals, from any of the myriad experimental investigations of regeneration by corals, gorgonians, bryozoans, sponges, and ascidians (see, for example, Bak *et al.* 1977; Palumbi & Jackson 1982, 1983; Wahle 1983*b*).

The second kind of evidence for lack of senescence comes from induced juvenilization of individuals by experimentally decreasing their size below threshold levels sufficient for reproduction (Wahle 1983*a*; Kojis & Quinn 1985; Szmant-Froelich 1985*b*). Small portions of large, reproductively mature colonies cease reproduction after isolation from the original 'parent' colonies which themselves continue to reproduce, as do isolates above the threshold size. In Szmant-Froelich's experiments, the parent colonies exceeded 2 m diameter and thus were minimally a century old, and probably much older (cf. Hughes & Jackson 1985). The small isolates from these retain their parental genetic age, but they came to behave reproductively like colonies less than three or four years old.

It would appear that reproductive value, in the sense of Fisher (1930), can increase indefinitely with increasing individual and genet size in clonal animals (Caswell 1985). The apparent absence of senescence in these organisms supports the idea that senescence evolves in response to age-specific rates of reproduction and mortality, rather than being an inevitable physiological necessity (Williams 1957; Bell 1984).

### Sexual versus asexual reproduction and recruitment

Reproduction and recruitment, defined simply as the appearance of new individuals in a population, occur sexually via larvae, the only means available to aclonal species. Asexually, they result from localized death of tissues, fragmentation, fission or budding, and asexual production of larvae. The relative contribution of these sexual versus asexual modes to recruitment of new individuals into populations of clonal animals ranges widely. In general, asexual processes appear to be far more successful.

Larval production and recruitment rates of clonal species vary from non-existent to very high values, especially among broadcasting species, but rarely approach those of aclonal species (Jackson 1983, 1985). Recruits derived from larvae are so rare among many abundant clonal species as to be undetectable, even in quite intensive searches (Bak & Engel 1979; Tunnicliffe 1981, 1983; Rylaarsdam 1983; Szmant-Froelich 1985a). In some cases, as for *Pocillopora damicornis* in the eastern Pacific, reproduction involving larvae may not normally occur (Richmond 1985).

In contrast, recruitment of asexually derived colonies is common among most clonal species, although their origin tends to vary with growth form, material, design, and habitat (Glynn *et al.* 1972; Hughes & Jackson 1980, 1985; Jackson & Winston 1981; Woodley *et al.* 1981; Bothwell 1982; Highsmith 1982; Winston 1983; Lasker 1984; Benayahu & Loya 1984; Farrant 1985; Heyward & Collins 1985; Jackson 1985). This has been true throughout the Phanerozoic (Jackson 1983; McKinney 1983). Among sessile species, fragmentation is the most common mechanism of propagation among branching species, whereas more massive organisms are more likely to divide through localized partial mortality (Highsmith 1982; Done & Fisk 1986; Hughes & Jackson 1985). Asexual production is typically most prolific for branching species, but even among encrusting and massive forms it commonly exceeds input from larvae. Budding and fission are most common among free-living animals like sea anemones, fungiid corals, and lunulitiform bryozoans (Marcus & Marcus 1962; Goreau & Yonge 1968; Francis 1973, 1979; Sebens 1982; Cook & Chimonides 1983).

Electrophoretic analysis suggests that asexual production of larvae is common among anthozoan cnidarians (Black & Johnson 1979; Stoddart 1983; Ayre 1984; Ayre & Resing 1986; Stoddart & Black 1985). This conclusion is based on studies of several species whose larvae are consistently genetically identical to their brood-parents, even at polymorphic loci. In contrast, larvae of other species regularly show non-parental genotypes consistent with expectations of sexual reproduction. These results raise questions of interpretation concerning gametogenesis and sexuality based solely on descriptive histological techniques in all other invertebrates for which fertilization has not actually been observed. Asexual production of larval-like individuals also commonly results from production of buds and gemmules in sponges, hibernaculae in bryozoans, and 'polyp bail-out' in corals (Ryland 1970; Bergquist 1978; Ayling 1970; Sammarco 1982; Richmond 1986).

DISPERSAL AND GENETIC COMPOSITION OF LOCAL POPULATIONS

One of the most striking features of clonal benthic animals is their characteristically short-distance (that is, philopatric) dispersal of both sexual and asexual propagules compared with aclonal species (Ryland 1981; Jackson 1985, 1986).

Both clonal and aclonal sea anemones may be transported tens to hundreds of metres by currents (Mackie 1974; Shick *et al.* 1979; Ayre 1984), and passive dispersal of sponges and corals or their fragments may take place over distances of 10–50 m during storms. Such dispersal has actually been observed for *Acropora cervicornis* and labelled rope-like sponges (Woodley *et al.* 1981; Wulff 1985). Comparable dispersal has also been inferred by mapping clonal boundaries by using tissue grafting (Bothwell 1982; Neigel & Avise 1983a,b; Hunter 1986) and by electrophoresis (Ayre 1984; Stoddart 1984a, b; Willis & Ayre 1985). Grafting is not entirely reliable in distinguishing genetically different clones (Curtis *et al.* 1982; Ayre & Resing 1986; Stoddart *et al.* 1985). Nevertheless, considerable evidence suggests that clonal dispersal of fragments up to 50 m or more is common on reefs over a few years time. The only limits to the process are inhospitable areas too broad for fragments to traverse in a single episode of dispersal (Ayre 1984).

All clonal ascidians, all but about 30 species of bryozoan, most sponges, hydrozoans, and octocorals, and an appreciable minority of clonal corals produce brooded, short-lived larvae that disperse much shorter distances than larvae of aclonal animals (Knight-Jones & Moyse 1961; Kott 1974; Ryland 1981; Jackson 1985, 1986). Many, perhaps most, of these brooded larvae are capable of settlement within minutes or seconds of their release from their parents, even though they commonly swim for much longer periods when confined in containers in the laboratory. The best data are Olson's (1985) for the clonal ascidian *Didemnum molle*. He managed to follow 14 larvae *in situ*, from release to settlement, all of which settled within 40–370 s of their release and within 2–12 m of their parents.

In addition to this general pattern, the great majority of species with *extremely* short-distance dispersal, often by benthic larvae that crawl less than 1 m from their brood parent, are clonal. Examples include demosponges (Ayling 1980), hydroids (Pyefinch & Downing 1949; Nishihira 1967; Williams 1965, 1976; Hughes 1977), hydrocorals (Ostarello 1970), scleractinians (Lewis 1974), alcyonaceans (Sebens 1983; Benayahu & Loya 1984; Benayahu 1986, personal communication; Farrant 1985), cyclostome bryozoans (C. McFadden, personal communication), and ascidians (van Duyl *et al.* 1981; R. Grosberg, personal communication). Brooded larvae of many aclonal species also disperse very short distances, but these are mostly very small or free-living animals (Gerodette 1981; Fadlallah & Pearse 1982a; Strathmann & Strathmann 1982). In contrast, the great majority of species with extremely long-lived (teleplanic) larvae are aclonal (Eckelbarger 1978; Scheltema 1971; Fadlallah & Pearse 1982b; Tranter *et al.* 1982; Scheltema & Williams 1983).

The most glaring exception to the pattern of philopatric larval dispersal by clonal animals is that of scleractinian reef-building corals, particularly in Australia where about two-thirds of all the species on the Great Barrier Reef synchronously broadcast gametes during one to three nights per year (Harrison *et al.* 1984). Even there, however, the proportion of brooding species with potentially short-distance dispersal is much higher than for major groups of associated aclonal animals like bivalves, barnacles, or echinoderms.

Frequent asexual reproduction by fragmentation and injury, asexual production of larvae, and the retention of the products of sexual and asexual reproduction near their site of origin should all promote a high degree of genetic relatedness in many clonal populations. Thus parents, siblings, and clonemates may all be mixed together within small areas. Patterns consistent with this view have been demonstrated by electrophoretic analyses of populations of an anemone (Ayre 1984), coral (Stoddart 1984a, b), bryozoan (Gooch & Schopf 1971), and ascidian (Sabbadin 1978). Moreover, to the extent that fertilization occurs among neighbouring individuals that have a high probability of being clonemates or relatives, considerable inbreeding must also occur. The fact that their progeny are likely to settle nearby will contribute to the process even more with every passing generation.

Various observations suggest that high levels of relatedness in clonal populations, and perhaps also the occurrence of inbreeding, are not just an accidental consequence of short-distance dispersal, but the result of natural selection for these characteristics (Jackson 1986). The evidence, most of which is circumstantial and equivocal, includes: evolution of decreased potential for larval dispersal among hydroids (Hyman 1940; Ryland 1981; Boero 1984), gymnolaemate bryozoans (Zimmer & Woollacott 1977), and clonal ascidians (Kott 1974); polyembryony in cyclostomes (Harmer 1893); spontaneous tissue degeneration and localized constrictions in the branches of gorgonians that suggest they are 'designed' to fragment (Lasker 1984), and sibling gregariousness of larvae at settlement (Keough 1984; R. Grosberg, personal communication). Of these, sibling gregariousness is the most important because it is a specific behaviour that has evolved polyphyletically (bryozoans and ascidians) and independently of the distance of larval dispersal, or of any simple predisposition for asexual reproduction.

## Evolutionary basis for short-distance dispersal

There are at least three possible kinds of evolutionary explanations for widespread philopatry among clonal animals, none of which are mutually exclusive. These concern theories of sexuality as an individual adaptation (Williams 1975; Maynard-Smith 1978), costs and benefits of forming chimeras (L. Buss and R. Grosberg, personal communication), and consequences of inbreeding (Shields 1982).

According to the strawberry-coral model, the cost of meiosis is offset by sexual production of very large numbers of genetically diverse propagules that are widely dispersed, thereby maximizing chances that some offspring will find and succeed in some other location (Williams 1975). These predictions describe well the characteristics of most aclonal animals like oysters and barnacles. They do not, however, describe most clonal animals which, on the contrary, tend to produce relatively few sexually derived propagules that commonly disperse quite short distances.

It is important to note, however, that among these contradictions to the strawberry-coral model, the probable proximity of siblings due to philopatry actually increases the probability that one of the most important assumptions in all of Williams' models for the explanation of sex as an individual adaptation is met, namely that of intense competition among sibling recruits to a local population (Williams & Mitton 1973). Moreover, among clonal animals living on hard substrata, environments change greatly over short distances due to heterogeneities in distributions of potential competitors for space, as well as other factors (Jackson 1977, 1979; Sebens & Thorne 1985), so that wide dispersal should not be necessary to encounter 'new' environments where the parental genotype is no longer optimal. There is, however, little evidence for sibling competition among postlarvae of clonal species which usually settle in low densities, and are more likely to suffer in conflicts with preexisting adults (Jackson 1977, 1979). On the other hand, the evidence I have presented here suggests that these adults could also be siblings. In short, Williams' arguments may be right for clonal animals, but only because their life histories differ in every important respect from those he assumed in the strawberry-coral model.

The most interesting aspect of the life histories of clonal animals is unquestionably the widespread retention of sexuality, with its potential 50% loss in fitness due to the 'cost of meiosis', when the sexual propagules are dispersed over such short distances. Moreover, the evolutionary trend has been to decrease the dispersal distance of sexual propagules, even though similar propagules can be produced asexually that disperse comparably short distances. This pattern is not a taxonomic artefact; it has evolved many times in many clonal clades, and thus must be adaptive. Exceptions like the majority of reef corals do not change this, but simply require an alternative explanation. Indeed, Australian corals seem to fit certain aspects of the strawberry-coral model quite nicely. Catastrophic mortality due to epidemic outbreaks of the coralivorous starfish *Acanthaster planci* or bleaching of tissues (possibly due to epidemic disease?) occur in a complex mosaic pattern all along the Great Barrier Reef every decade or so (Bradbury *et al.* 1985; Done & Fisk 1986). Production of vast numbers of gametes and larvae allows exploitation of distant refuges, although contrary to Williams' view, this could occur as easily by larvae derived from parthenogenetic as from fertilized eggs. The proportion of investment in sexual versus asexual dispersal should depend on the commonness or rarity of the parental habitat elsewhere (Hamilton & May 1977); if it is predictably and commonly distributed, asexual dispersal may be favoured.

An entirely different kind of explanation for philopatric sexual dispersal by clonal animals is related to postlarval fusion, a process that results in sudden size increase, and thus may increase survival at this vulnerable stage. Philopatry could decrease the potential genetic costs of fusion by helping to keep siblings together (Buss 1982, 1983; Buss *et al.* 1985). Fusion by postlarvae of most clonal groups has been observed in the laboratory (Duerden 1902; Boschma 1929; Stephenson 1931; Schijifsma 1935, 1939; Edmundson 1946; Knight-Jones & Moyse 1961; Harrigan 1972; Sabbadin 1978), often in cases where the participants were known to be siblings, or as we now realize, perhaps, clonemates (Stoddart 1983). The fusion hypothesis needs support from observations of postlarval fusion in the field. If this is common, it may also help to explain widespread fusion of allografts in grafting studies.

Another hypothesis is that philopatry is an adaptation to promote inbreeding as the means of reproduction that best duplicates genotypes over many generations, minimizing as it does the negative effects of both Muller's ratchet and the cost of meiosis (Shields 1982). Shields predicted that the advantages of inbreeding should be greatest for long-lived organisms with low fecundity living in stable environments, all characteristics of most clonal animals (Jackson 1977, 1985). The fit seems good and deserves investigation.

## Biogeography and evolution

Philopatry, inbreeding, and asexual reproduction should severely restrict gene flow between local populations, and thus increase chances of evolutionary adaptation to local environmental conditions and of speciation among clonal animals. Genetic and morphological differentiation consistent with this prediction have been observed for the cheilostome bryozoan *Schizoporella errata* sampled over a distance of 102 km along the southern shore of Cape Cod (Schopf & Dutton 1976). Significant morphological and genetic changes occur over distances as little as 11–13 km, a result consistent with the very short larval life of this species. In contrast to this pattern, however, between 80 and 89% of the electrophoretically sampled genome in this species is identical over a distance of 1000 km along the eastern coast of North America, including the Cape Cod region (Gooch & Schopf 1971). This relative homogeneity on a broad regional scale, in the face of marked local differentiation of populations, suggests some mode of dispersal must be acting other than the larval stage.

The sea anemone *Actinia tenebrosa* and coral *Pocillopora damicornis* also exhibit significant genetic differentiation of local populations consistent both with clonal proliferation and short-distance dispersal by larvae (Ayre 1984; Stoddart 1984a, b). However, these species of anemone presumably differ from *Schizoporella* in their ameiotic production of larvae, despite routine production of sperm and eggs (we say presumably because *Schizoporella* has not been examined for this). Although sexually produced larvae have not been observed in either species, their existence was inferred from high levels of variation observed by pooling local populations, existence of unique heterozygous genotypes, and the concordance of pooled clonal populations to Hardy–Weinberg expectations. Moreover, both authors have assumed that these presumed sexual propagules are the agents of long-distance dispersal and gene flow in these species. However, as pointed out by Ayre (1984), this latter assumption would be invalid if asexual propagules were found to disperse between local populations, and at least for *Pocillopora damicornis* there is strong evidence that this is true.

There is no evidence for or against intense local selection in any of these three species.

What is called *Pocillopora damicornis* is almost certainly a complex of species (Richmond 1986, personal communication). Local populations in southwestern Australia differ considerably in allelic frequencies over distances of a few kilometres (Stoddart 1984b), and populations in Enewetak, Hawaii and Panama differ strikingly in all life-history characteristics measured (Richmond 1986). These differences suggest considerable potential for allopatric speciation, as is supported by existence of reproductively isolated 'morphs' in the eastern Pacific (Richmond 1985, personal communication). On the other hand, this species is the most commonly rafted coral in the Pacific, occurring commonly on drifting pumice (Jokiel 1984), and it possesses a larva that is at least theoretically competent to survive the journey across the eastern Pacific barrier (Richmond 1982, 1986).

The examples of *Schizoporella* and *Pocillopora* are instructive for calling into question the kinds of assumptions that are commonly made regarding larval life and species distributions. In fact, there is no correlation between the presumed length of larval life and the geographic ranges of clonal species, including corals, bryozoans, and colonial ascidians (Kott 1974; Jackson et al. 1985; Jackson 1986). Species of clonal animals are, on average, distributed at least as widely as aclonal animals with long-lived planktonic larvae like barnacles and molluscs. The only explanation is dispersal of the normally sessile stage by some process such as rafting, as observed for *Pocillopora damicornis* (Jokiel 1984). Indeed, the primary role of larvae in promoting long-distance dispersal of clonal animals may be to get them onto and off rafts, although this might also occur simply by growth and fragmentation (many potential rafts like kelps and mats of sea grasses break away from the bottom after they have been colonized (Gerodette 1981; R. C. Highsmith, unpublished). In such cases, or in those in which larvae are produced asexually, the distributions of single clones may be enormous. Many genets of *Pocillopora damicornis* are probably alive today on both sides of the Pacific Ocean.

## MACROEVOLUTION

Two theories predict that the chances of speciation should be different for clonal and aclonal animals. The first derives from Stanley's (1975) observation that speciation is more likely to occur in sexual clades than in asexual clones, and Chapman's (1981) prediction that the probability of evolutionary divergence and speciation should increase with increasing potential for genetic recombination. The second theory relates the probability of speciation to generation time (Potts 1984). By this model, generation times of clonal animals like corals are so long as to approach the average intervals between catastrophic environmental change, so that processes leading to speciation and extinction cannot go to completion.

Species durations are, on average, a function of the probability of speciation and the probability of extinction. If the standing crop of species stays constant, high probability of speciation implies high probability of extinction, and thus short average durations of species. Likewise, if the standing crop of species is constant, low probability of speciation implies low probability of extinction, and thus long species durations. Thus both the Stanley–Chapman and Potts theories suggest that species durations should be longer for clonal species than for aclonal species.

We tested these predictions by examining the durations of 1381 species of scleractinian corals as compiled from 128 monographic publications of faunas ranging in age from mid-Triassic to Pleistocene. Holocene and Recent faunas were excluded to avoid biases due to presumably artificially short durations of unpreserved modern taxa (cf. Stanley 1979). (References to sources are available from A.G.C.) They cover a wide range of places and times, but inevitably over-represent American and Indo-European faunas compared with other less studied regions. Many published data were checked by reference to major museum collections, and obvious duplications were eliminated by using available synonymies, but no taxonomic revisions were attempted. There seems no reason to believe that the inevitable biases inherent in such a data set should be different for clonal or aclonal species.

All corals were classified as clonal if there was any evidence for budding or fragmentation of polyps, and as aclonal if there was not. Both species diversity and the relative proportions of clonal and aclonal corals have varied considerably since the Triassic, but without any consistent long-term trends (Coates & Jackson 1985). Data are present in table 1. The mean duration of clonal species is slightly but not significantly greater than that of aclonal species ($t = 2.00, p = 0.16$), despite the enormous sample size for both groups.

TABLE 1. MEAN SPECIES DURATIONS IN MILLIONS OF YEARS FOR CLONAL AND ACLONAL SCLERACTINIAN CORALS

|  | number of species | mean duration | standard deviation |
|---|---|---|---|
| aclonal corals | 376 | 9.06 | 5.51 |
| clonal corals | 1005 | 9.57 | 6.06 |

The corals were also classified by growth form (Coates & Jackson 1985), and their mean species duration compared by analysis of variance (table 2). Differences are highly significant ($F = 3.34, p = 0.001$); erect clonal species have evolved faster than all other growth forms, especially massive clonal species. Thus, to the extent that erect growth forms are more specialized than others (cf. Cheetham 1971; Cheetham & Thomsen 1981), table 2 suggests that

Table 2. Mean species durations in millions of years for different growth forms of scleractinian corals

| growth form | number of species | mean duration | standard deviation |
|---|---|---|---|
| solitary aclonal | 376 | 9.06 | 5.51 |
| solitary clonal | 15 | 9.53 | 5.15 |
| pseudocolonial | 157 | 9.21 | 5.40 |
| massive | 645 | 10.00 | 6.02 |
| erect | 188 | 8.40 | 6.63 |

more specialized species evolve more rapidly than less specialized species, whether clonal or not.

These data do not exclude the possibility that clonality influences evolutionary durations of coral species. However, they do demonstrate that should such an effect occur, it is of much smaller extent than those due to variations in life-history features or some other factors related to growth form. This should not be surprising, because the relative proportions of clonal and aclonal taxa have not obviously changed on the sea floor throughout the Phanerozoic (Jackson 1983).

N. Knowlton showed us that G. Williams was more likely to be right about clonal animals because the strawberry-coral model was wrong, and helped in many other ways. Discussions with L. W. Buss, T. P. Hughes, and S. Lidgard were important to the development of many of the ideas presented here.

## References

Adey, W. H. 1978 Coral reef morphogenesis: a multidimensional model. *Science, Wash.* **202**, 831–837.

Ayling, A. L. 1980 Patterns of sexuality, asexual reproduction and recruitment in some subtidal marine Demospongia. *Biol. Bull.* **158**, 271–282.

Ayre, D. J. 1984 The effects of sexual and asexual reproduction on geographic variation in the sea anemone *Actinia tenebrosa*. *Oecologia* **62**, 222–229.

Ayre, D. J. & Resing, J. M. 1986 Sexual and asexual production of planulae in reef corals. (In preparation.)

Babcock, R. C. 1984 Reproduction and distribution of two species of *Goniastrea* (Scleractinia) from the Great Barrier Reef province. *Coral Reefs* **2**, 187–204.

Bak, R. P., Brouns, J. J. & Heys, F. M. 1977 Regeneration and aspects of spatial competition in the scleractinian corals *Agaracia agaricites* and *Montastrea annularis*. *Proc. 3rd int. Coral Reef Symp.* **1**, 143–148.

Bak, R. P. M. & Engel, M. S. 1979 Distribution, abundance and survival of juvenile hermatypic corals (Scleractinia) and the importance of life history strategies in the parent coral community. *Mar. Biol.* **54**, 341–352.

Beklemishev, W. N. 1969 *Principles of comparative anatomy of invertebrates*, vol. 1. Promorphology. (Translated by J. M. MacLennon.) University of Chicago Press.

Bell, G. 1982 *The masterpiece of nature: the evolution and genetics of sexuality*. London: Croom Helm.

Bell, G. 1984 Evolutionary and nonevolutionary theories of senescence. *Am. Nat.* **124**, 600–603.

Benayahu, Y. 1986 Faunistic composition and patterns in the distribution of soft corals (Octocorallia Alcyonacea) along the coral reefs of Sinai peninsula. *Proc. 5th Int. Coral Reef Symp.* (In the press.)

Benayahu, Y. & Loya, Y. 1984 Substratum preferences and planula settling of two Red Sea alcyonarians: *Xenia macrospiculata* Gohar and *Parerythropodium fulvum* (Forskal). *J. exp. mar. Biol. Ecol.* **83**, 249–261.

Bergquist, P. R. 1978 *Sponges*. London: Hutchinson.

Bernstein, B. B. & Jung, N. 1979 Selective pressures and coevolution in a kelp canopy community in southern California. *Ecol. Monogr.* **49**, 335–355.

Black, R. & Johnson, M. S. 1979 Asexual viviparity and population genetics of *Actinia tenebrosa*. *Mar. Biol.* **53**, 27–31.

Boero, F. 1984 The ecology of marine hydroids and effects of environmental factors: a review. *Mar. Ecol.* **5**, 93–118.

Boschma, H. 1929 On the post-larval development of the coral *Meandra aereolata* (L.). *Publ. Carnegie Inst. Wash.* **391**, 129–147.

Bothwell, A. M. 1982 Fragmentation, a means of asexual reproduction and dispersal in the coral genus *Acropora* (Scleractinia: Astrocoeniida: Acroporidae) – a preliminary report. *Proc. 4th int. Coral Reef Symp.* **2**, 137–144.

Bradbury, R. H., Hammond, L. S., Moran, P. J. & Reichelt, R. E. 1985 Coral reef communities and the crown-of-thorns starfish: evidence for qualitatively stable cycles. *J. theor. Biol.* **113**, 69–80.

Buss, L. W. 1979 Habitat selection, directional growth, and spatial refuges: why colonial animals have more hiding places. In *Biology and systematics of colonial organisms* (ed. G. Larwood & B. R. Rosen), pp. 459–497. London: Academic Press.

Buss, L. W. 1982 Somatic cell parasitism and the evolution of somatic tissue compatibility. *Proc. natn. Acad. Sci. U.S.A.* **79**, 5337–5341.

Buss, L. W. 1983 Evolution, development, and the units of selection. *Proc. natn. Acad. Sci. U.S.A.* **80**, 1387–1391.

Buss, L. W., Moore, J. L. & Green, D. R. 1985 Autoreactivity and self-tolerance in an invertebrate. *Nature, Lond.* **313**, 400–402.

Caswell, H. 1985 The evolutionary demography of clonal reproduction. In *Population biology and evolution of clonal organisms* (ed. J. B. C. Jackson, L. W. Buss & R. E. Cook), pp. 187–224. New Haven: Yale University Press.

Chapman, R. W. 1981 Recombination potentials and evolutionary rates. *Am. Nat.* **118**, 384–393.

Cheetham, A. H. 1971 Functional morphology and biofacies distribution of cheilostome Bryozoa in the Danian Stage (Paleocene) of southern Scandinavia. *Smithson. Contr. Paleobiol.* **6**, 1–87.

Cheetham, A. H. & Thomsen, E. 1981 Functional morphology of arborescent animals: strength and design of cheilostome bryozoan skeletons. *Paleobiology* **7**, 355–383.

Coates, A. H. & Jackson, J. B. C. 1985 Morphological themes in the evolution of clonal and aclonal marine invertebrates. In *Population biology and evolution of clonal organisms* (ed. J. B. C. Jackson, L. W. Buss & R. E. Cook), pp. 67–106. New Haven: Yale University Press.

Connell, J. E. 1973 Population ecology of reef building corals. In *Biology and geology of coral reefs*, vol. 2 (ed. O. A. Jones & R. Endean), pp. 205–245. New York: Academic Press.

Cook, P. L. & Chimonides, P. J. 1983 A short history of the lunulite Bryozoa. *Bull. mar. Sci.* **33**, 566–581.

Crowell, S. 1953 The regression–replacement cycle of hydranths of *Obelia* and *Campanularia*. *Physiol. Zool.* **26**, 319–327.

Curtis, A. S. G., Kerr, J. & Knowlton, N. 1982 Graft rejection in sponges: genetic structure of accepting and rejecting populations. *Transplantation* **33**, 127–133.

Davis, L. V. 1971 Growth and development of colonial hydroids. In *Experimental coelenterate biology* (ed. H. M. Lenhoff, L. Muscatine & L. V. Davis), pp. 16–36. University of Hawaii Press.

Done, T. J. & Fisk, D. A. 1986 Effects of two *Acanthaster* outbreaks on coral community structure – the meaning of devastation. *Proc. 5th int. Coral Reef Symp.* (In the press.)

Duerden, J. E. 1902 Aggregated colonies in madreporarian corals. *Am. Nat.* **36**, 461–471.

Dustan, P. 1979 Distribution of zooxanthellae and photosynthetic chloroplast pigments of the reef-building coral *Montastrea annularia* Ellis and Solander in relation to depth on a West Indian reef. *Bull. mar. Sci.* **29**, 79–97.

Dyrynda, P. E. J. & Ryland, J. S. 1982 Reproductive strategies and life histories in the marine bryozoans *Chartella papyracea* and *Bugula flabellata*. *Mar. Biol.* **71**, 241–256.

Eckelbarger, K. J. 1978 Metamorphosis and settlement in the Sabellariidae. In *Settlement and metamorphosis of marine invertebrate larvae* (ed. F.-S. Chia & M. E. Rice), pp. 145–164. New York: Elsevier.

Edmondson, C. H. 1946 Behavior of coral planulae under altered saline and thermal conditions. *Occ. Pap. Bernice P. Bishop Mus.* **18**, 283–304.

Fadlallah, Y. H. & Pearse, J. S. 1982*a* Sexual reproduction in solitary corals: overlapping oogenic and brooding cycles, and benthic planulas in *Balanophyllia elegans*. *Mar. Biol.* **71**, 223–231.

Fadlallah, Y. H. & Pearse, J. S. 1982*b* Sexual reproduction in solitary corals: synchronous gametogenesis and broadcast spawning in *Paracyathus stearnsii*. *Mar. Biol.* **71**, 233–239.

Farrant, P. 1985 Reproduction in the temperate Australian soft coral *Capnella gaboensis*. *Proc. 5th int. Coral Reef Symp.* **4**, 319–324.

Fisher, R. A. 1930 *The genetical theory of natural selection*. Oxford University Press.

Foster, A. B. 1980 Phenotypic plasticity in the reef corals *Montastrea annularis* (Ellis & Solander) and *Siderastrea siderea* (Ellis & Solander). *J. exp. mar. Biol. Ecol.* **39**, 25–54.

Francis, L. 1973 Clone specific segregation in the sea anemone *Anthopleura elegantissima*. *Biol. Bull.* **144**, 64–72.

Francis, L. 1979 Contrast between solitary and clonal life styles in the sea anemone *Anthopleura elegantissima*. *Am. Zool.* **19**, 669–682.

Frost, T. M., de Nagy, G. S. & Gilbert, J. J. 1982 Population dynamics and standing biomass of the freshwater sponge *Spongilla lacustris*. *Ecology* **63**, 1203–1210.

Gerrodette, T. 1981 Dispersal of the solitary coral *Balanophyllia elegans* by demersal planular larvae. *Ecology* **62**, 611–619.

Glynn, P. W., Stewart, R. H. & McCosker, J. E. 1972 Pacific coral reefs of Panama: structure, distribution and predators. *Geol. Rdsch.* **61**, 483–519.

Gooch, J. L. & Schopf, T. J. M. 1971 Genetic variation in the marine ectoproct *Schizoporella errata*. *Biol. Bull.* **141**, 235–246.

Gordon, D. P. 1977 The aging process in bryozoans. In *Biology of bryozoans* (ed. R. M. Woollacott & R. L. Zimmer), pp. 335–376. New York: Academic Press.

Goreau, T. F. & Yonge, C. M. 1968 Coral community on muddy sand. *Nature, Lond.* **217**, 421–423.

Grosberg, R. K. 1982 Ecological, genetical and developmental factors regulating life history variation within a population of the colonial ascidian *Botryllus schlosseri* (Pallas) Savigny. Ph.D. dissertation. Yale University, New Haven, U.S.A.

Hamilton, W. D. & May, R. M. 1977 Dispersal in stable habitats. *Nature, Lond.* **269**, 578–581.

Harmer, S. F. 1893 On the occurrence of embryonic fission in cyclostomatous Polyzoa. *Q. Jl microsc. Sci.* **34**, 199–241.

Harrigan, J. F. 1972 The planula larva of *Pocillopora damicornis*: lunal periodicity of swarming and substratum selection behavior. Ph.D. dissertation. University of Hawaii, Honolulu, U.S.A.

Harrison, P. L., Babcock, R. C., Bull, G. D., Oliver, J. K., Wallace, C. C. & Willis, B. L. 1984 Mass spawning of reef corals. *Science, Wash.* **223**, 1186–1189.

Hartman, W. D. 1958 Natural history of the marine sponges of southern New England. *Peabody Mus. Nat. Hist. Yale Univ. Bull.* **12**, 1–155.

Harvell, D. 1984 Predator-induced defense in a marine bryozoan. *Science, Wash.* **224**, 1357–1359.

Hayward, P. J. 1973 Preliminary observations on settlement and growth in populations of *Alcyonidium hirsutum* (Fleming). In *Living and fossil Bryozoa* (ed. G. P. Larwood), pp. 107–113. London: Academic Press.

Heyward, A. J. & Collins, J. D. 1985 Fragmentation in *Montipora ramosa*: the genet and ramet concept applied to a reef coral. *Coral Reefs* **4**, 35–40.

Highsmith, R. C. 1982 Reproduction by fragmentation in corals. *Mar. Ecol. Prog. Ser.* **7**, 207–226.

Hughes, R. G. 1977 Aspects of the biology and life-history of *Nemertesia antennina* (L.) (Hydrozoa: Plumulariidae). *J. mar. Biol. Ass. U.K.* **57**, 641–657.

Hughes, T. P. 1984 Population dynamics based on individual size rather than age: a general model with a reef coral example. *Am. Nat.* **123**, 778–795.

Hughes, T. P. 1985 Life histories and population dynamics of early successional corals. *Proc. 5th int. Coral Reef Symp.* **4**, 101–106.

Hughes, T. P. & Jackson, J. B. C. 1980 Do corals lie about their age? Some demographic consequences of partial mortality, fission, and fusion. *Science, Wash.* **209**, 713–715.

Hughes, T. P. & Jackson, J. B. C. 1985 Population dynamics and life histories of foliaceous corals. *Ecol. Monogr.* **55**, 141–166.

Hunter, C. L. 1986 Assessment of clonal diversity and population structure of *Porites compressa* (Cnidaria, Scleractinia). *Proc. 5th int. Coral Reef Symp.* (In the press.)

Hyman, L. H. 1940 *The invertebrates*, vol. 1. *Protozoa through Ctenophora*. New York: McGraw-Hill.

Jackson, J. B. C. 1977 Competition on marine hard substrata: the adaptive significance of solitary and colonial strategies. *Am. Nat.* **111**, 743–767.

Jackson, J. B. C. 1979 Morphological strategies of sessile animals. In *Biology and systematics of colonial organisms* (ed. G. Larwood & B. R. Rosen), pp. 499–555. London: Academic Press.

Jackson, J. B. C. 1983 Biological determinants of present and past sessile animal distributions. In *Biotic interactions in recent and fossil benthic communities* (ed. M. Tevesz & P. W. McCall), pp. 39–120. New York: Plenum Press.

Jackson, J. B. C. 1985 Distribution and ecology of clonal and aclonal benthic invertebrates. In *Population biology and evolution of clonal organisms* (ed. J. B. C. Jackson, L. W. Buss & R. E. Cook, pp. 297–355. New Haven: Yale University Press.

Jackson, J. B. C. 1986 Dispersal and distribution of clonal and aclonal benthic invertebrates. *Bull. mar. Sci.* In the press.)

Jackson, J. B. C. & Hughes, T. P. 1985 Adaptive strategies of coral-reef invertebrates. *Am. Sci.* **73**, 265–274.

Jackson, J. B. C. & Wertheimer, S. P. 1986 Patterns of reproduction in five common species of Jamaican reef-associated bryozoans. In *Proc. 6th int. Conf. Bryozoa* (ed. G. P. Larwood & C. Nielsen). Fredensborg: Olsen & Olsen. (In the press.)

Jackson, J. B. C. & Winston, J. E. 1981 Modular growth and longevity in bryozoans. In *Recent and fossil Bryozoa* (ed. G. P. Larwood & C. Nielsen), pp. 121–126. Fredensborg: Olsen & Olsen.

Jackson, J. B. C., Winston, J. E. & Coates, A. G. 1985 Niche breadth, geographic range, and extinction of Caribbean reef-associated cheilostome Bryozoa and Scleractinia. *Proc. 5th int. Coral Reef Symp.* **4**, 151–158.

Jokiel, P. L. 1984 Long distance dispersal of reef corals by rafting. *Coral Reefs* **3**, 113–116.

Keough, M. J. 1984 Kin-recognition and the spatial distribution of larvae of the bryozoan *Bugula neretina* L. *Evolution* **38**, 142–147.

Knight-Jones, E. W. & Moyse, J. 1961 Intraspecific competition in sedentary marine animals. *Symp. Soc. exp. Biol.* **15**, 72–95.

Kojis, B. L. & Quinn, N. J. 1985 Puberty in *Goniastrea favulus* age or size limited? *Proc. 5th int. Coral Reef Symp.* **4**, 289–294.

Kott, P. 1974 The evolution and distribution of Australian tropical Ascidiacea. *Proc. 2nd int. Coral Reef Symp.* **1**, 405–423.

Lasker, H. R. 1984 Asexual reproduction, fragmentation, and skeletal morphology of a plexaurid gorgonian. *Mar. Ecol. Prog. Ser.* **19**, 261–268.

Leslie, P. H. 1945 On the use of matrices in certain population mathematics. *Biometrika* **33**, 183–212.

Lewis, J. B. 1974 The settlement behaviour of planulae larvae of the hermatypic coral *Favia fragum* (Esper). *J. exp. mar. Biol. Ecol.* **15**, 165–172.

Lutaud, G. 1961 Contribution a l'étude du bourgeonnement et de la croissance des colonies chez *Membranipora membranacea* (Linné), bryozoaire chilostome. *Annls Soc. r. Zool. Belg.* **91**, 157–300.

Mackie, G. O. 1974 Locomotion, flotation, and dispersal. In *Coelenterate biology* (ed. L. Muscatine & H. M. Lenhoff), pp. 313–357. New York: Academic Press.

Marcus, E. & Marcus, E. 1962 On some lunulitiform Bryozoa. *Bolm. Fac. Filos. Cienc. Sao Paulo, Zool.* **24**, 281–312.

McKinney, F. K. 1983 Asexual colony multiplication by fragmentation: an important mode of genet longevity in the Carboniferous bryozoan *Archimedes*. *Paleobiology* **9**, 35–43.

Neigel, J. E. & Avise, J. C. 1983 *a* Clonal diversity and population structure in a reef-building coral, *Acropora cervicornis*: self-recognition analysis and demographic interpretation. *Evolution* **37**, 437–453.

Neigel, J. E. & Avise, J. C. 1983 *b* Histocompatibility bioassays of population structure in marine sponges. *J. Hered.* **74**, 134–140.

Nishihira, M. 1967 Dispersal of the larvae of a hydroid, *Sertularella miurensis* in nature. *Bull. mar. Biol. Stat. Asamushi* **13**, 49–56.

Olson, R. R. 1985 The consequences of short-distance larval dispersal in a sessile marine invertebrate. *Ecology* **66**, 30–39.

Ostarello, G. L. 1976 Larval dispersal in the subtidal hydrocoral *Allopora caifornica* Verrill (1866). In *Coelenterate ecology and behavior* (ed. G. O. Mackie), pp. 331–337. New York: Plenum.

Palumbi, S. R. & Jackson, J. B. C. 1982 Ecology of cryptic coral reef communities. II. Recovery from small disturbance events by encrusting Bryozoa: the influence of 'host' species and lesion size. *J. exp. mar. Biol. Ecol.* **64**, 103–115.

Palumbi, S. R. & Jackson, J. B. C. 1983 Aging in modular organisms: ecology of zooid senescence in *Steginoporella* sp. (Bryozoa: Cheilostomata). *Biol. Bull.* **164**, 267–278.

Potts, D. C. 1984 Generation times and the Quarternary evolution of reef-building corals. *Paleobiology* **10**, 48–58.

Potts, D. C., Done, T. J., Isedale, P. J. & Fisk, D. A. 1985 Dominance of a coral community by the genus *Porites* (Scleractinia). *Mar. Ecol. Prog. Ser.* **23**, 79–84.

Pyefinch, K. A. & Downing, F. S. 1949 Notes on the general biology of *Tubularia larynx* Ellis & Solander, *J. mar. Biol. Ass. U.K.* **28**, 21–44.

Resing, J. M. 1986 The usefulness of the tissue grafting bioassay as an indicator of clonal identity in scleractinian corals (Great Barrier Reef – Australia). *Proc. 5th int. Coral Reef Symp.* (In the press.)

Richmond, R. H. 1982 Energetic considerations in the dispersal of *Pocillopora damicornis* (Linnaeus) planulae. *Proc. 4th int. Coral Reef Symp.* **2**, 153–156.

Richmond, R. H. 1986 Variations in the population biology of *Pocillopora damicornis* across the Pacific. *Proc. 5th int. Coral Reef Symp.* (In the press.)

Ryland, J. S. 1970 *Bryozoans*. London: Hutchinson University Library.

Ryland, J. S. 1979 Structural and physiological aspects of coloniality in Bryozoa. In *Biology and systematics of colonial organisms* (ed. G. Larwood & B. R. Rosen), pp. 211–242. London: Academic Press.

Ryland, J. S. 1981 Colonies, growth and reproduction. In *Recent and fossil Bryozoa* (ed. G. P. Larwood & C. Nielsen), pp. 221–226. Fredensborg: Olsen & Olsen.

Sabbadin, A. 1978 Genetics of the colonial ascidian, *Botryllus schlosseri*. In *Marine organisms: genetics, ecology, and evolution* (ed. B. Battaglia & J. A. Beardmore), pp. 195–209. New York: Plenum.

Sammarco, P. W. 1982 Polyp bail-out: an escape response to environmental stress and a new means of reproduction in corals. *Mar. Ecol. Prog. Ser.* **10**, 57–65.

Sarukhán, J. & Gadgil, M. 1974 Studies on plant demography: *Ranunculus repens* L., *R. bulbosus* L., and *R. acris* L. III. A mathematical model incorporating multiple modes of reproduction. *J. Ecol.* **62**, 921–936.

Scheltema, R. S. 1971 Larval dispersal as a means of genetic exchange between geographically separated populations of shallow-water benthic marine gastropods. *Biol. Bull.* **140**, 284–322.

Scheltema, R. S. & Williams, I. P. 1983 Long-distance dispersal of planktonic larvae and the biogeography and evolution of some Polynesian and western Pacific mollusks. *Bull. mar. Sci.* **33**, 545–565.

Schijifsma, K. 1935 Observations on *Hydractinia echinata* (Flem.) and *Eupagarus behardus* (L.). *Arch. Neerl. Zool.* **1**, 261–314.

Schijifsma, K. 1939 Preliminary notes on early stages in the growth of colonies of *Hydractinia echinata* (Flem.). *Arch. Neerl. Zool.* **4**, 93–102.

Schopf, T. J. M. & Dutton, A. R. 1976 Parallel clines in morphologic and genetic differentiation in a coastal zone marine invertebrate: the bryozoan *Schizoporella errata*. *Paleobiology* **2**, 255–264.

Sebens, K. P. 1982 Asexual reproduction in *Anthopleura elegantissima* (Anthozoa: Actiniaria): seasonality and spatial extent of clones. *Ecology* **63**, 434–444.

Sebens, K. P. 1983 The larval and juvenile ecology of the temperate octocoral *Alcyonium siderium* Verrill. II. Fecundity, survival, and juvenile growth. *J. exp. mar. Biol. Ecol.* **72**, 263–285.

Sebens, K. P. & Thorne, B. 1985 Coexistence of clones, clonal diversity, and the effects of disturbance. In *Population biology and evolution of clonal organisms* (ed. J. B. C. Jackson, L. W. Buss & R. E. Cook), pp. 357–398. New Haven: Yale University Press.

Shick, J. M., Hoffman, R. J. & Lamb, A. N. 1979 Asexual reproduction, population structure, and genotype-environment interactions in sea anemones. *Am. Zool.* **19**, 699–713.

Shields, W. M. 1982 *Philopatry, inbreeding, and the evolution of sex.* Albany: State University, New York Press.

Stanley, S. M. 1975 Clades versus clones in evolution: why we have sex. *Science* **190**, 382–383.

Stanley, S. M. 1979 *Macroevolution: pattern and process.* San Francisco: W. H. Freeman.

Stephenson, T. A. 1931 Development and formation of colonies in *Pocillopora* and *Porites*. *Sci. Rep. Great Barrier Reef Exped.* **3**, 113–134.

Stoddart, J. A. 1983 Asexual production of planulae in the coral *Pocillopora damicornis*. *Mar. Biol.* **76**, 279–284.

Stoddart, J. A. 1984a Genetical structure within populations of the coral *Pocillopora damicornis*. *Mar. Biol.* **81**, 19–30.

Stoddart, J. A. 1984b Genetical differentiation amongst populations of the coral *Pocillopora damicornis* off southwestern Australia. *Coral Reefs* **3**, 149–156.

Stoddart, J. A., Ayre, D. A., Willis, B., Heyward, A. J. 1985 Self-recognition in sponges and corals? *Evolution* **39**, 461–463.

Stoddart, J. A. & Black, R. 1985 Cycles of gametogenesis and planulation in the coral *Pocillopora damicornis* from south western Australia. *Mar. Ecol. Prog. Ser.* **23**, 153–164.

Strathmann, R. R. & Strathmann, M. F. 1982 The relationship between adult size and brooding in marine invertebrates. *Am. Nat.* **119**, 91–101.

Sugimoto, K. & Nakauchi, M. 1974 Budding, sexual reproduction, and regeneration in the colonial ascidian, *Symplegama reptans*. *Biol. Bull.* **147**, 213–226.

Szmant-Froelich, A. 1985a Reproductive ecology of Caribbean reef corals. *Coral Reefs*. (In the press.)

Szmant-Froelich, A. 1985b The effect of colony size on the reproductive ability of the Caribbean coral *Montastrea annularis* (Ellis and Solander). *Proc. 5th int. Coral Reef Symp.* **4**, 295–300.

Tranter, P. R. G., Nicholson, D. N. & Kinchington, D. 1982 A description of spawning and post-gastrula development of the cool temperate coral, *Caryophyllia smithi*. *J. mar. Biol. Ass. U.K.* **62**, 845–854.

Tunnicliffe, V. 1981 Breakage and propagation of the stony coral *Acropora cervicornis*. *Proc. natn. Acad. Sci. U.S.A.* **78**, 2427–2431.

Tunnicliffe, V. 1983 Caribbean staghorn populations: pre-Hurricane Allen conditions in Discovery Bay, Jamaica. *Bull. mar. Sci.* **33**, 132–151.

van Duyl, F. C., Bak, R. P. M. & Sybesma, J. 1981 The ecology of the tropical compound ascidian *Trididemnum solidum*. I. Reproductive strategy and larval behaviour. *Mar. Ecol. Prog. Ser.* **6**, 35–42.

Wahle, C. M. 1983a The roles of age, size and injury in sexual reproduction among Jamaican gorgonians. *Am. Zool.* **23**, 961.

Wahle, C. M. 1983b Regeneration of injuries among Jamaican gorgonians: the roles of colony physiology and environment. *Biol. Bull.* **165**, 778–790.

Wilbur, H. M., Tinkle, D. W. & Collins, J. P. 1974 Environmental certainty, trophic level, and resource availability in life history evolution. *Am. Nat.* **108**, 805–817.

Williams, G. B. 1965 Observations on the behaviour of the planulae larvae of *Clava squamata*. *J. mar. Biol. Ass. U.K.* **45**, 257–273.

Williams, G. B. 1976 Aggregation during settlement as a factor in the establishment of coelenterate colonies. *Ophelia* **15**, 57–64.

Williams, G. C. 1957 Pleiotropy, natural selection, and the evolution of senescence. *Evolution* **11**, 398–411.

Williams, G. C. 1975 *Sex and evolution.* Princeton University Press.

Williams, G. C. & Mitton, J. B. 1973 Why reproduce sexually? *J. theor. Biol.* **39**, 545–554.

Willis, B. L. & Ayre, D. J. 1985 Asexual reproduction and genetic determination of growth form in the coral *Pavona cactus*: biochemical genetic and immunogenic evidence. *Oecologia* **65**, 516–525.

Winston, J. E. 1983 Patterns of growth, reproduction and mortality in bryozoans from the Ross Sea, Antarctica. *Bull. mar. Sci.* **33**, 688–702.

Woodley, J. D., Chornesky, E. A., Clifford, P. A., Jackson, J. B. C., Kaufman, L. S., Knowlton, N., Lang, J. C., Pearson, M. P., Porter, J. W., Rooney, M. C., Rylaarsdan, K. W., Tunnicliffe, V. J., Wahle, C. M., Wulff, J. L., Curtis, A. S. G., Dallmeyer, M. D., Jupp, B. P., Koehl, M. A. R., Neigel, J. & Sides, E. M. 1981 Hurricane Allen's impact on Jamaican reefs. *Science, Wash.* **214**, 749–755.

Wulff, J. L. 1985 Dispersal and survival of fragments of coral reef sponges. *Proc. 5th int. Coral Reef Symp.* **5**, 119–124.

Yoshioka, P. M. 1982a Role of planktonic and benthic factors in the population dynamics of the bryozoan *Membranipora membranacea*. *Ecology* **63**, 457–468.

Yoshioka, P. M. 1982b Predator-induced polymorphism in the bryozoan *Membranipora membranacea*. *J. exp. mar. Biol. Ecol.* **61**, 233–242.

Zimmer, R. L. & Woollacott, R. M. 1977 Metamorphosis, ancestrulae, and coloniality in bryozoan life cycles. In *Biology of bryozoans* (ed. R. M. Woollacott & R. L. Zimmer), pp. 91–142. New York: Academic Press.

# Metabolic implications of modularity: studies on the respiration and growth of *Electra pilosa*

By D. J. Hughes and R. N. Hughes

*School of Animal Biology, University College of North Wales, Bangor, Gwynedd LL57 2UW, U.K.*

The mass-specific respiration rate of *Electra pilosa* is independent of colony size and therefore there is no allometric metabolic constraint on colonial growth rate (modular iteration). Budding, however, is confined to peripheral zooids and so the amount of meristem per unit area of colony declines as the colony grows. Hence the rate of zooid production per individual is a decreasing function of colony size. *E. pilosa* partly compensates for this, first by increasing the budding rate of peripheral zooids as the colony grows and second by expanding the peripheral meristem into lobes.

Among colonial invertebrates, a modular construction frees colonies from metabolic allometry and if modules retain their capacity for replication these may accumulate exponentially until restrained by extrinsic factors. Encrusting forms with strictly two-dimensional growth, however, are constrained by the peripheral location of the budding zone. This may be alleviated by faster budding, perhaps as a result of nutritional subsidies as the colony grows, and by departure from a circular shape.

## Introduction

The negative allometric relationship between metabolic rate and body mass is a fundamental property of most organisms that increase in size volumetrically (Hemmingsen 1961). As a rule, therefore, larger organisms metabolize more slowly than smaller ones and this might be expected to influence their growth and reproductive rates. Mass-specific metabolic rate has an allometric exponent of about $-0.25$ over the several orders of magnitude in body mass covered by interspecific comparisons and whereas intraspecific values may differ significantly from the norm, the exponent is always negative. Organisms expanding volumetrically, therefore, are destined to grow progressively more slowly once beyond the embryonic or juvenile stages (Bertalanffy 1960).

Modular colonial organisms such as hydroids, corals and bryozoans present an exception. As a result of fission or budding, genetically identical modules are added to the colony, increasing its biomass. Each module (polyp or zooid) is probably of a size optimizing its functional capacities. Consequently the variance of modular size within a colony is likely to be small: among Caribbean corals, for example, the coefficient of variation for polyp diameter within colonies is usually less than 10% (Lehman & Porter 1973); bryozoans may adjust the number of tentacles on the lophophore in response to microenvironmental conditions (Thorpe *et al.* 1986) but this represents only a slight variation on zooidal biomass. Yet by modular iteration, the colonial organism can increase its total biomass far beyond the constraints operating on the modules themselves (Hughes & Cancino 1985).

Many colonial benthic invertebrates live on ephemeral substrata or in situations of intense competition for space (Connell 1973; Ryland 1976, 1981; Buss 1979; Jackson 1979), so mechanisms avoiding constraints on the rate of modular iteration could be selectively advantageous. It is of interest to know, therefore, whether modular metabolic rate is an allometric function of colonial biomass, following the trend of most non-modular organisms, or whether it is independent of colonial biomass, without constraint on growth rate. To investigate these questions, the respiration rate and the rate of zooidal accumulation were measured in different sized colonies of the encrusting cheilostome bryozoan *Electra pilosa* (L.), a modular colonial animal with monomorphic zooids (Ryland & Hayward 1977) which normally buds only distally and distolaterally, producing on plane surfaces a simple two-dimensional structure (proximal and proximolateral buds can occur following lesion of the colony).

## Materials and methods

Larvae of *Electra pilosa* were allowed to settle naturally on glass microscope slides held in a Perspex rack suspended in the Menai Straits. Superfluous colonies and other fouling organisms were periodically scraped away. Colonies brought into the laboratory, from January to July, 1984, were acclimatized to 15 °C over two days and the encrusted slides cleaned of small fouling organisms and detritus. To measure colonial respiration rate, a circular Perspex chamber, 5 mm deep, was fitted round the downward-facing colony and sealed to the microscope slide with silicone grease (figure 1). Chambers of different diameters were used for different sized colonies,

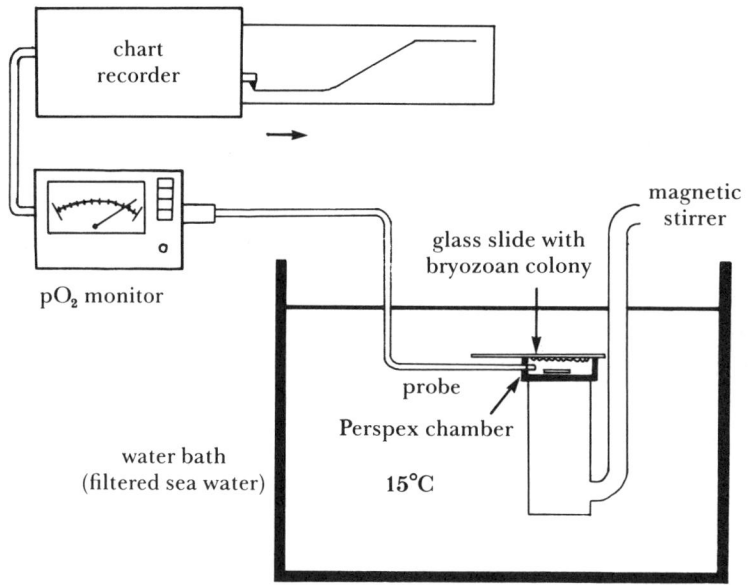

Figure 1. Diagram showing the apparatus used in the respirometry experiments.

keeping the working volume to a minimum in all cases. Circulation within the chamber was achieved with a magnetic stirrer and the oxygen tension of the initially saturated, filtered seawater was monitored with a Searle oxygen electrode, passing through a 2 mm hole in the wall of the chamber, and connected via a $pO_2$ monitor to a chart recorder. As a control, the procedure was repeated after moving the respiration chamber to an area of the slide adjacent to the colony. The colony was subsequently scraped from the slide, dried at 60 °C for 24 h, weighed, ashed at 500 °C for 6 h and reweighed.

To measure colonial growth rates during June 1983, 76 colonies on microscope slides were individually placed in a chamber of recirculating, cooled seawater, mounted on the stage of a Wild dissecting microscope (Cancino 1983) and drawn by using a camera lucida. The colonies were redrawn after replacement in the sea for 14 days. Above a size of 1000 zooids, only the perimeter of each colony was measured, the total number of zooids being predicted from a regression of zooid number on colony area (figure 3$a$). Perimeters and areas were measured from the drawings by using a microcomputer with digitizer.

## Results

### Respiration

Colonies of *Electra pilosa* consumed oxygen more slowly as it was depleted from the water, but at oxygen tensions above 90% saturation colonies consumed 1.77 ($\pm$1.40 s.d.) $\times 10^{-3}$ μl $O_2$ per zooid per hour. By comparison, Mangum & Schopf (1967) reported the consumption of $4.5 \times 10^{-4}$ μl $O_2$ per zooid per hour for *Bugula turrita* (Desor) at 20 °C and constant oxygen tension.

The regression of ln (mass-specific respiration rate) on ln (colonial biomass) was not statistically significant (figure 2), giving no grounds for refuting the hypothesis that modular respiration rate is independent of colony size.

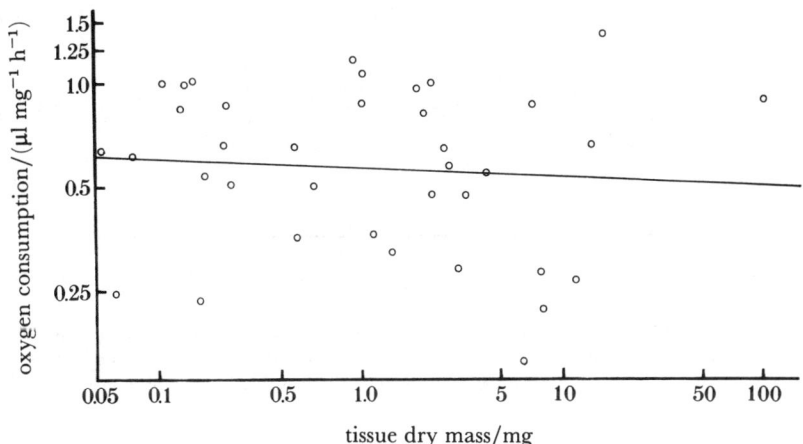

FIGURE 2. Mass-specific rate of oxygen consumption of colonies of *Electra pilosa* plotted as functions of colonial ash-free dry mass. Regression equation fitted to data: $y = -0.030x - 0.574$, standard error of coefficient = 0.008.

### Growth

The total number of zooids (figure 3$a$) and total biomass (figure 3$b$) were linear functions of colonial area, the mean biomass per zooid remaining at 3.70 ($\pm$1.79 s.d.) μg throughout all stages of colonial growth. The rate of zooid production per zooid was a decreasing function of colony size (figure 4$a$), whereas the rate of zooid production per unit of colonial perimeter was an increasing function of colony size (figure 4$b$). Since only peripheral zooids undergo budding, it was inferred that their budding rate increased as colony size increased.

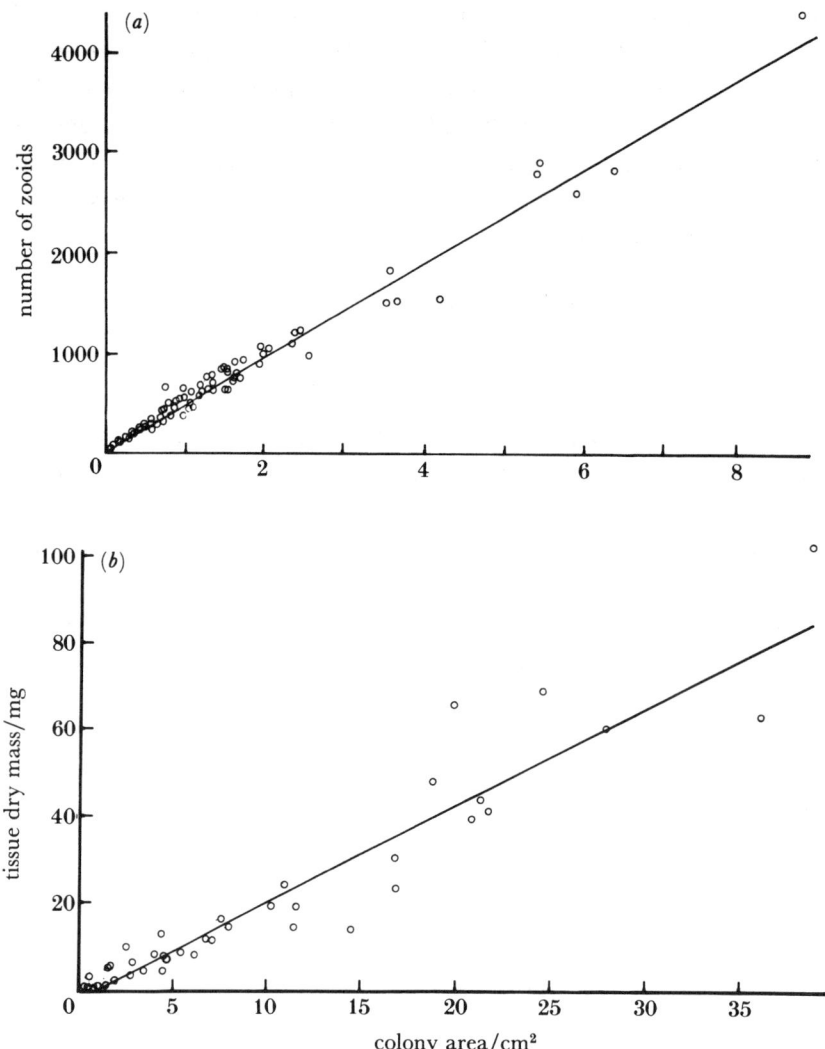

FIGURE 3. (a) Number of zooids per colony plotted as a function of colony area. Regression equation fitted to data: $y = 472x + 13.7$, standard error of coefficient = 5.5. (b) Ash-free dry mass of colony plotted as a function of colony area. Regression equation fitted to data: $y = 2.22x - 1.85$, standard error of coefficient = 0.014.

## DISCUSSION

### Respiration

The isometric relation between colonial respiration rate and biomass of *Electra pilosa* suggests that compartmentalization into modules avoids the metabolic allometry normally associated with volumetric somatic growth. Intra- and interspecific metabolic isometry is also found among small metazoans of up to about 1 mg wet mass, whereas in protozoans and larger metazoans metabolism is usually allometric (Hemmingsen 1961; Zeuthen 1970). Metabolic rate might be determined by the aggregative surface area of cells and organelles (Hemmingsen 1961; Zeuthen 1970; Peters 1983). Among unicellular organisms this surface area is confined by limits to the development of organelles, but the cellular construction of small metazoans increases their total surface area of membranes sufficiently to compensate for the decreasing ratio of

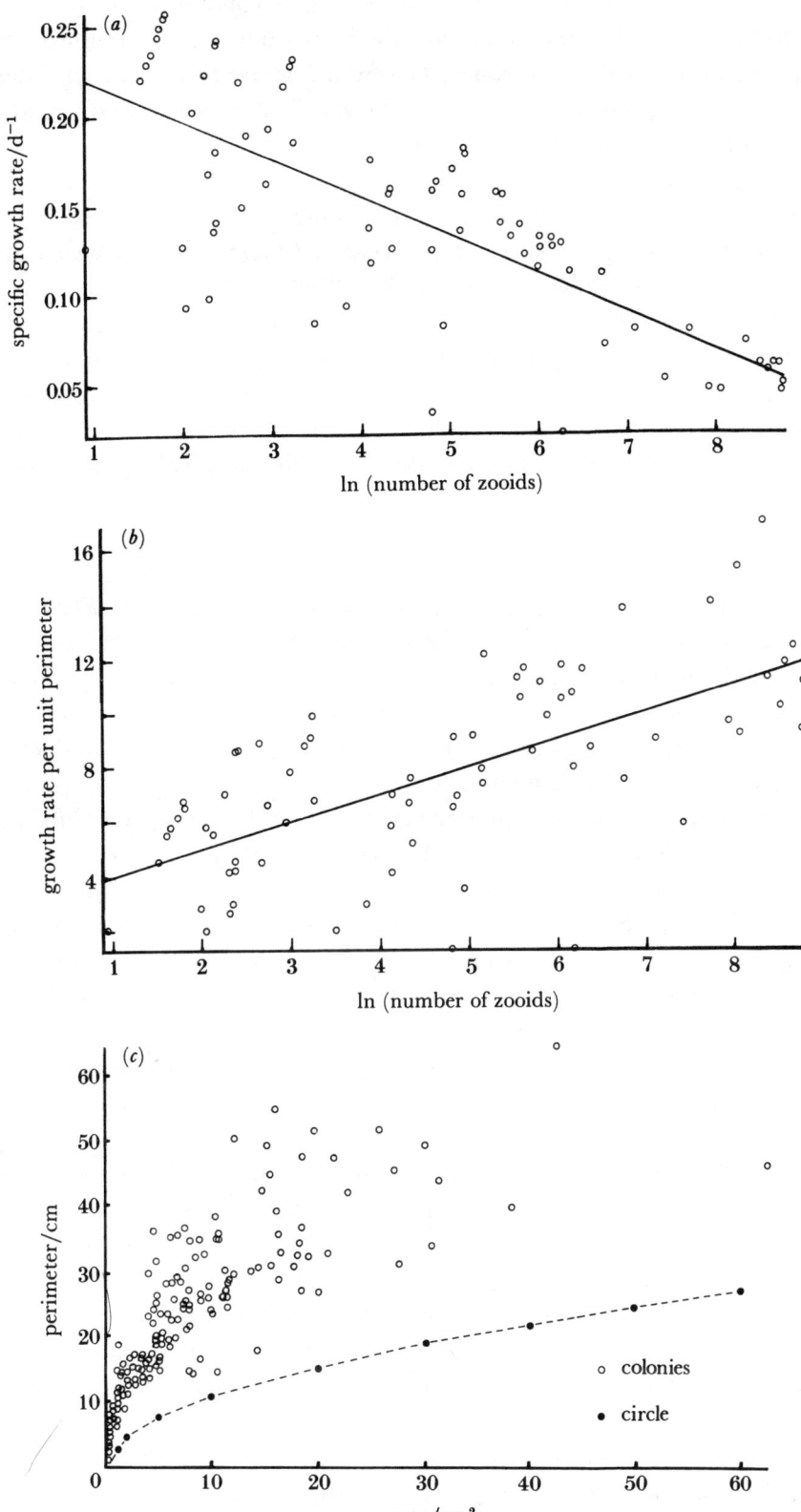

FIGURE 4(a) Specific growth rate (instantaneous rate of zooid production per individual, calculated as $(\ln n_2 - \ln n_1)/14$, where $n_1$ = number of zooids at start, $n_2$ = number after 14 days)) plotted as a function of ln (colony size) (where size is expressed as the geometric mean of the initial and final number of zooids present in the colony). (b) Finite growth rate per unit of perimeter (calculated as $(n_2 - n_1)/14\sqrt{(p_1 p_2)}$, where $p_1$ = length of perimeter at start, $p_2$ = length after 14 days) plotted as a function of ln (colony size). (c) Length of colonial perimeter plotted as a function of colony area.

somatic surface area to volume. With increases in biomass beyond about 1 mg, the degree of compensation declines and the allometric metabolic relationship is resumed. Metabolic allometry is preserved even in slices of tissue taken from *Littorina littorea* (L.) of different sizes but is obliterated by homogenizing these tissues (Newell & Pye 1971), partly, although not unequivocally, corroborating the surface area hypothesis.

The parallel between the small metazoans and modular colonial animals is striking; whereas the former maintain a high ratio of surface area to volume by somatic division into cells, the latter do it by division into polyps or zooids. The exposed surfaces of each zooid are apparently sufficient to maintain adequate gaseous exchange by diffusion (Ryland 1967).

Of course, the lack of allometry does not imply that modules are metabolically independent of one another; gymnolaemate bryozoan zooids are cytologically interconnected (Ryland 1979), and the translocation of metabolites has been demonstrated in *Membranipora membranacea* (L.) (Best & Thorpe 1985). Such metabolic interactions allow the nourishment of specialized, non-feeding zooids and, in certain cheilostomes, the electrophysiological coordination of zooids (Ryland 1979).

*Growth*

Despite the isometric relation between metabolic rate and biomass in *Electra pilosa*, the specific rate of zooidal budding declines as colonies grow. A constant specific rate of budding would cause an exponential increase in the number of zooids throughout colonial growth. However, the specific growth rate will remain constant only if there is no constraint on the number of successive multiplications that each module can undergo. *Electra pilosa* does not satisfy this requirement, since after establishment of the ancestrula, growth quickly changes from a linear to a two-dimensional pattern, producing, on an unlimited plane substratum, an approximately circular colony in which only the peripheral zooids have room to bud. This constraint also applies to other bryozoan species in which exponential growth has been reported, for example *Plumatella repens* (Bushnell 1966), in which each zooid has only a limited number of budding sites. For an encrusting, two-dimensional colony, therefore, the number of zooids will increase as a quadratic function (Kaufmann 1970; Wass & Vail 1978) and the number of peripheral zooids as a linear function of time.

*Electra pilosa*, however, shows two properties that partly compensate for the constraint on budding imposed by two-dimensional growth. First, peripheral zooids, although remaining constant in size (figure 3a,b), bud faster as the colony grows (figure 4b), perhaps being subsidized by metabolites translocated from the non-replicating zooids accumulating within. Second, the periphery becomes expanded into lobes, increasing the perimeter, and hence the number of replicating zooids. The colony perimeter rapidly becomes about twice that of a circle of the same area, and remains so throughout later growth (figure 4c). Frontal budding is potentially another method by which the constraints set by two-dimensional growth can be avoided, but in species such as *Celleporella hyalina* (L.) this is usually concerned with the production of non-replicating sexual zooids, and growth remains essentially two-dimensional (Cancino 1983).

In conclusion, a modular construction frees colonies from the influence of metabolic allometry, with the result that the specific rate of modular iteration need not decline with increasing colonial biomass. If all modules retain their capacity for replication they may accumulate exponentially until influenced by external constraints such as resource depletion, competition, architectural instability or environmental change. This, however, is only possible with three-dimensional growth, as occurs in some branching hydroids and bryozoans or in mound-forming massive corals. Two-dimensional colonial growth restricts replication to the peripheral modules, but the development of a non-circular perimeter and nourishment of peripheral modules by translocation allow modular accumulation to exceed the quadratic function of time appropriate for a circular colony expanding at a constant rate.

We thank Dr J. Davenport and Mr A. Woolmington for providing the advice and equipment necessary for the respirometry.

## References

Bertalanffy, L. von 1960 Principles and theory of growth. In *Aspects of normal and malignant growth* (ed. W. W. Nowinski), pp. 137–259. Amsterdam: Elsevier.

Best, M. A. & Thorpe, J. P. 1985 Autoradiographic study of feeding and the colonial transport of metabolites in the marine bryozoan *Membranipora membranacea*. *Mar. Biol.* **84**, 295–300.

Bushnell, J. H. 1966 Environmental relations of Michigan Ectoprocta, and dynamics of natural populations of *Plumatella repens*. *Ecol. Monogr.* **36**, 95–123.

Buss, L. W. 1979 Bryozoan overgrowth interaction: the interdepence of competition for space and food. *Nature, Lond.* **281**, 475–477.

Cancino, J. M. 1983 *Demography of animal modular colonies*. Ph.D. thesis, University of Wales.

Connell, J. H. 1973 Population ecology of reef building corals. In *Biology and geology of coral reefs, vol. II, biol. 1.* (ed. O. A. Jones & R. Endean), pp. 205–245. New York: Academic Press.

Hemmingsen, A. M. 1960 Energy metabolism as related to body size and respiratory surfaces, and its evolution. *Rep. Steno Meml Hosp. Nordisk Insulin Lab.* **9**, 1–110.

Hughes, R. N. & Cancino, J. M. 1985 An ecological overview of cloning in Metazoa. In *Population biology and evolution of clonal organisms* (ed. J. B. C. Jackson, L. W. Buss & R. E. Cook), pp. 153–186. New Haven: Yale University Press.

Jackson, J. B. C. 1979 Overgrowth competition between encrusting cheilostome ectoprocts in a Jamaican cryptic reef environment. *J. anim. Ecol.* **48**, 805–823.

Kaufmann, K. W. 1970 A model for predicting the influence of colony morphology on reproductive potential in the Phylum Ectoprocta. *Biol. Bull.* **139**, 426.

Lehman, J. T. & Porter, J. W. 1973 Chemical activation of feeding in the Caribbean reef-building coral *Montastrea cavernosa*. *Biol. Bull.* **145**, 140–149.

Mangum, C. P. & Schopf, T. J. M. 1967 Is an ectoproct possible? *Nature, Lond.* **213**, 264–266.

Newell, R. C. & Pye, V. I. 1971 Quantitative aspects of the relationship between metabolism and temperature in the winkle *Littorina littorea* (L.). *Comp. Biochem. Physiol.* **38B**, 635–650.

Peters, R. H. 1983 *The ecological implications of body size*. Cambridge University Press.

Ryland, J. S. 1967 Polyzoa. *Oceanogr. Mar. Biol. Ann. Rev.* **5**, 343–369.

Ryland, J. S. 1976 Physiology and ecology of marine bryozoans. *Adv. Mar. Biol.* **14**, 285–443.

Ryland, J. S. 1979 Structural and physiological aspects of coloniality in Bryozoa. In *Biology and systematics of colonial organisms* (ed. G. Larwood & B. R. Rosen), pp. 211–242. London and New York: Academic Press.

Ryland, J. S. 1981 Colonies, growth and reproduction. In *Recent and fossil bryozoa* (ed. G. P. Larwood & C. Nielsen), pp. 221–226. Fredensborg: Olsen & Olsen.

Ryland, J. S. & Hayward, P. J. 1977 *British anascan bryozoan: Cheilostomata: Anasca. Synopses of the British fauna no. 10* (ed. D. M. Kermack). London, New York and San Francisco: Academic Press.

Thorpe, J. P., Clarke, D. R. K. & Best, M. A. 1985 Natural variation in tentacle number in marine bryozoans and the possible effects of intraspecific and interspecific ecological competition for food. In *Bryozoa: Ordovician to Recent. Proc. Sixth int. Conf. int. Bryozool. Ass.* (ed. C. Nielsen & G. P. Larwood). Fredensborg: Olsen & Olsen.

Wass, R. E. & Vail, L. L. 1978 Encrusting bryozoa exhibit linear growth. *Search* **9**, 42–44.

Zeuthen, E. 1970 Rate of living as related to body size in organisms. *Pol. Arch. Hydrobiol.* **17**, 21–30.

# Some life-history consequences of modular construction in plants

By A. R. Watkinson[1] and J. White[2]

[1] School of Biological Sciences, University of East Anglia, Norwich NR4 7TJ, U.K.
[2] Department of Botany, University College Dublin, Dublin 4, Ireland

The nature and life-history consequences of modular construction in plants are discussed with particular reference to growth, reproduction and survival. Plants grow by the iteration of modular units and as a consequence growth can be described in terms of the population dynamics of these structural units. Changes in size, whether positive or negative, depend on the birth and death rates of modules; however, if the births continue to exceed the deaths, plants then have the capability of attaining enormous sizes, especially if they are clonal. The population nature of plant growth also means that plants of the same age may show large variation in individual size if individuals differ in their relative growth rates. Correlations between age and size are often, therefore, very weak.

Constraints on the allocation of resources accumulated during growth have important implications for the reproductive schedules of plants, but the analysis of constraint functions has so far revealed little about the actual detail of these schedules. All the meristems of semelparous plants are involved in or die at reproduction and as a consequence death of the genet follows reproduction. For iteroparous plants, however, there are fundamental differences between the reproductive schedules of plants with a single shoot module and those with many shoot modules. The former demonstrate a relatively constant rate of reproduction from year to year following maturity whereas the latter show a continual increase in fecundity with size and age. The reproductive schedules of clonal plants are further discussed in relation to the allocation of meristems to either growth or reproduction.

The pattern of mortality is examined at both the level of the module and the genet. Particular attention is focused on the survival and senescence of leaves and shoots; there is no equivalent regular shedding of organs in unitary organisms. Whereas genet senescence and death are coincident with shoot module death in semelparous plants, there is no evident relation between them in iteroparous plants. The life span of the genet reflects the birth and death rates of its modules and both aclonal and clonal plants that are iteroparous may achieve considerable longevity. The longevity of aclonal plants often seems to be restricted by the accumulation of dead material and the problems of being large. Clonal plants are, in contrast, potentially immortal. It is questionable whether the genets of iteroparous plants show senescence as defined for unitary organisms since there is no separation of germ plasm from soma and since apical meristems do not appear to senesce. Insofar as they retain the capacity for rejuvenescence from apical meristems, genets of modular organisms do not senesce; it is only the constituent organs that show senescence, death and decay.

## 1. Introduction

The many and varied life histories of plants have traditionally been subdivided by the duration of life of the genetic individual (genet) into annual, biennial and perennial. Recent demographic research has increasingly cast doubt on the distinctiveness of the biennial life cycle. Very few plants are obligately biennial; most 'biennials' appear to be facultative (Kelly 1985). They may either, like *Verbascum thapsus*, have annual, biennial or triennial geographical races (Reinhartz 1984) or, like *Digitalis purpurea* and *Senecio jacobaea*, exhibit biennial or perennial life cycles in response to environmental circumstances (van der Meijden & van der Waals-Kooi 1979). Most biennials are therefore probably best considered as short-lived perennials that may complete their life cycles within two years, given optimal conditions (Harper 1977).

A more recent subdivision of life histories by demographers has used another criterion, recognizing two fundamentally different patterns of reproduction as a function of age. These were named semelparous and iteroparous life histories by Cole (1954). The former has an obscure etymology, but refers to those organisms whose reproduction is confined to a single age class and followed by death. Iteroparous organisms reproduce (or are physiologically capable of reproducing) more than once. Botanists have traditionally referred to plants with such life histories as monocarpic and polycarpic, respectively. In seed plants a further useful distinction may also be made between shoots that have only one terminal and more or less synchronous reproductive episode, and shoots with more persistent and indeterminate vegetative expansion, bearing lateral reproductive structures over an indefinite period, often years.

A description of the life history of a plant in terms of the actively growing phase of the life cycle is, of course, only a partial description. Plant populations exist in two parts: the growing plants and the dormant seeds. Most species possess some form of seed dormancy and it is only at the time of germination, which is the formal equivalent of birth in animals (Harper & White 1974), that the embryo acquires independence from the parental tissue. The dormancy of seeds has three important consequences. First, by delaying germination it slows down the rate of increase of the population: only when one generation follows on directly from the next is the growth rate of the population maximized. Second, the longevity of seeds confers perenniality on many annual plants and produces populations in which the generations may overlap. In a sense annual plants with long-lived seeds can be considered as long-lived semelparous perennials. The dormancy of at least a fraction of the seed crop also provides a safeguard against unsuitable conditions in a variable environment (León 1985).

Among life-history phenomena it is the relationship of reproductive activity to age that has received most attention from evolutionary theorists, by consideration of the role of natural selection in moulding the forms of life histories, as described by the age-specific survival probabilities and relations of fecundity with age (Law 1979; Charlesworth 1980). Attention has been focused in particular on selection for iteroparity as opposed to semelparity and on such questions as the effects of selection on the age of first reproduction, reproductive effort and senescence. All of these studies have treated unitary and modular organisms as if they were equivalent. It is the aim of this paper to examine whether there are any life-history features that are particularly associated with the repetitive, modular construction characteristic of plants. And since the life history of an organism reflects the pattern of allocation of resources between reproduction, maintenance and growth with time, we shall consider some of the consequences of modular construction in plants as they affect growth, reproduction and survival.

## 2. Growth

### (a) A demographic approach to growth

Populations of unitary organisms are composed of individuals, the genotypes of which specify a unitary morphology and a life cycle that proceeds remorselessly from the zygote through juvenility and a reproductive phase to senility and death (Harper & Bell 1979). In contrast populations of *modular organisms* are composed of individual genets (individuals that develop from zygotes) each of which is made up of a collection of unit tissues (above the level of the cell, a distinction necessary unless all multicellular organisms are to be considered modular). In plants the zygote develops into an organism in which one or more structural tissue units are iterated by one or (usually) many growing points, themselves capable of self-perpetuation. For various types of morphological analysis the structural unit chosen may be a leaf with its axillary meristem, a bud, a metamer, a shoot, a branch system or ramet, indeed any structural tissue unit that is iterated. This is a very broad definition of modular growth which encompasses all vascular plants.

There is a more restricted definition: the term *module* is used by some morphologists (Hallé *et al.* 1978) to refer to a monopodial shoot which is produced by a single apical meristem and terminated by a reproductive structure. Thus, a talipot palm (*Coryphya umbraculifera*), nearly 20 m tall with a single apical meristem, would be regarded as a single module; but in our terms it shows modular growth or construction, growing by the repeated, sequential iteration of structural tissue units. We shall refer to modules in this stricter sense as *shoot modules*, but otherwise use the term unqualified for reiterated structural units.

White (1984) has used the term metamer for a structural unit below the level of shoot module and the phrase 'modular growth' may be more or less synonymous with 'metameric growth'; but the equivalence, so evident in plants, cannot be taken for granted for animals: not all metameric (segmented) animals are necessarily modular, particularly if they lack the capacity for proliferation of 'somatic replicas produced by budding or fission' (Hughes 1983), one of the most characteristic features of modular organisms. The taxonomic limit of modular construction in animals remains a moot point.

Unlike a unitary organism, the product of a zygote in a modular organism has population properties of its own. This is because the growth of a genet can be described in terms of the number of modules which in turn depends on the rate at which modules are born and the rate at which they die. The modular growth of a genet can be described by the equation

$$\eta_{t+1} = \eta_t + B - D, \tag{1}$$

where $\eta$ is the number of modules, $B$ is the number of module births and $D$ is the number of module deaths (Noble *et al.* 1979; Harper 1981). While this equation takes no account of whether the modules are vegetative or reproductive it nevertheless provides a basis for the growth of individual plants to be studied at a demographic level (Bazzaz & Harper 1977). More complex models of individual plant growth involving matrices are described by Maillette (1982) and McGraw & Antonovics (1983). Clearly if the growth of an individual genet can be described in terms of the population dynamics of modules then the modules of each genet will have an age structure together with other population attributes such as expectation of life and age-specific mortality. As the birth and death rates of modules can be expected to vary with the conditions in which a genet finds itself then so can the age structure. For example the leaf population of *Linum usitatissimum* grown in deionized water for 100 days shows an age structure dominated by leaves that are less than 20 days old whereas the leaves live much longer on plants grown in a full nutrient solution (Harper 1981). Leaf senescence here is related to the withholding of nutrients.

Applying the concepts of population growth to a population of modules allows the rate of growth of a genet to be calculated from one census to the next by subtracting $\eta_t$ from both sides of (1) to give

$$\Delta\eta = B - D. \quad (2)$$

If $\Delta\eta$ is positive the population of modules will increase in number, whereas if $\Delta\eta$ is negative the number of modules will decline. Assuming that the number of module births and deaths does not change either with time or with the number of modules, then the genet will increase to an infinite size if $\Delta\eta$ is positive and die if $\Delta\eta$ is negative. Yet for plants where the modules remain part of a physiologically integrated whole and contact of the shoots with the soil is through a single axis (for example, most trees and annual herbs) an exponential phase of growth is typically followed by a period in which the rate of iteration of new modules and accumulation of biomass declines until a maximum size is reached (Franco 1985). The rate of production of new modules is in this case either dependent on time or the number of modules already present. If the latter, then competition between modules for a limited supply of resources around the base of the plant or the effect of overlapping leaves on each others' activity presumably results in either a decrease in the birth rate of modules, an increase in mortality or both. There is a clear analogy here with density-dependent population growth in unitary organisms. Alternatively, constructional constraints may well impose a limit on the size that an individual genet can attain. Interestingly, the branch modules of a tree get progressively smaller as the number of modules increases (Borchert & Honda 1984). This may reflect the most efficient way that a branching system can support a photosynthetic canopy. If the modules did not get smaller as the branching system developed it is likely that the whole plant (if dependent on a single trunk) would eventually collapse from biomechanical limitations.

In contrast clonal plants that spread laterally and root at the nodes have at least the potential capability (depending on the branching system) of increasing in size indefinitely; new resources can always be tapped as the genet forages into new areas. Clonal growth thus allows plants to escape from some biomechanical size constraints. Clones of bracken fern (*Pteridium aquilinum*) have commonly been observed up to 300 m in diameter (estimated to be about 700 years old) and a few clones reach almost 500 m across, probably about 1400 years old (Oinonen 1967). Various estimates have been made of the extent of clonal growth in quaking aspen (*Populus tremuloides*): two clones of 10.1 and 43.3 ha, containing about 15000 and 47000 ramets, respectively, have been described by Kemperman & Barnes (1976), who also suggest that some clones may approach 81 ha in area. A few other examples of the extent of clonal growth are given by Cook (1983). For these plants there is no decrease in the size of modules as the number of modules increases. Those genets that space their modules far apart can spread over an area quickly ('guerrilla' growth form) whereas those with short internodes will have a 'phalanx' growth form, where the modules are tightly packed (Lovett Doust 1981). Clearly the way in which the modules of a genet are packed and whether or not they remain connected can have major effects on the way a genet exploits environmental resources and interacts with neighbours (Harper 1981).

*(b) Variation in plant size*

Individual plant genets may attain enormous sizes but they may also show immense variation in size. Such plasticity is largely a consequence of modular growth and the number of modules iterated (Harper 1981). Small individuals are those in which few modules have been iterated or many have aborted or died. In contrast a high birth rate of modules or a low rate of modular abortion and death, or both, results in vigorous growth and large individuals. Even individuals of the same age within a monoculture consist of a size hierarchy of individuals with a few large dominants and many relatively small individuals. Such hierarchies develop soon after seedling emergence and the inequality between individuals appears to become even greater as seedlings grow. Simple exponential growth is sufficient to cause such a shift from a symmetrical to a highly asymmetrical distribution with large size differentials if individuals differ slightly in their relative growth rates (Koyama & Kira 1956). As such it is tempting to suggest that greater size inequalities might develop in clonal plants where the phase of exponential growth might be expected to last for longer than in those plants with a single stem, especially if the individuals are widely spaced. But other factors may play a role. In high density stands where interference occurs between individuals asymmetric or one-sided competition may further exaggerate the difference between individuals leading to dominance and suppression (Aikman & Watkinson 1980). As reproductive output is generally highly correlated with plant size large individuals will typically be more fecund while small individuals will usually have higher mortality. Consequently, in sharp contrast with unitary organisms, there will often be a very weak correlation between age and the reproductive performance and mortality of many plants.

Dominance and suppression may lead to a decline in the size of some individuals since the death rate of modules ($D$) may exceed the birth rate ($B$), but it is not the only factor. Grazing and pathogen activity may also result in a decline in module number. With grazing the death rate of modules can be increased by animals removing them completely or by removing parts of them: the latter while not resulting in immediate module death does result in an increase in the death rate (Dirzo 1984). Where animals continue to remove modules at a rate greater than the birth rate the plant will decrease in size and perhaps die whereas if grazing is at a level such that $B > D$ the plant will continue to iterate and accumulate new modules. Animals that feed on seedlings may often cause death (Crawley 1983) but established plants are much more resistant to herbivory because most herbivores remove only parts of the plant or tap resources leaving other plant parts that are capable of regeneration through the iteration of new modules. Frequently the heavy grazing of a plant that results in a decrease in the number of modules may be followed by a recovery period in which new modules can be iterated as long as there are some undamaged meristems. For example, the cinnabar moth *Tyria jacobaeae* often defoliates ragwort *Senecio jacobaea* such that there is an apparent seed loss of 100%. On fertile soils the plants can compensate partly for this defoliation by the iteration of new modules and the production of a second crop of flowers (Islam & Crawley 1983). Seed yield is decreased and seed maturity delayed in comparison with ungrazed plants but the longevity of the plant may be increased because defoliation tends to increase the number of rosettes as damaged plants produce new rosettes from root buds and from the crown of the root stock.

Again the shrinkage of plants as a result of grazing and their potential for regrowth means that the age and size of a genet are poorly correlated. Since the fecundity and survival of plants is often much more closely related to size than age, a number of authors (for example, Werner & Caswell 1977; Kirkpatrick 1984) have argued that it is better to classify the life history characteristics of plants by size rather than age which is the most often used classification for unitary organisms. Population models based on size-related parameters that incorporate shrinkage and fragmentation, as well as the more familiar processes of growth, reproduction and death, are described by Hughes (1984). Others have argued for a classification based on age-state (Sarukhán & Gadgil 1974) which comprises a somewhat arbitrary classification of plants into a number of categories (for example, seed, juvenile, young reproductive, senile) based partly on age and size. It would of course be preferable to monitor the fate of individual plants according to both their age and size (Law 1983) but this has seldom been attempted (see Werner & Caswell 1977). Such large samples and such detailed monitoring are required to calculate all the necessary transition probabilities between the various age and size classes that the labour involved is daunting. Nevertheless an understanding of the variance in life history characteristics that occurs both in relation to size and age is essential if one is to understand the evolution of life history patterns (Lacey *et al.* 1983).

## 3. Reproduction

### (a) *Constraints on the allocation of resources*

How does an organism allocate the resources that it accumulates during growth? If there were no constraints on the design of organisms one would expect natural selection to optimize growth, reproduction and survival irrespective of the age of the plant. In this case the organism with the highest absolute fitness would be immortal, start to produce progeny almost immediately after it was born and continue to produce large numbers of offspring at frequent intervals as it grew older; it would be a 'Darwinian demon' (Law 1979). No unitary organism of this type exists and the assumption that there is no relation between present reproduction and future survival or reproduction is clearly incorrect. There are constraints on the resources that can be accumulated over a given time interval and constraints on the way that these resources can be divided up between the competing demands of maintenance, growth and reproduction (Law 1979). Although a greater allocation of resources to reproduction might be expected to increase the number of offspring produced, fewer resources to maintenance might increase the risk of mortality and fewer resources to growth might reduce both survival and the potential for reproduction later in life. It seems reasonable to suppose, therefore, that reproduction early in life will vary inversely with reproduction and the chance of survival later in life.

The importance of the relation between present reproduction and subsequent reproduction and survival for the evolution of life histories has been extensively explored by using the notion of reproductive value (see Charlesworth 1980). By using this concept it has been shown that a knowledge of the form of the constraint functions between reproduction, growth and survival together with a knowledge of the factors affecting them is central to an understanding of the evolution of reproductive schedules. Unfortunately we have very little information on the nature of these constraint functions in plants except for a few plants such as *Agave* (Schaffer & Schaffer 1977), *Astrocaryum* (Piñero *et al.* 1982) and *Dipsacus* (Caswell & Werner 1978), which have one or few shoot modules. In *Poa annua* (Law 1979), which has an indeterminate pattern of growth, the variance in the data is so great that it is impossible to judge the exact nature of the constraint functions. There are too few data yet to generalize but this may well typify the problem in defining constraint functions for plants with many shoot modules and especially for clonal plants.

In iterating new modules plants are producing shoot systems that are partly capable of paying their own carbon costs (Watson 1984). This applies not only to vegetative modules but also to flowers and fruits (Bazzaz *et al.* 1979). These factors compound the problems of measuring reproductive effort and the determination of survival costs (Tuomi *et al.* 1983). In addition, developmental constraints may influence the way in which apical meristems (considered as a resource) are allocated to either reproduction or growth (Armstrong 1982; Watson 1984) in a way quite unlike that in unitary organisms, where there is no separation of the germ plasm from the soma. A meristem may either generate new meristems allowing further growth or it may become a flower primordium that produces seed and dies (Harper 1981); once committed to flowering it is precluded from generating new meristems. Thus growth form, which reflects the sequential iteration of modules, imposes constraints on growth and reproduction. The exact significance of this for the reproductive schedule of a plant can be fully appreciated only if growth form and modular construction are taken into account in investigations of resource allocation.

(b) *Reproductive schedules*

(i) *Semelparous plants*

All the meristems of a semelparous or monocarpic plant are involved in (or die at) reproduction and as an inevitable consequence death of the genet follows (figure 1a). Some authors (for example, Charlesworth 1980) regard all annual plants as semelparous, in that they reproduce within a single season, but many annuals (for example, *Poa annua*, *Senecio vulgaris*) have indeterminate growth and continue to form new shoots as long as conditions are favourable. Each individual shoot may be monocarpic but because all shoots do not flower synchronously the genet as a whole may be considered polycarpic or iteroparous. Indeed Kirkendall & Stenseth (1984) have questioned whether any annual plant is semelparous, on the somewhat restrictive criterion that organisms with true semelparity cluster their progeny in one, condensed reproductive phase. Most annual plants by their argument are uniseasonally iteroparous or polycarpic; this is due either to a succession of monocarpic shoots (for example, *Senecio vulgaris*) or to the persistent reproductive activity of polycarpic shoots (for example, *Cakile maritima*). Meusel (1955) had earlier made a distinction between annual plants that produce their shoots in one or more growth cycles, and there is undoubtedly a wide variety of

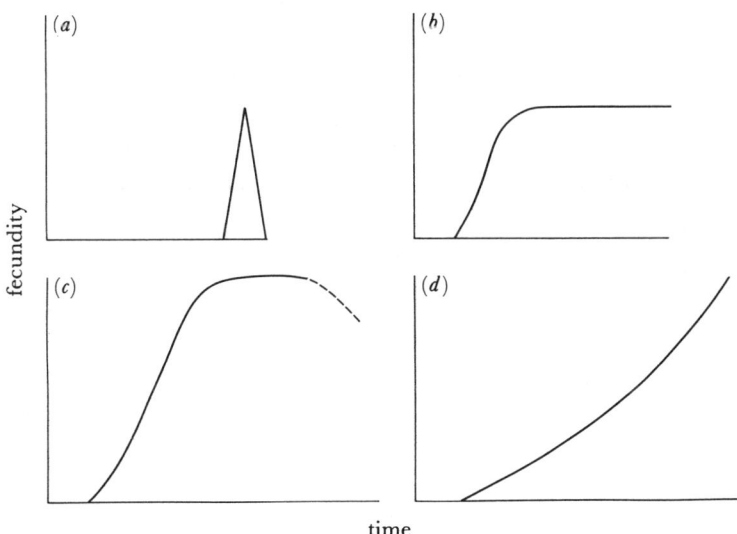

FIGURE 1. Schematic diagrams for the reproductive schedules of semelparous plants (a) and iteroparous plants with a single shoot module (b) or multiple shoot modules which are in turn either aclonal (c) or clonal (d).

reproductive schedules among plants traditionally referred to as annuals. Relatively synchronous reproduction of shoots or meristems of a genet may perhaps be characteristic of some winter annuals (for example, *Vulpia fasciculata, Erophila verna*), but it is perhaps more conspicuous in many facultative biennials (for example, *Daucus carota*) once a critical size threshold for flowering is achieved (Gross 1981). These 'biennials' scarcely (if at all) propagate themselves vegetatively and often have only a single vegetative axis; this restricts the time span of genet reproduction. Maturation of the primary, secondary and tertiary umbels of *Pastinaca sativa*, for example, takes place within three or four weeks (Hendrix 1984). Such relatively synchronous monocarpy of meristems is, it seems invariably, soon followed by senescence and death. There are relatively few examples of semelparous plants among long-lived perennials. These include some bamboos, *Agave* spp., *Strobilanthes* spp. (Janzen 1976), *Frasera* spp. (Inouye & Taylor 1980; Threadgill *et al.* 1981), *Lobelia telekii* (Young 1984), and a few tropical trees (Veillon 1971; Tomlinson & Soderholm 1975; Foster 1977). The large majority of perennial plants are iteroparous, since after some shoots flower each genet retains the capacity for sequential reproduction by the formation of new meristems, themselves eventually capable of reproduction.

(ii) *Iteroparous plants*

Consideration of the reproductive schedules of iteroparous species shows that there are fundamental differences between plants with a single apical meristem and those with many apical meristems (figure 1*b–d*). The latter may in turn be divided into those where the shoot modules remain part of a physiologically integrated whole (aclonal) and those with shoot modules that may become physiologically independent (clonal). The first category is typified by the coconut palm (*Cocos nucifera*) which has a single apical meristem and trunk. No visible trunk is formed until the palm is several years old and the apical meristem has attained its full diameter. At this point the diameter of the stem remains relatively constant and the trunk is gradually built up by the accumulation of modular units. An adult palm consists of 25–35 leaves; one leaf is usually shed as a new leaf unfurls, thus maintaining a constant leaf area and number. Flowering commences at 6–12 years of age and a typical plantation palm has 8–10 leaves from whose axils fruit bunches have been harvested, 10–14 leaves supporting fruit bunches in various stages of development and 10–12 opened leaves with axillary spadices in different stages of growth (Purseglove 1972). The seed production of a palm may therefore remain relatively constant over a large number of years (figure 2), since although the total mass of the tree is increasing the number of leaves remains relatively constant. *Astrocaryum mexicanum* similarly shows a relatively constant seed production for a given individual once it reaches reproductive maturity although there may be large year-to-year variation (Piñero & Sarukhán 1982). There appeared to be no decline in fecundity with age in natural populations of this understory palm, which if knocked flat by falling branches or trees has a remarkable capacity to recover by turning erect at the stem apex and continuing its growth (Sarukhán *et al.* 1985). Coconut trees in commercial plantations do, however, show a decline in reproductive capacity after about 60 years, although this may vary greatly between individuals and between various selected cultivars (Child 1964).

The fecundity of a plant with numerous apical meristems may generally be expected to increase over a much longer period of time after the onset of reproduction, as the number of shoot modules continues to increase. In those (aclonal) plants where modules remain part of

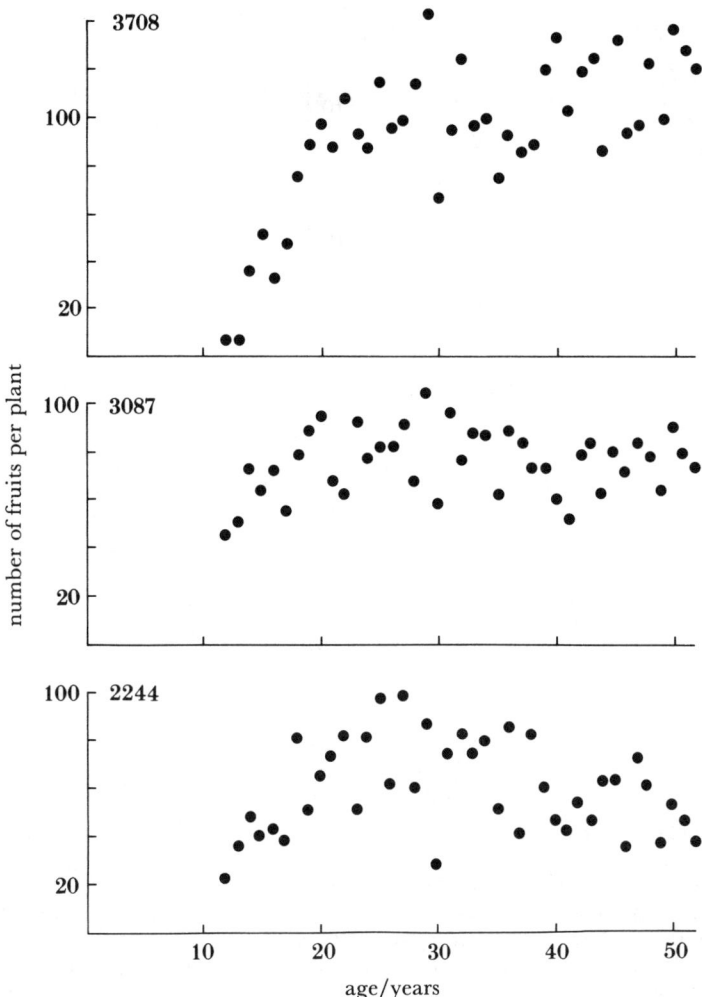

FIGURE 2. Fecundity as a function of age in *Cocos nucifera* (coconut): the reproductive output of three separate trees (cv. West Coast Tall) which were planted in 1917 and began fruiting in 1928. The total numbers of fruits during the period of observation are indicated for each tree; note year-to-year variation. (From data kindly supplied by E. V. Nelliat, Central Plantation Crops Research Institute, Kerala, India.)

a physiologically integrated whole (at least for some resources such as water, if not for others such as carbon (Watson & Casper 1984)), a maximum level or plateau of fecundity may be expected, followed by a decline as a result of disease, damage or senescence (figure 1c). Where shoot modules of a (clonal) genet achieve physiological independence, typically by making their own roots, genet fecundity may be expected to match the growth of the modular population and increase indefinitely, even exponentially (figure 1d). Under optimal conditions for modular expansion there is no reason to expect fecundity to decline. But the nature of gene flow between individuals may, however, change as the clone expands and this may affect fecundity, especially if genets are self-incompatible (Handel 1985).

Increasing fecundity with increasing plant size has been recorded frequently by plant demographers in recent years. Two general patterns are apparent, as Kohyama (1982) indicated for trees. The first shows an increase in reproductive output with size or age, followed by a plateau at maturity and perhaps a later decline. So far this pattern is only known for

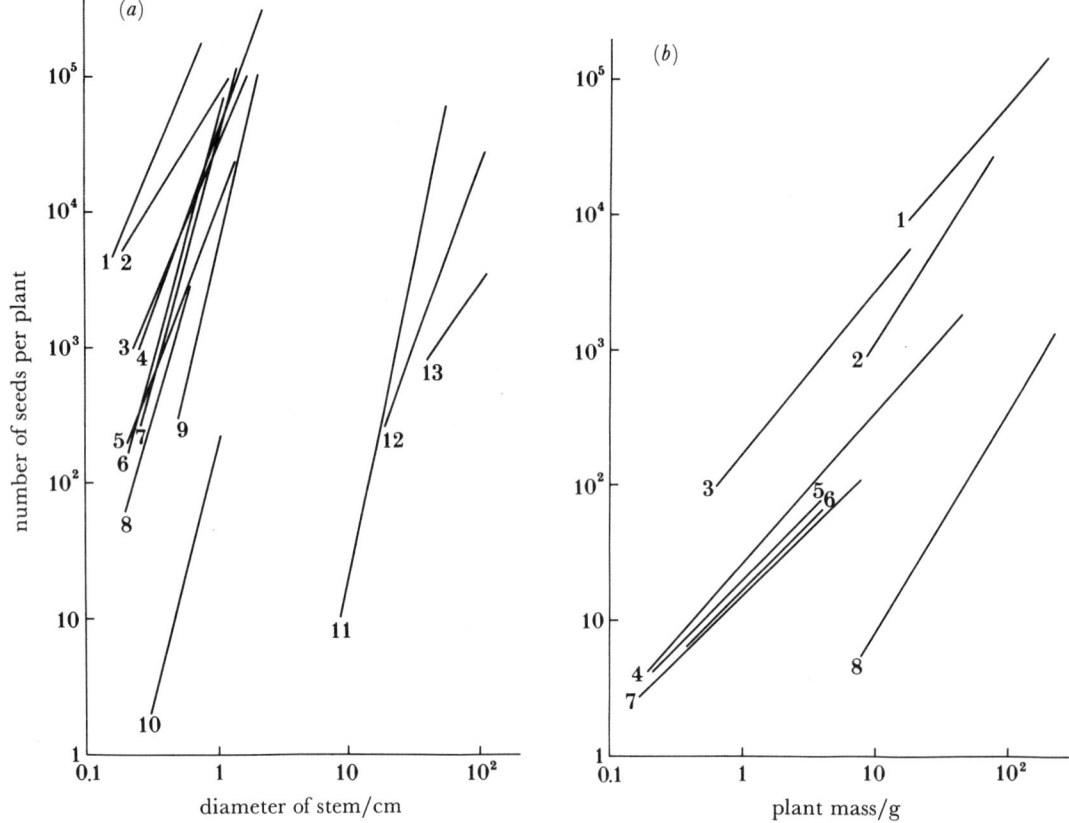

FIGURE 3. Fecundity as a function of size in seed plants. (a) Size expressed as basal stem diameter for herbs (1–9) and a shrub (10) or diameter at breast height for trees (11–13). (1) *Erigeron canadensis* (Hayashi & Numata 1968); (2) *E. annuus* and *E. strigosus* (Hayashi 1984a); (3) *Oenothera parviflora* (Hayashi & Numata 1968); (4) *Chenopodium album* (Fukuda & Hayashi 1982); (5) *Polygonum persicaria* (Hayashi 1984a); (6) *Chenopodium album* (Hayashi & Numata 1984); (7) *Artemisia princeps* (Hayashi 1984a); (8) *Daucus carota* (Holt 1972); (9) *Lactuca virosa* (Boorman & Fuller 1984); (10) *Symphoricarpos occidentalis* (Pelton 1953); (11) *Bursera simaruba* (Hubbell 1980); (12) *Quercus crispula* (Kanazawa 1982); (13) *Pentaclethra macrolobata* (Hartshorn 1975).

(b) Size expressed as above-ground plant mass, for perennial (1–3) and annual (4–8) herbs. (1) *Solidago sempervirens* (Cartica & Quinn 1982); (2) *Rumex crispus* and *R. obtusifolius* (Weaver & Cavers 1980); (3) *Plantago asiatica* (Kawano & Matsuo 1983; figure 12); (4) *Bromus sterilis* (Firbank et al. 1984); (5) *Vulpia fasciculata* (Watkinson 1982); (6) *Setaria faberi* (Kawano & Miyake 1983); (7) *S. pallide-fusca* (Kawano & Miyake 1983); (8) *Ambrosia trifida* (Abul-Fatih et al. 1979). The lines for *Setaria* spp. and *Plantago asiatica* are samples from a wider range of data given by the authors.

Note that the slopes of the lines in (a) and (b) are a function of the size parameter given in the source literature; but since plant mass ($w$) is typically related to stem diameter ($d$) by the power function $w \propto d^\alpha$, where $\alpha$ is approximately 2.5 for both trees (White 1981) and herbs (Hayashi 1984b), the difference between them is more apparent than real.

(aclonal) trees and for the shrub *Staavia dodii* (Moll & Gubb 1981). There is reasonably sound evidence for it in *Pinus ponderosa*, *Abies concolor* (Fowells & Schubert 1956) and *Abies veitchii* (Kohyama 1982), though some other reports, such as those for *Quercus* spp. (Downs & McQuilkin 1944; see Silvertown 1982), are based on inadequate evidence.

Far more general is the second pattern of reproductive output as a function of size or age: a sustained exponential increase, truncated only by death of the plant (figure 3). Most of the evidence for this comes not from clonal plants, where one might expect best to find it, but from aclonal plants. These examples (which include several semelparous plants) provide botanical counterparts of the size- and age-related reproductive output of the gorgonian coral illustrated by Harper & Bell (1979). The data on which figure 3 is based sometimes have a high variance (as for *Bursera simarouba*), but the general trend in all cases is an exponential increase with size. The exponential relationships recorded for trees (figure 3a: 11–13) are not likely to be maintained indefinitely, if they live long enough to become senescent through disease or disintegration. But it is questionable how often trees ever become senescent through physiological decline, before they are simply destroyed by natural forces such as windthrow (Harper 1977; Ogden 1985). It is clear, however, that the largest (and possibly, sometimes, chronologically the oldest) individuals in a population contribute disproportionately to reproductive output and have a preponderant influence on the recruitment of the young, as Harper (1977) suggested. The genetic basis of such differential reproduction remains almost unknown, though there is recent evidence for it in *Pinus ponderosa* (Linhart *et al.* 1979). But it seems that long-lived modular organisms whose reproductive output not only remains undiminished, but *increases* with age, may have much more profound and lasting influence on the genetic structure of local populations than long-lived unitary organisms.

The fruit production of *Opuntia* species on the Galapagos Islands also increases exponentially with plant size, since fecundity is determined by the number of flowers per pad (flattened shoot module); reproduction begins when plants have accumulated about 20 pads (Racine & Downhower 1974). This particular example is an instructive illustration of the real morphological basis of the fecundity patterns for various species illustrated in figure 3: plant size is a derived parameter which hides this basis, the number of modules of which the plant is constructed. This may be further exemplified by the data for *Lactuca virosa*: the exponential relationship between fecundity and stem basal diameter (figure 3a) can be expressed also as an exponential relationship between fecundity and the number of nodes (or metamers), since node number and basal diameter are almost isometrically related (Boorman & Fuller 1984). Several other investigators have tried to relate fecundity to morphological parameters directly: for example, 'degree of branchiness' in *Xanthium strumarium* (Lechowicz 1984); branch number in *Salicornia* (Jefferies *et al.* 1983) and *Staavia dodii* (Moll & Gubb 1981); node number in *Ceratiola ericoides* (Johnson 1982).

The preceding discussion shows that most iteroparous plants demonstrate an increase in fecundity with size and age. This contrasts strongly with the reproductive schedules of iteroparous unitary organisms as illustrated, for example, by *Drosophila* and man where different design constraints are involved. For these species there is a rapid rise in fecundity to a maximum early in life once reproduction has been initiated followed by a slow decline (Charlesworth 1980). This pattern is not exhibited by plants. Other unitary organisms such as many small birds and mammals that maintain a constant size following reproductive maturity show a relatively constant rate of reproduction from year to year. This pattern of allocation is similar to that in plants with a single apical meristem. It is perhaps the cold-blooded vertebrates that continue to grow in size throughout their reproductive life and show a positive correlation between fecundity and age that are most similar to the large majority of plants.

### (c) *Meristem allocation in clonal plants*

Although the life cycles and reproductive systems of clonal plants are becoming well understood (see, for example, Kawano 1985) their reproductive schedules remain almost unknown. It is nevertheless possible to examine how modular construction in clonal plants

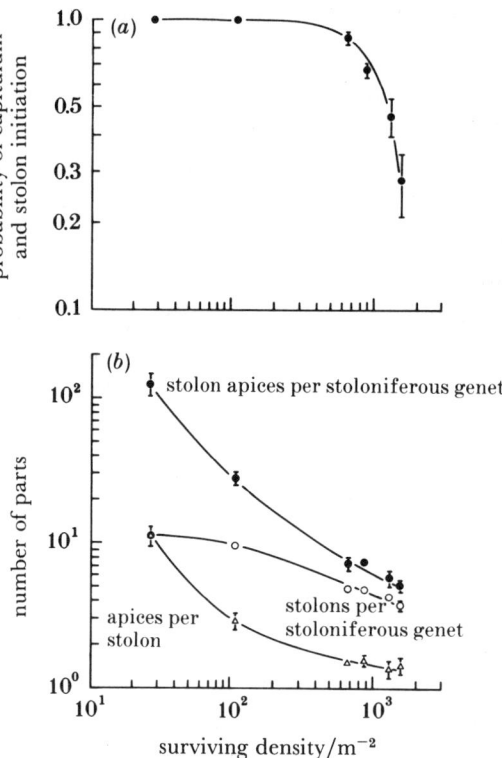

FIGURE 4. (a) Probability of a genet of *Hieracium pilosella* forming a capitulum and stolons in relation to surviving genet density. (b) The number of apices and stolons per genet and the number of apices per stolon in relation to surviving genet density. (From Bishop & Davy 1985.)

imposes constraints on reproduction, through the allocation of meristems to either reproduction or growth. Three recent studies on *Eichhornia crassipes* (Watson 1984), *Hieracium pilosella* (Bishop & Davy 1985) and *Potentilla anserina* (Eriksson 1985) demonstrate the power of this approach. For example, *Hieracium pilosella* is a rosette-forming stoloniferous herb with a close physiological coupling between the evocation of the terminal inflorescence bud and the development of one or more axillary buds into stolons. Only after floral evocation will one or more of the axillary buds develop into stolons and even if the capitulum then aborts or is grazed off stolon development continues. Despite the close coupling of flower and stolon production there is nevertheless considerable scope for plasticity in the allocation of meristems to reproduction and clonal growth since flowers may abort, and the number of stolon apices depends on the number of stolons initiated and the number of stolon branches. How then does intraspecific competition affect apical meristem allocation? Bishop & Davy (1985) showed that the number of reproductive and stolon apices per genet declined with density but that the form of the plastic response was rather different (figure 4). Moreover, the number of apices allocated to reproduction on a per unit area basis rose to a peak and then declined as the density of genets increased, a finding that contradicts the linear models of Williams (1975) and Abrahamson (1980) that predict a shift from growth to reproduction with increasing population density. Neither does this finding support the model of Armstrong (1982, 1984) which predicts a constant reproductive effort irrespective of conditions. Certainly the geometry of growth strongly influences the allocation of resources to growth and reproduction in *Hieracium pilosella* but so do the conditions in which the plants find themselves.

Moreover, the growth form of *H. pilosella* has important implications for its demography, as shown by Bishop & Davy (1984) in a study on a nutrient-poor grassland. There the exclusion of rabbits from the sward had little impact on the rosette density of *H. pilosella* compared with grazed plots but it significantly reduced the rate of turnover of rosettes resulting from clonal growth. The death of rosettes was related to rabbit grazing in a rather complex way since only 11% of losses could be attributed directly to rabbit activity. In fact the single most important cause of death in both the grazed and ungrazed areas was senescence following flower initiation, since *H. pilosella* rosettes are semelparous. The probability of flower initiation was, however, higher in the grazed areas and as a consequence mortality was greater too. The higher mortality in the grazed areas was thus inextricably linked to the higher rates of flower initiation as was the recruitment of rosettes, since clonal growth is also coupled to flower initiation. This example beautifully illustrates the complex constraints that growth form may impose not only on the reproductive schedules of modular organisms but also on their population dynamics.

### 4. Senescence and survival

There are no qualitative differences between the survivorship curves of genets in populations of semelparous plants and those of organisms with unitary construction (see, for example, Watkinson 1981). However, the modular construction of plants calls for a more detailed demography of the constituent parts of each genet: growth by accretion of modular units may be offset by their death and loss. There is no equivalent of regular shedding of organs in unitary organisms. What influence then has the senescence and loss of organs such as leaves and shoots on the survival of genets?

#### (a) The senescence of leaves and semelparous plants

The pattern of mortality has now been examined both at the level of the modular unit and the genet for a large number of plant species. Particular attention has been focused on the survival of leaves. Leaves often show sequential or asynchronous senescence: each leaf has only a limited life span so that as the shoot continues to grow the older leaves senesce and die progressively. This pattern is well illustrated by the survivorship schedules of overlapping leaf cohorts in evergreen trees such as *Pseudotsuga menziesii* (Mitchell 1974) and herbaceous plants such as *Ammophila arenaria* (Huiskes & Harper 1979). In contrast there is a relatively synchronous annual senescence of leaves at the onset of winter in some deciduous forest trees (Kikuzawa 1983) and many hemicryptophyte herbs (for example, *Mercurialis perennis*, *Urtica dioica*). Whatever the causes of leaf senescence and whether these are controlled by factors internal or external to the plant it is clear that it is a tightly controlled process at the modular level and that the sequence of events is usually highly ordered until the terminal stages are underway (Sexton & Woolhouse 1984).

In perennial, iteroparous plants with multiple shoot modules, whether aclonal (for example, trees) or clonal, leaf senescence and abscission are not directly associated with genet senescence. Indeed, by permitting the recovery of nutrients from leaves that have become shaded and suppressed by prolonged modular growth of the shoot system or damaged by herbivores and pathogens, leaf senescence may have benefits for genet survival: photosynthetically inefficient organs are abscised as new ones are formed (Leopold 1961). This contrasts with aclonal plants which have one or few monocarpic shoot modules where the terminal stages of leaf senescence are often coincident with death of the whole genet (characteristically annuals and 'biennials').

Many studies of senescence in plants have been conducted with such semelparous plants where senescence is typically associated with flower and fruit formation. The association of flowering and seed-filling with senescence in semelparous plants has led to the suggestion that death results from the mobilization of nutrients by developing seeds and exhaustion of the rest of the plant as resources are diverted to maximize the production of seed. This view of senescence is, however, too simple and the process of senescence in semelparous plants remains poorly understood (Sexton & Woolhouse 1984). Even among semelparous plants most attention has been focused on annuals, biennials and facultative biennials which are by far the most common forms of plant that demonstrate synchronous reproduction. No attention has been given to the senescence of semelparous plants that are long-lived perennials. These include those with one or a few shoot modules (for example, most *Agave* spp.), aclonal plants with numerous shoot modules (for example, *Tachigalia versicolor*) and clonal plants (for example, some bamboos, *Strobilanthes* spp.). Rather than a physiological interpretation of senescence, predator satiation has sometimes been advanced (Janzen 1976) as the reason for the evolution of semelparity in long-lived, clonal perennials; whatever the reason it would appear that it is a life history restricted to plants as we are not aware of any animals that demonstrate it.

### (b) Shoot dynamics and the longevity of aclonal plants

Whereas genet senescence and death are coincident with shoot module death in semelparous plants, there is no evident relationship between them in iteroparous plants. Among aclonal plants individual shoot modules may be monocarpic (for example, species of *Aesculus*, *Magnolia*, *Rhododendron* and *Rhus*), but within the genet they are markedly asynchronous in reproductive activity, with well defined annual flushes of flowering of some but not all shoots. Formation of new shoots occurs by the continued modular growth of the genet for an indefinite and sometimes prolonged period of hundreds of years. Monocarpic shoots may be incorporated into the permanent framework of the plant (for example, *Rhododendron*) or abscised (for example, *Pinus*). Here shoot senescence and abscission enhances the structural properties of the crown by allowing new shoots to be accommodated: this is especially true in trees where crown architecture may change considerably during plant ontogeny (for example, Borchert & Honda 1984).

Shoot death and abscission appear to have no detrimental effect on the survival probability of the genet in iteroparous aclonal species (Millington & Chaney 1973). Rather death of the genet in adult plants seems to be associated with the accumulation of dead material and the problems of being large. Many trees and bushes exhibit considerable longevity. For example, *Pinus aristata* may attain ages of approximately 4900 years while *Quercus robur* may reach an age of 1200 years (Harper & White 1974). There is nevertheless considerable variation between species. Among trees conifers generally appear to attain greater ages than angiosperms while among bushes some species may live only a few years (for example, *Daphne mezereum*) and others over 200 years (for example, *Salix arctica*).

### (c) Shoot dynamics and the longevity of clonal plants

Even more spectacular are the age of clonal plants whether they are trees, shrubs or herbs. Estimates for the longevity of clonal plants (see Cook 1983) include *Larrea tridentata* (11000+ years), *Populus tremuloides* (10000+ years), *Convallaria majalis* (670+ years) and *Pteridium aquilinum* (1400 years). The causes of death in old genets are many and various. Among extrinsic factors fire, disease and competition may all cause the death of plants but a noticeable feature of clonal plants where the shoot modules lose physiological contact is that the death risk of the genet decreases with increasing clonal proliferation (Cook 1983) as a result, for example, of a reduction in the risk of systemic infection by pathogens.

The life span of the genet reflects the birth and death rates of its modules. Hence if the birth rate of modules continues to exceed the death rate the genet is potentially immortal. Our knowledge of the shoot module dynamics of clonal perennials is now considerable (Harper 1977; Silvertown 1982) but remains piecemeal: monocotyledonous herbs have received particular attention (see, for example, Noble *et al.* 1979; Callaghan 1984). For example, Noble *et al.* (1979) found that the shoots of *Carex arenaria*, which remain connected by rhizomes, show an increase in the probability of death with time. The life expectancy of a shoot varies depending on the time of recruitment into the population and the phase of sand dune development in which it is found, but most shoots are typically biennial. The rhizome segments associated with the shoots are, however, much longer lived, resulting in age structures of above and below ground parts that are quite different. While older parts of the rhizome eventually die, resulting in fragmentation of the genet, the continuous iteration of new shoot modules means that the genet may be extremely long-lived and potentially immortal. This conjunction of short life-span of shoot modules and long-life or even potential immortality of the genet is a noticeable feature of clonal plants. The life expectancy of individual tillers of *Eriophorum vaginatum* is about seven years under favourable conditions (Fetcher & Shaver 1983) but the estimated ages of mature tussocks range from 122 to 187 years (Mark *et al.* 1985). For *Ranunculus repens* most shoot modules have a half-life of only one year while the half-life of genets has been estimated at approximately eight-and-a-half years (Soane & Watkinson 1979). Much more dramatic, however, are the individual, non-fragmented clumps of creosote bush (*Larrea tridentata*) of seedling origin which may achieve ages of 60 years (and exceptionally up to 90 years) compared with the clones into which they eventually fragment, which may live for several thousand years (Vasek 1980).

We have established that the 'partial senescence' (Turner 1950) or 'asynchronous senescence' (Palumbi & Jackson 1983) of modules has no necessary effect on the senescence or death risk of the genet. This is a striking feature of modular organisms, not confined to plants (Palumbi & Jackson 1983; Potts 1984; Hughes & Jackson 1985) and calls into question Hamilton's (1966) argument that 'senescence is an inevitable consequence of the working of natural selection'. Hamilton asserted that the concept of a non-senescing organism demanded a mortality risk that does not change with age and a fertility schedule which increases exponentially as the organism grows older. Both of these criteria may be met by modular organisms, where senescence is *localized* within the genet and does not necessarily involve the entire organism. There is undoubtedly evidence that the probability of genet death is independent of age among established plants for some species (for example, Antonovics 1972) and we have already shown that fecundity may be expected to increase with both age and size for many plants.

### (d) Senescence in iteroparous plants

What role does senescence play in the death of iteroparous species with modular construction? Plants are usually ignored or mentioned only in passing in general considerations of senescence (Comfort 1979; Charlesworth 1980), while specific discussions of plant senescence consider only the senescence of modules and semelparous plants (see, for example, Sexton & Woolhouse 1984).

Most theories of senescence include either evolutionary or physiological explanations (Bell 1984) and definitions of senescence reflect these two viewpoints. For example, Charlesworth (1980) regards senescence as the tendency for age-specific survival probabilities and age-specific fecundities to decline with increasing age for individuals of sufficiently advanced age. This definition is then qualified by the statement that the senescent decline in fecundity and survival should reflect the decline in performance of many different physiological functions with age. This implies that senescence is an inherent characteristic of the plant and yet the increased probability of death with age in plants is usually connected with extrinsic factors. The increased probability of wind-fall and lightning strikes in trees that results from an increase in size, death from interference in a successional environment, and the increased susceptibility to disease that results from mechanical damage surely cannot be regarded as reflections of senescence.

Certainly there are examples of a decline in reproductive output with age, as in *Staavia dodii* (Moll & Gubb 1981), and a decline in life expectancy with age has been reported for *Astrocaryum mexicanum* (Sarukhán 1980). Declines in the physiological activity of plants with age in general reflect the accumulating burden of respiratory tissue, vascular transport problems associated with ever-increasing distances that water and nutrients need to be moved, and increased susceptibility to pathogens and herbivores. These declines in physiological performance relate predominantly to aclonal plants, however, and are primarily side consequences of being large (Harper 1977). Whether the genets of clonal plants show senescence is not at all clear. There is a continuous turnover of shoot modules and in contrast with aclonal plants there are no tissues as old as the genet. It has been suggested that the decline in vigour of *Ammophila arenaria* in the later stages of sand-dune succession represents senescence (Eldred & Maun 1982) but it can also be interpreted in terms of interference (Watkinson *et al.* 1979). Similarly, examples of senescence among clonally propagated crop plants or forest trees ('senile degeneration') are probably in most cases due to more or less simultaneous epidemic infection, probably by viruses (Bijhouwer 1931; Wangermann 1965). Indeed it has been suggested that 'immunity to mosaic virus is the key to immortality in the potato' (Salaman, quoted by Bijhouwer 1931).

Although there may be a decline in physiological activity of genets, particularly if shoot modules remain physiologically connected, it must be questioned whether apical meristems senesce at all. They may cease to function in shoot extension (a process referred to as parenchymatization) or differentiate in their growth potential (Hallé *et al.* 1978), but they seem to show little evidence of senility. The continued and undiminished potential for vegetative propagation of distinctive cultivars of fruit trees has been cited as evidence to rebut the notion of senility in meristems (Schaffalitzky de Muckadell 1959). In genets with pronounced differences in juvenile and mature growth expression (for example, *Citrus*, *Eucalyptus*, *Hedera*), some meristems may 'age', but other meristems may remain undifferentiated and 'juvenile', even in mature plants. The continued capacity of meristems for rejuvenescence of the genet is especially well recorded in trees, where, somewhat paradoxically perhaps, those at the base of the trunk, the oldest part of the genet, retain the capacity to reiterate new shoots; this may be related to their proximity to the root system (Nozeran *et al.* 1982). Insofar as it retains the capacity for rejuvenescence from apical meristems, a genet of a modular organism does not show senescence, despite the senescence, death and decay of constituent organs. A clone of *Lemna*, for example, does not age. Individual fronds and the meristems they bear age and die, but rejuvenation takes place during clonal growth, as a result of which the average physiological age of the clone remains constant (Wangermann 1965). Advancing age of a parent frond affects bud size but has little or no effect on the lifespan or budding capacity of the offspring (Claus 1972).

Thus in modular organisms senescence of some organs is associated with rejuvenescence and continued genet survival: partial senescence of the genet has replaced the senescence and death of unitary organisms (Turner 1950). Charlesworth (1980) argued that senescence was 'almost certain to evolve whenever there is separation of soma and germ plasm'. Modular organisms, however, show no such separation of soma and germ plasm (Buss 1983) and many remain capable of rejuvenescence indefinitely. Moreover, where fecundity increases with age there should be relatively low rates of senescence with respect to survival since increasing fecundity tends to reduce the selective differential between different age classes. These characteristics are particularly evident in iteroparous plants with numerous shoot modules. For these plants fecundity increases with age and size and the apical meristems retain the capacity for somatic maintenance and expansion by modular growth. Since there is no separation of germ plasm from soma this also allows the potential for the individual genet to evolve in relation to changing conditions. Somatic mutations in the apical meristem may, if they confer higher fitness, be incorporated in all subsequent modular units that are iterated thus allowing selection to occur between different modules of the same genet (Whitham *et al.* 1984). It is still quite unclear how important somatic mutation is in the evolution of modular organisms. In this volume Hardwick argues that the organization of meristems may be such as to weed out mutant cell lines. Most observed somatic mutations in higher plants appear to be cytoplasmic or sometimes due to virus infections. Great somaclonal variation is displayed in plants regenerated from potato meristems but in normal cultivation potato varieties seem to show high stability. If somatic mutation of the genome is a frequent feature in modular organisms, natural selection is to be expected *within* genets and between modules. We still lack the evidence from which to argue this case.

### References

Abrahamson, W. G. 1980 Demography and vegetative reproduction. In *Demography and evolution in plant populations* (ed. O. T. Solbrig), pp. 89–106. Oxford: Blackwell Scientific Publications.

Abul-Fatih, H. A., Bazzaz, F. A. & Hunt, R. 1979 The biology of *Ambrosia trifida* L. III. Growth and biomass allocation. *New Phytol.* **83**, 829–838.

Aikman, D. P. & Watkinson, A. R. 1980 A model of growth and self-thinning in even-aged monocultures of plants. *Ann. Bot.* **45**, 419–427.

Antonovics, J. 1972 Population dynamics of the grass *Anthoxanthum odoratum* on a zinc mine. *J. Ecol.* **60**, 351–365.

Armstrong, R. A. 1982 A quantitative theory of reproductive effort in rhizomatous perennial plants. *Ecology* **63**, 679–686.

Armstrong, R. A. 1984 On the quantitative theory of reproductive effort in clonal plants: refinements of theory, with evidence from goldenrods and mayapples. *Oecologia* **63**, 410–417.

Bazzaz, F. A., Carlson, R. W. & Harper, J. L. 1979 Contribution to reproductive effort by photosynthesis of flowers and fruits. *Nature, Lond.* **279**, 554–555.

Bazzaz, F. A. & Harper, J. L. 1977 Demographic analysis of the growth of *Linum usitatissimum*. *New Phytol.* **78**, 193–208.

Bell, G. 1984 Evolutionary and nonevolutionary theories of senescence. *Am. Nat.* **124**, 600–603.

Bijhouwer, A. P. C. 1931 Old and new standpoints on senile degeneration. *J. Pomol. Hort. Sci.* **9**, 122–144.

Bishop, G. F. & Davy, A. J. 1984 Significance of rabbits for the population regulation of *Hieracium pilosella* in Breckland. *J. Ecol.* **72**, 273–284.

Bishop, G. F. & Davy, A. J. 1985 Density and the commitment of apical meristems to clonal growth and reproduction in *Hieracium pilosella*. *Oecologia* **66**, 417–422.

Boorman, L. A. & Fuller, R. M. 1984 The comparative ecology of two sand dune biennials: *Lactuca virosa* L. and *Cynoglossum officinale* L. *New Phytol.* **69**, 609–629.

Borchert, R. & Honda, H. 1984 Control of development in the bifurcating branch system of *Tabebuia rosea*: a computer simulation. *Bot. Gaz.* **145**, 184–195.

Buss, L. W. 1983 Evolution, development, and the units of selection. *Proc. natn. Acad. Sci. U.S.A.* **80**, 1387–1391.

Callaghan, T. V. 1984 Growth and translocation in a clonal southern hemisphere sedge, *Uncinia meridensis*. *J. Ecol.* **72**, 529–546.

Cartica, R. J. & Quinn, J. A. 1982 Resource allocation and fecundity of populations of *Solidago sempervirens* along a coastal dune gradient. *Bull. Torrey Bot. Club* **109**, 299–305.

Caswell, H. & Werner, P. A. 1978 Transient behaviour and life history analysis of teasel (*Dipsacus sylvestris* Huds.). *Ecology* **59**, 53–66.

Charlesworth, B. 1980 *Evolution in age-structured populations*. Cambridge University Press.

Child, R. 1964 *Coconuts*. London: Longmans.

Claus, W. D. 1972 Lifespan and budding potential of *Lemna* as a function of age of the parent – a genealogic study. *New Phytol.* **71**, 1081–1095.

Cole, L. C. 1954 The population consequences of life history phenomena. *Q. Rev. Biol.* **29**, 103–137.

Comfort, A. 1979 *The biology of senescence*, 3rd edn. Edinburgh: Churchill Livingstone.

Cook, R. E. 1983 Clonal plant populations. *Am. Sci.* **71**, 244–253.

Crawley, M. J. 1983 *Herbivory: the dynamics of animal–plant interactions*. Oxford: Blackwell Scientific Publications.

Dirzo, R. 1984 Herbivory: a phytocentric viewpoint. In *Perspectives on plant population ecology* (ed. R. Dirzo & J. Sarukhán), pp. 141–165. Sunderland, Massachusetts: Sinauer Associates.

Downs, A. A. & McQuilkin, W. E. 1944 Seed production of Southern Appalachian oaks. *J. For.* **42**, 913–920.

Eldred, R. A. & Maun, M. A. 1982 A multivariate approach to the problem of decline in vigour of *Ammophila*. *Can. J. Bot.* **60**, 1371–1380.

Eriksson, O. 1985 Reproduction and clonal growth in *Potentilla anserina* L. (Rosaceae): the relation between growth form and dry weight allocation. *Oecologia* **66**, 378–380.

Fetcher, N. & Shaver, G. R. 1983 Life histories of tillers of *Eriophorum vaginatum* in relation to tundra disturbance. *J. Ecol.* **71**, 131–147.

Firbank, L. G., Manlove, R. J., Mortimer, A. M. & Putwain, P. D. 1984 The management of grass weeds in cereals, a population biology approach. *Proc. 7th int. Symp. Weed Biol., Ecol. System.*, pp. 375–384. Paris: E.W.R.S.

Foster, R. B. 1977 *Tachigalia versicolor* is a suicidal neotropical tree. *Nature, Lond.* **268**, 624–626.

Fowells, H. A. & Schubert, G. H. 1956 Seed crops of forest trees in the pine region of California. *U.S. Dept. Agric. Tech. Bull.* **1150**, 48 pages.

Franco, M. 1985 A modular approach to tree production. In *Studies on plant demography: a Festschrift for John L. Harper* (ed. J. White), pp. 257–271. London: Academic Press.

Fukuda, H. & Hayashi, I. 1982 Ecology of dominant plant species of early stages in secondary succession: on *Chenopodium album* L. *Jap. J. Ecol.* **32**, 517–526.

Gross, K. L. 1981 Prediction of fate from rosette size in four 'biennial' plant species: *Verbascum thapsus*, *Oenothera biennis*, *Daucus carota* and *Tragopogon dubius*. *Oecologia* **48**, 209–213.

Hallé, F., Oldeman, R. A. A. & Tomlinson, P. B. 1978 *Tropical trees and forests: an architectural analysis*. Berlin: Springer-Verlag.

Hamilton, W. D. 1966 The moulding of senescence by natural selection. *J. theoret. Biol.* **12**, 12–45.

Handel, S. N. 1985 The intrusion of clonal growth systems on plant breeding systems. *Am. Nat.* **125**, 367–384.

Harper, J. L. 1977 *Population biology of plants*. London: Academic Press.

Harper, J. L. 1981 The concept of population in modular organisms. In *Theoretical ecology: principles and applications*, 2nd edn, (ed. R. M. May), pp. 53–77. Oxford: Blackwell Scientific Publications.

Harper, J. L. & Bell, A. D. 1979 The population dynamics of growth form in organisms with modular construction. In *Population dynamics* (ed. R. M. Anderson, B. D. Turner & L. R. Taylor), pp. 29–52. Oxford: Blackwell Scientific Publications.

Harper, J. L. & White, J. 1974 The demography of plants. *A. Rev. Ecol. Syst.* **5**, 419–463.

Hartshorn, G. S. 1975 A matrix model of tree population dynamics. In *Tropical ecological systems: trends in terrestrial and aquatic research* (ed. F. B. Golley & E. Medina), pp. 41–51. New York: Springer-Verlag.

Hayashi, I. 1984a Secondary succession of herbaceous communities in Japan: seed production of successional dominants. *Jap. J. Ecol.* **34**, 375–382.

Hayashi, I. 1984b Secondary succession of herbaceous communities in Japan: quantitative features of the growth form of successional dominants. *Jap. J. Ecol.* **34**, 47–53.

Hayashi, I. & Numata, M. 1968 Ecology of pioneer species of early stages in secondary succession. II. The seed production. *Bot. Mag., Tokyo* **81**, 55–66.

Hendrix, S. D. 1984 Variation in seed weight and its effects on germination in *Pastinaca sativa* L. (Umbelliferae). *Am. J. Bot.* **71**, 795–802.

Holt, B. F. 1972 Effect of arrival time on recruitment, mortality and reproduction in successional plant populations. *Ecology* **53**, 668–673.

Hubbell, S. P. 1980 Seed predation and the coexistence of tree species in tropical forests. *Oikos* **35**, 214–229.

Hughes, R. N. 1983 Evolutionary ecology of colonial reef-organisms, with particular reference to corals. *Biol. J. Linn. Soc.* **20**, 39–58.

Hughes, T. P. 1984 Population dynamics based on individual size rather than age: a general model with a reef coral example. *Am. Nat.* **123**, 778–795.

Hughes, T. P. & Jackson, J. B. C. 1985 Population dynamics and life histories of foliaceous corals. *Ecol. Mon.* **55**, 141–166.

Huiskes, A. H. L. & Harper, J. L. 1979 The demography of leaves and tillers of *Ammophila arenaria* in a dune sere. *Oecol. Plant.* **14**, 435–446.

Inouye, D. W. & Taylor, O. R., Jr 1980 Variation in generation time in *Frasera speciosa* (Gentianaceae), a long-lived perennial monocarp. *Oecologia* **47**, 171–174.

Islam, J. & Crawley, M. J. 1983 Compensation and regrowth in ragwort (*Senecio jacobaea*) attacked by cinnabar moth (*Tyria jacobaeae*). *J. Ecol.* **71**, 829–843.

Janzen, D. H. 1976 Why bamboos wait so long to flower. *Ann. Rev. Ecol. Syst.* **7**, 347–391.

Jefferies, R. L., Jensen, A. & Bazely, D. 1983 The biology of the annual *Salicornia europaea* agg. at the limits of its range in Hudson Bay. *Can. J. Bot.* **61**, 762–773.

Johnson, A. F. 1982 Some demographic characteristics of the Florida rosemary *Ceratiola ericoides* Michx. *Am. Midl. Nat.* **108**, 170–174.

Kanazawa, Y. 1982 Some analyses of the reproduction process of a *Quercus crispula* Blume population in Nikko. 1. A record of acorn dispersal and seedling establishment for several years at three natural stands. *Jap. J. Ecol.* **32**, 325–331.

Kawano, S. 1985 Life history characteristics of temperate woodland plant in Japan. In *Handbook of vegetation science 3. The population structure of vegetation* (ed. J. White), pp. 515–549. Dordrecht: Dr W. Junk.

Kawano, S. & Matsuo, K. 1983 Studies on the life history of the genus *Plantago*. I. Reproductive energy allocation and propagule output in wild populations of a ruderal species, *Plantago asiatica* L., extending over a broad altitudinal gradient. *J. Coll. Lib. Arts Toyama Univ.* (*Nat. Sci.*) **16**, 85–112.

Kawano, S. & Miyake, S. 1983 The productive and reproductive biology of flowering plants. X. Reproductive energy allocation and propagule output of five congeners of the genus *Setaria* (Gramineae). *Oecologia* **57**, 6–13.

Kelly, D. 1985 On strict and facultative biennials. *Oecologia* **67**, 292–294.

Kemperman, J. A. & Barnes, B. V. 1976 Clone size in American aspens. *Can. J. Bot.* **54**, 2603–2607.

Kikuzawa, K. 1983 Leaf survival of woody plants in deciduous broad-leaved forests. 1. Tall trees. *Can. J. Bot.* **61**, 2133–2139.

Kirkendall, L. R. & Stenseth, N. C. 1985 On defining 'breeding once'. *Am. Nat.* **125**, 189–204.

Kirkpatrick, M. 1984 Demographic models based on size, not age, for organisms with indeterminate growth. *Ecology* **65**, 1874–1884.

Kohyama, T. 1982 Studies on the *Abies* population of Mt Shimagare. II. Reproductive and life history traits. *Bot. Mag., Tokyo* **95**, 167–181.

Koyama, H. & Kira, T. 1956 Intraspecific competition among higher plants. VIII. Frequency distribution of individual plant weight as affected by the interaction between plants. *J. Inst. Polytech. Osaka City Univ.* **7**, 73–94.

Lacey, E. M., Real, L., Antonovics, J. & Heckel, D. G. 1983 Variance models in the study of life histories. *Am. Nat.* **122**, 114–131.

Law, R. 1979 Ecological determinants in the evolution of life histories. In *Population dynamics* (ed. R. M. Anderson, B. D. Turner & L. R. Taylor), pp. 81–103. Oxford: Blackwell Scientific Publications.

Law, R. 1983 A model for the dynamics of a plant population containing individuals classified by age and size. *Ecology* **64**, 224–230.

Lechowicz, M. J. 1984 The effects of individual variation in physiological and morphological traits on the reproductive capacity of the common cocklebur, *Xanthium strumarium* L. *Evolution* **38**, 833–844.

León, J. A. 1985 Germination strategies. In *Evolution: essays in honour of John Maynard Smith* (ed. P. J. Greenwood, P. H. Harvey & M. Slatkin), pp. 129–142. Cambridge University Press.

Leopold, A. C. 1961 Senescence in plant development. *Science, Wash.* **134**, 1727–1732.

Linhart, Y. B., Mitton, J. B., Bowman, D. M., Sturgeon, K. B. & Hamrick, J. L. 1979 Genetic aspects of fertility differentials in ponderosa pine. *Genet. Res., Camb.* **33**, 237–242.

Lovett Doust, L. 1981 Population dynamics and local specialization in a clonal perennial (*Ranunculus repens*). I. The dynamics of ramets in contrasting habitats. *J. Ecol.* **69**, 743–755.

Maillette, L. 1982 Structural dynamics of silver birch. II. A matrix model of the bud population. *J. appl. Ecol.* **19**, 219–238.

Mark, A. F., Fetcher, N., Shaver, G. R. & Chapin III, F. S. 1985 Estimated ages of mature tussocks of *Eriophorum vaginatum* along a latitudinal gradient in central Alaska, U.S.A. *Arct. Alp. Res.* **17**, 1–5.

McGraw, J. B. & Antonovics, J. 1983 Experimental ecology of *Dryas octopetala* ecotypes. II. A demographic model of growth, branching and fecundity. *J. Ecol.* **71**, 899–912.

Meijden, E. van der & Waals-Kooi, R. E. van der 1979 The population ecology of *Senecio jacobaea* in a sand dune system. I. Reproductive strategy and the biennial habit. *J. Ecol.* **67**, 131–153.

Meusel, H. 1955 Über Wuchsform und Wuchsdauer mediterraner Einjahrspflanzen. *Wiss. Zeit. Univ. Halle, Math.-Nat.* **4**, 643–649.

Millington, W. F. & Chaney, W. R. 1973 Shedding of shoots and branches. In *Shedding of plant parts* (ed. T. T. Kozlowski), pp. 149–204. New York: Academic Press.

Mitchell, R. G. 1974 Estimation of needle populations on young, open-grown Douglas-fir by regression and life table analysis. *U.S. Dept. Agric. For. Res. Pap.* **PNW-181**, 14 pp.

Moll, E. J. & Gubb, A. A. 1981 Aspects of the ecology of *Staavia dodii* in the South Western Cape of South Africa. In *The biological aspects of rare plant conservation* (ed. H. Synge), pp. 331–342. Chichester: John Wiley & Son.

Noble, J. C., Bell, A. D. & Harper, J. L. 1979 The population biology of plants with clonal growth. I. The morphology and structural demography of *Carex arenaria*. *J. Ecol.* **67**, 983–1008.

Nozeran, R., Ducreux, G. & Rossignol-Bancilhon, L. 1982 Réflexions sur les problèmes de rajeunissement chez les végétaux. *Bull. Soc. Bot. Fr.* **129**, Lettres bot., 107–130.

Ogden, J. 1985 Past, present and future: studies on the population dynamics of some long-lived trees. In *Studies on plant demography: a Festschrift for John L. Harper* (ed. J. White), pp. 3–16. London: Academic Press.

Oinonen, E. 1967 The correlation between the size of Finnish bracken (*Pteridium aquilinum* (L.) Kuhn) clones and certain periods of site history. *Acta Forest. Fenn.* **83 (2)** (51 pages.)

Palumbi, S. R. & Jackson, J. B. C. 1983 Aging in modular organisms: ecology of zooid senescence in *Steginoporella* sp. (Bryozoa; Cheilostomata). *Biol. Bull.* **164**, 267–278.

Pelton, J. 1953 Studies on the life-history of *Symphoricarpos occidentalis* Hook. in Minnesota. *Ecol. Mon.* **23**, 17–39.

Piñero, D. & Sarukhán, J. 1982 Reproductive behaviour and its individual variability in a tropical palm, *Astrocaryum mexicanum*. *J. Ecol.* **70**, 461–472.

Piñero, D., Sarukhán, J. & Alberdi, P. 1982 The costs of reproduction in a tropical palm, *Astrocaryum mexicanum*. *J. Ecol.* **70**, 473–481.

Potts, D. C. 1984 Generation times and the Quaternary evolution of reef-building corals. *Paleobiology* **10**, 48–58.

Purseglove, J. W. 1972 *Tropical crops: monocotyledons*. London: Longman.

Racine, C. H. & Downhower, J. F. 1974 Vegetative and reproductive strategies of *Opuntia* (Cactaceae) in the Galapagos Islands. *Biotropica* **6**, 175–186.

Reinhartz, J. A. 1984 Life history variation of common mullein (*Verbascum thapsus*). I. Latitudinal differences in population dynamics and timing of reproduction. *J. Ecol.* **72**, 897–912.

Sarukhán, J. 1980 Demographic problems in tropical systems. In *Demography and evolution in plant populations* (ed. O. T. Solbrig), pp. 161–188. Oxford: Blackwell Scientific Publications.

Sarukhán, J. & Gadgil, M. 1974 Studies on plant demography: *Ranunculus repens* L., *R. bulbosus* L. and *R. acris* L. III. A mathematical model incorporating multiple modes of reproduction. *J. Ecol.* **62**, 921–936.

Sarukhán, J., Piñero, D. & Martínez-Ramos, M. 1985 Plant demography: a community-level interpretation. In *Studies on plant demography: a Festschrift for John L. Harper* (ed. J. White), pp. 17–31. London: Academic Press.

Schaffalitzky de Muckadell, M. 1959 Investigations on aging of apical meristems in woody plants and its importance in silviculture. *Forstl. Forsgsvaesen Danmark* **25**, 309–455.

Schaffer, W. M. & Schaffer, M. W. 1977 The adaptive significance of variations in reproductive habit in the Agavaceae. In *Evolutionary ecology* (ed. B. Stonehouse & L. Perrins), pp. 261–276. London: Macmillan.

Sexton, R. & Woolhouse, H. W. 1984 Senescence and abscission. In *Advanced plant physiology* (ed. M. B. Wilkins), pp. 469–497. London: Pitman.

Silvertown, J. W. 1982 *Introduction to plant population ecology*. London: Longman.

Soane, I. D. & Watkinson, A. R. 1979 Clonal variation in populations of *Ranunculus repens*. *New Phytol.* **82**, 557–573.

Threadgill, P. F., Baskin, J. M. & Baskin, C. C. 1981 The ecological life cycle of *Frasera caroliniensis*, a long-lived monocarpic perennial. *Am. Midl. Nat.* **105**, 277–289.

Tomlinson, P. B. & Soderholm, P. K. 1975 The flowering and fruiting of *Corypha elata* in south Florida. *Principes* **19**, 83–99.

Tuomi, J., Hakala, T. & Haukioja, E. 1983 Alternative concepts of reproductive effort, cost of reproduction, and selection in life-history evolution. *Am. Zool.* **23**, 25–34.

Turner, C. L. 1950 The reproductive potential of a single clone of *Pelmatohydra oligactis*. *Biol. Bull.* **99**, 285–299.

Vasek, F. C. 1980 Creosote bush: long-lived clones in the Mojave Desert. *Am. J. Bot.* **67**, 246–255.

Veillon, J. M. 1971 Une Apocynacée monocarpique de Nouvelle Calédonie *Cerberiopsis candelabrum* Vieill. *Adansonia* (N.S.) **11**, 625–639.

Wangermann, E. 1965 Longevity and ageing in plants and plant organs. *Encyl. Pl. Physiol.* **15**, 1026–1057.

Watkinson, A. R. 1981 The population ecology of winter annuals. In *The biological aspects of rare plant conservation* (ed. H. Synge), pp. 253–264. Chichester: John Wiley.

Watkinson, A. R. 1982 Factors affecting the density response of *Vulpia fasciculata*. *J. Ecol.* **70**, 149–161.

Watkinson, A. R., Huiskes, A. H. & Noble, J. C. 1979 The demography of sand dune species with contrasting life cycles. In *Ecological processes in coastal environments* (ed. R. L. Jefferies & A. J. Davy), pp. 95–112. Oxford: Blackwell Scientific Publications.

Watson, M. 1984 Developmental constraints: effect on population growth and patterns of resource allocation in a clonal plant. *Am. Nat.* **123**, 411–426.

Watson, M. & Casper, B. 1984 Morphogenetic constraints on pattern of carbon allocation in plants. *Ann. Rev. Ecol. Syst.* **15**, 233–258.

Weaver, S. E. & Cavers, P. B. 1980 Reproductive effort in two perennial weed species in different habitats. *J. appl. Ecol.* **17**, 505–513.

Werner, P. A. & Caswell, H. 1977 Population growth rates and age vs. stage distribution models for teasel (*Dipsacus sylvestris* Huds.). *Ecology* **58**, 1103–1111.

White, J. 1981 The allometric interpretation of the self-thinning rule. *J. theor. Biol.* **89**, 475–500.

White, J. 1984 Plant metamerism. In *Perspectives on plant population ecology* (ed. R. Dirzo & J. Sarukhán), pp. 15–47. Sunderland, Massachusetts: Sinauer Associates.

Whitham, T. G., Williams, A. G. & Robinson, A. M. 1984 The variation principle: individual plants as temporal and spatial mosaics of resistance to rapidly evolving pests. In *A new ecology: novel approaches to interactive systems* (ed. P. W. Price, C. N. Slobodchikoff & W. S. Gaud), pp. 15–51. New York: John Wiley.

Williams, G. C. 1975 *Sex and evolution*. Princeton University Press.

Young, T. P. 1984 The comparative demography of semelparous *Lobelia telekii* and iteroparous *Lobelia keniensis* on Mount Kenya. *J. Ecol.* **72**, 627–650.

# Growth and form in modular animals: ideas on the size and arrangement of zooids

By J. S. Ryland[1] and G. F. Warner[2]

[1] *Department of Zoology, University College of Swansea, Swansea SA2 8PP, U.K.*
[2] *Department of Pure and Applied Zoology, University of Reading, Reading RG6 2AJ, U.K.*

Modular (colonial) invertebrates are mostly aquatic, sessile, active or passive suspension feeders. This paper proposes and discusses some generalizations concerning form that apparently are related to the sessile colonial mode of life. In contrast to the size of related unitary forms, the modules are small, maximizing feeding surface relative to metabolic mass and favouring production of a high energy surplus. Increasing colonial integration in ascidians and hydroids is associated with decreasing module size but in Bryozoa, with the lophophore as index, with some increase in size. The smallest lophophores are found in species with apparently primitive, near-linear branching. Among bryozoans with compact encrusting colonies, however, species with larger lophophores can outcompete abutting neighbours with smaller lophophores. Lophophore size may then be a compromise between energetic advantage and competitive disadvantage. Whereas internal filterers tend to have modules grouped to produce larger exhalant openings, favouring stronger discharge flow, in Bryozoa it appears advantageous to attain the maximum coverage of expanded lophophores. In Cheilostomata, lophophores are generally close packed, except at excurrent chimneys, and zooid size and shape are then directly linked to the dimensions of the lophophore. Bryozoa Cyclostomata, however, have evolved away from close-packed lophophores and quincuncial zooids towards fasciculated arrangements, possibly providing structural excurrent channels in a group that lacks the colonial coordination to maintain non-skeletal chimneys.

Variations in colony form are related to mode of growth, the disposition of modules to maximize filtration, and interactions with environmental factors. Increasing surface area leads to increased drag imposed by water movements. This may place constraints on growth and form, or may be exploited to augment filtration. Passive filterers often produce erect, branching, planar colonies oriented normal to directional currents. Bilaterally symmetrical, dish-shaped colonies with downstream zooids may occur in unidirectional flow. Erect bryozoan colonies more commonly are irregularly tufted or regularly branched in three dimensions, being then adapted to flows that vary in direction or velocity, or both.

## Introduction

The modular invertebrates are those that form colonies of replicated units. The units arise, in the first instance, from a zygote but are thereafter effectively reproduced by non-sexual (mitotic) processes. Moreover, as formed, they remain in physical and physiological communication, thereby imparting to the colony some degree of individuality greater than that represented solely by an aggregation (clone) of otherwise comparable units. Such colonial invertebrates include many hydrozoan and anthozoan coelenterates, calyssozoans, bryozoans, pterobranchs, and some of the tunicates. Following Huxley (1851), the module of an invertebrate colony is a definite entity, the zooid. Sponges generally function ecologically as colonies and arguably are constructed on a modular basis, but the modules comprise the oscula together with their aquiferous canals; unlike zooids, they are labile in form through time.

The useful concept of modular construction was introduced to biology by Harper (1977). Unlike zooids, modules are not exclusive to animals and provide a term equally applicable to an analogous organization in plants. Also, modules can be arbitrarily defined according to need or purpose. In animal colonies the primary modules are the zooids, but zooids themselves are often grouped in replicated patterns, variously 'systems' or 'cormidia' (Beklemishev 1969), which constitute modules of the second order. Further, colonies may replicate entire fronds, or divide into similar daughter colonies or 'cormomeres' (Oka & Usui 1944), these constituting modules of a third order.

Modularity and coloniality represent two viewpoints on the same phenomenon. 'Colony' in the modular sense is perhaps an unfortunate application of the word, for the Latin *colonia* was a settlement at a distance from the parent body, as in later human geographical colonies. It was applied in zoology first to the societies of Hymenoptera and communally nesting birds, and only in the late 19th century was it coined (Nicholson 1872) in what is here termed the modular sense. Haeckel (1866) had by then already appropriated 'cormus', from the Greek for a tree trunk, for a colonial body, but the word seems not to have found favour among English-speaking biologists, and persists mainly in the derivative cormidia. The meaning and applications of the term 'coloniality' have been considered recently by Rosen (1979) in a postscript introduction to a number of specialized contributions on that topic (Larwood & Rosen 1979).

Workers on colonial invertebrates were stimulated greatly by the appearance, first in German and then in English (1969), of Beklemishev's *Principles of comparative anatomy of invertebrates*. In this work Beklemishev identified certain universal trends which he associated with increased integration within colonies. His synthesis led to a reexamination of many of the major colonial taxa by specialists, resulting in some important publications on coloniality in sponges, Hydrozoa, corals, Octocorallia, Bryozoa, etc. (Mackie 1963; Boardman *et al.* 1973; Larwood & Rosen 1979). The increasing use of scuba as a research tool led contemporaneously to the discovery and study of animal communities in which sessile, colonial invertebrates dominated (Jackson 1977). These approaches converged with the introduction of the modularity concept and its development with reference to organization, genetic bases, evolutionary consequences, and community ecology (Chapman & Stebbing 1980; Chapman 1981; Jackson *et al.* 1985).

One of the most frequently remarked, though not invariable, characters of modular invertebrates is the small size of the modules (zooids) compared with their unitary relatives. The difference may be seen in such paired groupings as Scleractinia (corals) and Actiniaria (anemones), Bryozoa and Brachiopoda (lophophorates), and the 'simple' and 'compound' Ascidiacea. The reasons for small size, and the extent to which size varies within colonial taxa, however, have not been well explored. It might be expected that module size would further vary according to the manner of arrangement within colonies or with the degree of integration. Thus a superficial review of coral genera, exemplifying one of the more familiar modular taxa, suggested that massive forms budding by fission (intratentacular budding) often have large corallites. Many arborescent forms have much smaller corallites and, in many, greater integration implied by the ramose pattern and, in *Acropora*, differentiation of apical zooids. We have accordingly compiled some data and reviewed some of the literature dealing with three other modular groups, Ascidiacea, thecate Hydroida and Bryozoa, to see whether the inverse correlation of size and colonial integration occurs elsewhere. In view of a possibly causal relation between integration and module size, and to avoid circular argument, some definition of integration seems necessary.

Beklemishev's trends that indicate increasing levels of integration are (i) the weakening individuality of zooids, (ii) the intensifying individuality of colonies, and (iii) the development of cormidia ('colonies within colonies'). The first two were exemplified by a number of usually parallel developments (Beklemishev 1969, p. 483). The concept of colonial integration in this way was attributed to Herbert Spencer. The source appears to be *First principles* (1862). Spencer perceived integration as the process of aggregating the diffuse into a whole, during either evolution (for example, of arthropods from annelids) or ontogeny. Among 'compound animals... integration is displayed not within the limits of an individual only but by the union of many individuals' (he cited salps and *Pyrosoma*). He commented on unifying features such as the colonial tunic in Botryllinae, the 'common system of nutrition' in Hydrozoa and corals, and noted what we now term somatic polymorphism in Siphonophora and their approximation to a 'single organism' (or 'superorganism' of Mackie (1963)).

The definition of integration in the *Shorter English dictionary* (edn 3, 1973) is: 'the making up of a whole by adding together or combining the separate parts or elements: a making whole or entire'. As applied to colonies, the degree of integration is usually inferred from comparative morphological series (see Boardman & Cheetham 1973; Cook 1979), although physiological criteria are applied when possible (see, for example, Mackie 1963; Ryland 1979). There seems to be no suggestion that change in size *per se* is associated with degree of integration. Indeed, as we show in this paper, the correlations with size are contradictory.

### The physiological implications of module size

The significant biological manifestations of size are mass (or volume), $W$, proportional to the cube of the linear dimensions, and surface area, $A$, which varies as the square of the linear dimensions. Hence,

$$A \propto W^{\frac{2}{3}}.$$

The consequences of this relation are well-known in growth terms, and inevitably were introduced by D'Arcy Thompson (1942) very early in his treatise *On growth and form*. Obviously, if surface is proportional to the two thirds power of volume, the surface:volume ratio of an animal decreases as it grows. In many modular animals (but excepting corals with intratentacular budding and some ascidians) the module does not continue to grow through life. If modules were small to maximize their surface:volume ratio, evolution would favour budding methods which did not first involve increasing the size of the parent module. However, growth of colonies tends to be essentially two-dimensional, avoiding disproportionate increase in mass.

The pattern of growth in individual organisms (including modular forms: see, for example, Kaufmann (1981)) may be approximated by one or more descriptive models, which describe a flexed curve decaying to some asymptotic upper limit, for example, the von Bertalanffy, Gompertz, and logistic (Verhulst–Pearl) equations. The von Bertalanffy growth equation is descriptive and derivable empirically (Gulland 1983); but as formulated (von Bertalanffy 1957), following the idea of Pütter (1920), it had a physiological basis related to the acquisition and utilization of energy, the former related to surface area and the latter to volume. Thus, the instantaneous rate of change in mass is given by the difference between surface-related anabolism and volume-related catabolism,

$$dW/dt = aW^b - kW^c,$$

where $a$ and $k$ are coefficients respectively of anabolism and catabolism, and the exponents $b = 0.67$ and $c = 1$ when growth is isometric.

The modular invertebrates discussed here are sedentary, usually sessile, and either suspension feeders *sensu stricto* or 'passive' suspension feeders, in both cases wholly dependent on a dilute food supply brought to them by water currents (Chapman & Stebbing 1980). Von Bertalanffy's $b$ is therefore related to the surfaces available for particle capture. In passive feeders such as hydroids, antipatharians and gorgonians, the ability to capture particles will be directly proportional to the surface area opposed to the water flow; in true suspension feeders, whether unitary or modular, particle capture depends on the area of ciliated surface used to generate flow. Any ability to use dissolved organic matter (see, for example, Best 1985) would also be dependent on surface area. The exponent $c$ is a function of processes occurring everywhere, and thus more nearly related to mass or volume.

Von Bertalanffy's equation can be shown as a pair of curves plotted against increasing mass (figure 11.19 in Jones 1976). One, representing energy utilization (or cost of living), based on mass to the power 1, would be linear; the other, representing the energy derived from food intake would be a decaying exponential based on $W^{0.67}$. The vertical distance between the two lines, energy surplus, $E_s$, decreases with increasing mass until the two lines intercept at $W_\infty$, the maximum attainable. In unitary organisms $E_s$ has been termed the scope for growth (Warren & Davis 1966), where growth is both somatic and gametic. In the modular context, $E_s$ expresses the energy available for gamete production within the zooid plus that transmitted to the extending, budding or regenerating zones of the colony.

If the exponent $c$ departs from unity, the lower line will be curvilinear. In fish, $b$ seems generally close to the theoretical 0.67 but $c$, from experiments on both mass loss and oxygen consumption, is nearer 0.7–0.8 than 1 (Zeuthen 1953; Jones 1976). Perhaps more obviously relevant are some figures for unitary suspension feeders, from Jørgensen (1975). In three bivalves, *Argopecten irradians*, *Mytilus edulis*, and *Cardium edule*, in which water transport is proportional to $W^b$, $b = 0.52$, 0.60 and 0.58. For oxygen consumption, proportional to $W^c$, $c = 0.68$ (10 °C) or 0.87 (18 °C), 0.75 and 0.77, respectively. The ratio $b/c$ decreases with increasing body size, as does growth efficiency. Sebens (1982a) cites a number of other values of $c$ in the range 0.84–1.05. Another appropriate data set refers to anemones, several of which are clonal though not colonial. Oxygen consumption varies considerably between the expanded state in water and the contracted state in air. A re-plot of data from Shick *et al.* (1979) provides an approximate value of 0.8 for $c$ in *Anthopleura elegantissima*. This may be compared with the results of Sebens (1979), for three anemones, *A. elegantissima*, *A. xanthogrammica* and *M. senile*. For these, tentacle surface area increased as the 0.45, 0.71, and 0.73 power of mass, prey capture as the 0.60, 0.64, and 0.60 power, and metabolic costs derived from starvation as the 0.77, 0.98, and 0.80 power, respectively. Thus for a range of poikilotherms, values have been derived that evaluate $b$ as 0.45–0.73 and $c$ as 0.68–0.98. While different values, of course, have implications for particular energy budgets and life styles, none affects the generalization (for $b < c$) that small organisms generate proportionally the greatest energy surplus. It has, of course, been recognized for some time that among filter feeders small species have the highest clearance rates (per unit dry mass) and that in bryozoan colonies it is zooid size and not colony size that is relevant (Bullivant 1968).

Sebens' (1979) simulations generated important conclusions, particularly that large size, for example, of solitary anemones, was optimal where prey were large and infrequent, and that a clone or colony of small, fixed-size modules was optimal where prey were small and frequent (the assumed pattern for modular coelenterate suspension feeders such as zoanthids and octocorals). Between these extremes, and optimal for intermediate prey size and frequency, are clones or colonies with modules of intermediate unit size (for example, fissile anemones). Presumably (Sebens 1982a), binary fission producing half-sized daughter modules is advantageous only if small prey are being used at that time and capture success does not suffer.

The general applicability of Sebens' (1979, 1982a) deductions seem clearly supported by the ascidian, hydroid and lophophorate data presented here. The unitary examples, by their very growth, are constantly reducing the relative area of their food-capturing surfaces, and therefore their energetic efficiency. Their specific growth rate therefore slows as size increases; or, the attainment of large size requires an allometric increase in the area of the collecting surface. Thus in Stolidobranchia, the ascidian suborder in which unitary form is most highly developed, a pleated branchial sac with four to eight folds per side has evolved, increasing the number of meshes. Concentric or spiral stigmata, which have evolved in both Phlebobranchia (Corellidae) and Stolidobranchia (Molgulidae), would not seem generally to increase the pore:filament ratio but maximize the length of the ciliated interface. Extreme examples are seen in *Corynascidia* in which the meshes have a spider's web appearance (Herdman 1882, plate 25) and in *Bostrychobranchus* in which they bulge as spiralled infundibula (Van Name 1945). (It should, however, be recorded that some solitary phlebobranchs with unmodified branchial sacs are large and some molgulids are small.) Comparably, in brachiopods, the evolution of spirolophes and plectolophes (Rudwick 1970) has facilitated lophophores bearing hundreds of filaments (that is, tentacles).

In contrast to the above examples, the modular Ascidiacea and lophophorates (Bryozoa) maintain small, determinate zooids; their growth strategy, as recognized by Sebens (1982a), is directed towards increasing the colonial surface area, whether encrusting or variously erect, so maximizing the number of modules. Thus volume does not increase to the detriment of relative collecting surface, and allometric adjustment is not necessary.

If it is accepted that having small modules confers energetic advantages on the colony, but that there are probably undetermined constraints on the lower limit, might increasing integration facilitate some depression of that limit? The following three sections examine three taxa for evidence that better colonial integration is accompanied by a decrease in module size.

## Module size and degree of colonial integration in Tunicata Ascidiacea

The ancestor of present-day ascidians may have been solitary or loosely colonial (Millar 1966). Evolutionary pathways have led to the development of both complex unitary forms (for example, Molgulidae) and highly integrated modular genera (for example, Didemnidae among Aplousobranchia, Botryllinae among Stolidobranchia). Some of the usual criteria for degree of integration are not applicable: for example, there is no zooidal polymorphism among ascidians, hence no cormidia, although there may be groupings of identical zooids as 'systems' and in some didemnids an apparently well regulated process of colonial fission (Ryland *et al.* 1984), an important adaptation to maintain high specific rates of growth. Moreover, although zooids are propagated by blastogenesis, the morphology of budding, the tissues involved, and the consequences to the parent zooid vary considerably (Berrill 1950; Millar 1966; Kott 1982).

Although these seemingly diverse methods of budding may have evolved separately, Millar (1966) argues a strong case for regarding the more complex methods of aplousobranch blastogenesis as derived from a primitive kind, in which the buds arise on a stolon incorporating an extension of the ventral blood vessel (*Clavelina, Perophora*). In Aplousobranchia a tissue known as epicardium becomes involved and the site of blastogenesis is moved from the stolon to the body. Diazonid and polyclinid budding is accomplished by strobilation of the abdomen, with the rest of the zooid often degenerating during the process. Didemnids generally have very flat, spreading colonies with minute zooids. Blastogenesis is clearly of a derived type with a pair of buds originating in the pyloric region; the parent zooid divides into thorax and abdomen, with the buds providing another abdomen or thorax as appropriate. The process can be continual and the zooids functional for most of the time.

Colonies occur in only one stolidobranch family, Styelidae, and whatever the ascidian ancestry it is difficult to envisage this as other than separately evolved. The method is termed pallial and involves outgrowth from the body wall incorporating ampullae of the test vessels. At first (Polyzoinae) this provides a method of cloning, for the short vascular stolons often degenerate as the daughter zooids develop (*Eusynstyela*), although they may form a continuous crust as in certain growth forms of *Distomus*. In the Botryllinae, however, the test vessels of developing buds merge with others to produce a truly colonial system.

The ascidian branchial sac consists basically of transverse rows of linear stigmata. In Phlebobranchia and Stolidobranchia inner longitudinal vessels are present, which effectively divide the stigmata into groups. It is impossible to use a standardized index of size of the branchial sac throughout the class, because taxonomic works usually state the number of transverse rows of stigmata for Aplousobranchia, while for the other two suborders they indicate the number of longitudinal vessels. In any case, use of a unidimensional index to describe the surface area of a three-dimensional structure will minimize differences. In figure 1 the number of rows or number of longitudinal vessels (mainly from Berrill (1950) and Millar (1970)) characterizing genera have been plotted according to whether the habit is solitary, colonial, or colonial with systems. The use of a logarithmic ordinate tends visually to conceal quite large differences, and the branchial sacs of unitary species are often an order of magnitude bigger, even in the single dimension.

The arrangement of zooids in systems around a common cloaca, instead of being irregularly disposed on stolons or a spreading base, is the most obvious indicator of the degree of integration. As shown in figure 1, particularly in stolidobranchs, the forms with systems (Botryllinae) have obviously smaller branchial sacs. Botryllines are the most highly integrated of all colonial ascidians, not merely through morphological systems but physiologically through the colonial vascular system. Budding and gonad maturation are in this way synchronized throughout the colony. The size reduction accompanying the evolution of systems is less marked in aplousobranchs unless there is a further breakdown within the category 'with systems'. In figure 1 polyclinids are distinguished from didemnids. It may be argued that the derived method of budding (though perhaps inextricably associated with reduction in size) is an advanced feature;

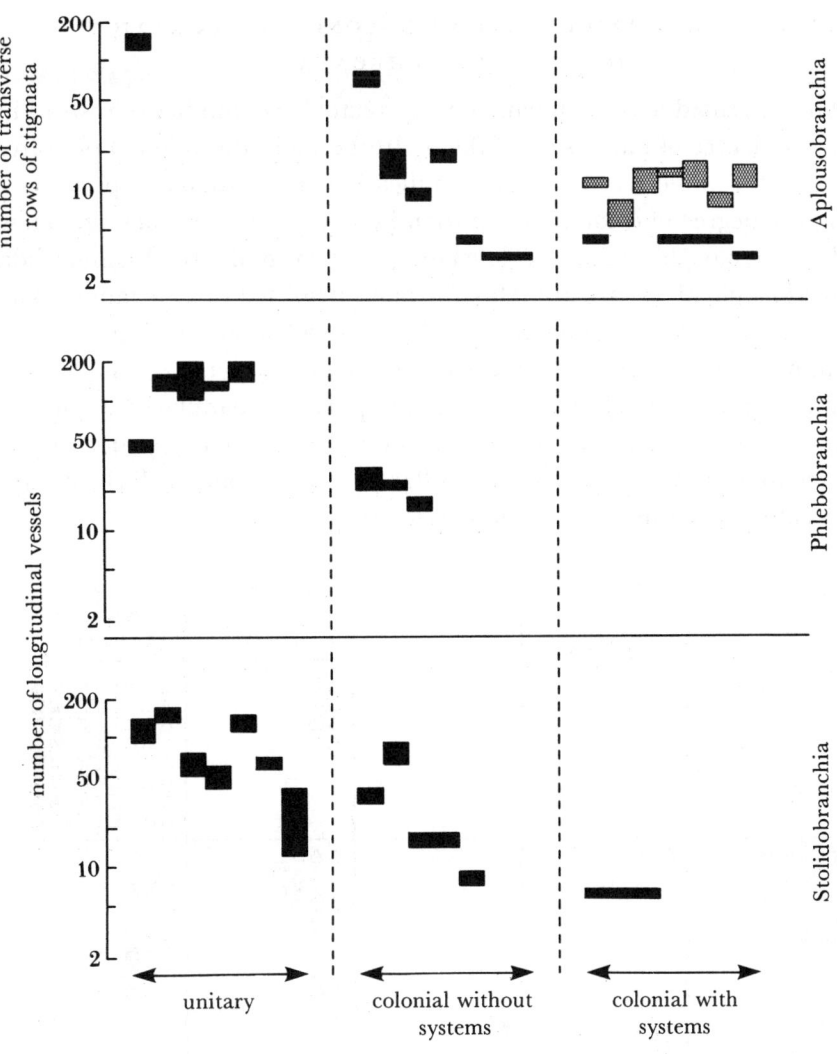

FIGURE 1. Correlation of the size of the branchial sac with habit in Ascidiacea. The index of size in Aplousobranchia is the number of transverse rows of stigmata; in Phlebobranchia and Stolidobranchia it is the number of longitudinal vessels. Each column is based on one taxon, with measurements taken from the literature. In Aplousobranchia with systems, hatched columns represent polyclinid genera, black columns didemnid genera.

and that although the didemnid colony is characteristically a flat, spreading one (in Herbert Spencer's terms less integrated than the compact blobs of polyclinids), it is ecologically advantageous and in fact an advanced feature. Moreover, there is clearly colonial control over the regular fission and movement described in tropical didemnids (Ryland et al. 1984), although the mechanism is unknown. Overall we conclude that the ascidians provide a rather clear example of modular size decreasing as colonial integration increases. An important consequence of reduced zooid size, and possibly the chief advantage, is an accelerated rate of replication and colony growth.

## MODULE SIZE AND DEGREE OF COLONIAL INTEGRATION IN HYDROIDA THECATA

Naumov (1969) presented a table in which he correlated the number of rows of hydranths per branch with the degree of immersion of the hydrothecae in the colonial perisarc. He also indicated the degree of complexity of colonial ramification. This tabulation provides a suitable base for assessing the degree of colonial integration independently of zooid size. However, on grounds of level of polymorphism and the development of cormidia, the Plumulariidae (taken here as including aglaophenines) are arguably the best integrated thecate family. They do not display multiple rows of hydranths along the branches. Instead of Naumov's plotted correlation, therefore, type of hydrotheca was compared with branching pattern. A sample of 50 species in each of three groupings, namely Campanulinidae plus Campanulariidae plus Lafoeidae, Sertulariidae, and Plumulariidae, which together comprise a large proportion of all thecates, was taken from the literature (Naumov 1969; Millard 1975; Gibbons & Ryland 1986). Their hydrothecal and colonial characters are summarized in figure 2.

| colony \ hydrothecae | pedicellate | sessile | under 50% adnate or immersed | over 50% adnate or immersed |
|---|---|---|---|---|
| monopodial growth point with ordered branching: pinnate, whorled, dichotomous or helicoid | | ▼ △ | △□ △□ △□ △□ △□ | △△△△□□□■■ △△△△□□□■■ △△△△□□□■■ △△△△□□□■■ △△ □□□■■ △△ □□□■■ △△ □□□■■ △△ □□□□■ |
| monopodial growth point with single stem or irregularly arranged planar or arborescent branching | △ | △ | △△△□ △△△□ △△△□ △△ □ △△ □ △△ △△ | △△□ △△□ △△□ △△□ △△■ △△■ △△■ |
| sympodial or monopodial with terminal hydranth: irregular branching with hydranths in single row | △○○ ▼○○ ▼○○ ▼⊗○ ▼⊗○ ▼⊗○ ▼⊗ ⊗ | ▼ | ▼ | |
| stolonial | ⊗○○▼ ⊗○○▼ ⊗○○▼ ⊗○○▼ ⊗○○ ⊗○○ ⊗○ ⊗○ | ⊗▼ ⊗▼ | | ○ Campanulinidae and Campanulariidae ⊗ with medusa ▼ Lafoeidae △ Sertulariidae □ Plumulariidae ■ with phylactocarps or corbulae |

FIGURE 2. Correlation of colony form and type of branching with the morphology of the hydrotheca in 150 species of Hydroida Thecata.

This diagrammatic tabulation shows clearly how simple colony forms and branching patterns are correlated both with pedicellate hydrothecae and the presence of a medusa: these are characteristic of campanulinids, campanulariids and lafoeids. Adnate or immersed hydrothecae are correlated with colonies that have monopodial growing points and ordered branching patterns: they are characteristic of sertulariids and plumulariids. However, the latter not only are characterized by nematothecae and cormidia but, as indicated, frequently have special colonial reproductive structures (phylactocarps and corbulae). Taking all criteria together, this suggests an integrational series (Campanulinidae+Campanulariidae+Lafoeidae) < Sertulariidae < Plumulariidae. It is not suggested that the three families united by parentheses form a 'natural' group.

For assessing hydranth size, two indices seemed possible, neither entirely satisfactory. The

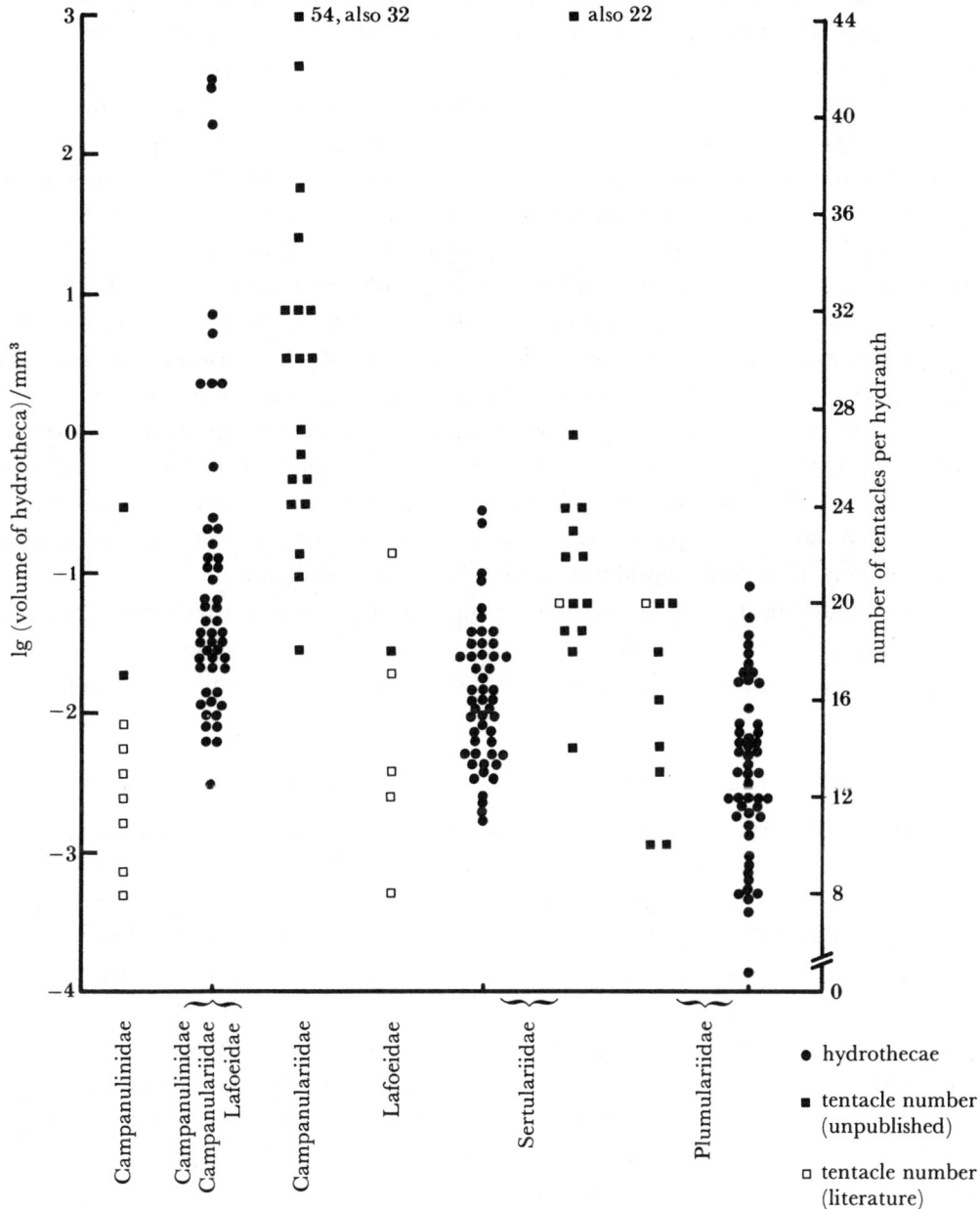

FIGURE 3. Indices of hydranth size in families of Hydroida Thecata. Solid circles (left ordinate) are estimates of hydrothecal volume based on published measurements for 150 species (as in figure 2). Squares (right ordinate) are mean tentacle numbers. Open squares represent numbers taken from the literature; solid squares are counts made by P. F. S. Cornelius or J. S. Ryland. Normally each species is represented by one point, but if two widely separated counts have been obtained for any nominal species, both are included.

first was an estimate, for the same 150 species, of mean hydrothecal volume, taken from published measurements and making appropriate allowances for geometry. One problem is that in certain thecates the hydranth does not fit into the hydrotheca. This particularly affects the Haleciidae, for which reason the family was excluded from the study, but it applies to some extent to the Plumulariidae. The second possibility was to use tentacle number. Unfortunately, contrary to the situation in Bryozoa, tentacle number is rarely included in taxonomic descriptions; when it is, the values may be inaccurate (P. F. S. Cornelius, personal communication) and, one suspects, sampling biased towards the lower end. This deficiency in the literature has been recognized by Cornelius (personal communication), who has been collecting data which have been made available and incorporated with our own into figure 3.

The results for hydrothecal volume appear unequivocal, the three ranges being well separated (Mann–Whitney $U$-test, $p \ll 0.01$). The very high values in the first group are attributable to a few species of *Bonneviella* (Bonneviellidae) also included. The tentacle counts appear more ambiguous. The three families in the first grouping have been separated and low numbers in the literature characterize Campanulinidae and Lafoeidae. For the Campanulariidae, Sertulariidae, and Plumulariidae, based mainly on original data, the distribution of numbers looks consistent with that of hydrothecal volumes. In *Aglaophenia*, one of the most advanced plumulariids, the number is always 10 (A. Svoboda, personal communication).

One other caution should be mentioned. The Plumulariidae are essentially a warm water, even tropical, family. A greater proportion of hydrothecal measurements was derived from Millard's (1975) southern African and Gibbons & Ryland's (1986) Fijian collections. It is well-known that the zooids in some colonial taxa are smaller in warm water; furthermore, water at 5 °C is 2.5 times as kinematically viscous as at 35 °C (Vogel 1981); a smaller hydranth in tropical waters, therefore, would not involve lower Reynolds numbers.

It is concluded that, with the possible exception of a few primitive forms, mean hydranth size in thecate hydroids does decrease as the level of integration increases. That this may incorporate a temperature-related effect in the Plumulariidae cannot be excluded.

### Module size and degree of colonial integration in Bryozoa

In marine Bryozoa (freshwater Bryozoa are rather different and not treated here) the dimensions of the lophophore provide the indicator of size most appropriate for the present discussion. Several observers have counted the tentacles in expanded lophophores and tentacle number is well correlated with size in terms of bell diameter and height (Ryland 1975).

Tentacle number in marine Bryozoa ranges from 8 to nearly 40. Winston (1977) has reviewed feeding and her figure 2 summarizes the data on tentacle number for 212 species as histograms presented separately for five major taxa: orders Cyclostomata (class Stenolaemata), Ctenostomata Stolonifera, Ctenostomata Carnosa, Cheilostomata Anasca, and Cheilostomata Ascophora (all class Gymnolaemata). This breakdown to subordinal level precludes consideration of tentacle number in relation to states of integration within these major taxa, leaving us only zooid form and their degree of association. The most primitive zooid shape in these marine taxa, as in Phylactolaemata, appears to be cylindrical. This shape is preserved to the greatest extent in Cyclostomata and Ctenostomata Stolonifera, in the former often combined with crustose colonies but in the latter more commonly associated with spreading stolons or more compact mats of tall zooids. In these two taxa the lophophores are almost universally small, commonly 8–10 tentacles and rarely exceeding 13. It seems reasonable to assume that this represents the primitive situation in marine bryozoans, and it is likely that eight tentacles represents the minimum from which an effective funnel can be constructed.

The carnose (commonly encrusting) Ctenostomata present a strikingly different range of numbers, with rather large lophophores of 14–25 tentacles predominating. The two cheilostome suborders display a wide range, with a rather flat mode of 11–16 in Anasca and 11–19 in Ascophora. The data pose a number of questions, some at present intractable; but we believe that there may be an explanation for the difference in size so dramatically apparent in these groups compared with Stolonifera. Many Carnosa and many Cheilostomata typically form compact encrusting colonies, which are rounded in outline, at least when young. Many such bryozoans live on substrata that become excessively crowded, with acute competition for space, with colonies being stopped or even overgrown by competitors (Gordon 1972; Stebbing 1973; Jackson 1977). There are no consistent winners in these communities (Stebbing 1973; Buss & Jackson 1979; Russ 1982; Rubin 1982).

An important contribution towards understanding mechanisms that might, in Bryozoa, be involved in determining the outcome of spatial encounters between species was made by Buss (1979a). He studied two anascan species, *Onychocella alula* (17 tentacles, funnel diameter 0.79 mm) and *Antropora tincta* (12 tentacles, 0.41 mm diameter). Buss was able to show that, at the interface between contiguous colonies of the two species, larger lophophores effectively interfered with the smaller, depriving them of food. This study has recently been developed by Best (1985), looking at the species inhabiting the *Fucus serratus* community earlier studied by Stebbing (1973). With carnosans *Flustrellidra hispida* (28–30 tentacles, funnel diameter 1.2 mm) and *Alcyonidium hirsutum* (17–18 tentacles, 0.77 mm diameter), the larger lophophores again overshadow the smaller. The capture of water by the marginal bells of *F. hispida* starves the immediately adjacent zooids of *A. hirsutum*, although there is enhanced feeding at a distance of four to five zooids from the edge, where the lophophores evidently benefit from the enhanced flow of water imperfectly cleared by *F. hispida*. These two investigations indicate that success in one kind of competition depends on having the larger lophophore. Clearly other factors are involved; for example, growth rate, in many cases, may be greater in species with smaller zooids (Buss 1979a). The angle at which opposing growing edges approach can be decisive (Rubin 1982), while longevity is another significant variable (Jackson & Winston 1981).

We consider that the stolonate life form is the more primitive in Ctenostomata, quick to colonize but less persistent (Buss 1979b; Jackson 1979): the guerilla strategy of Harper & Bell (1979). Small zooids are advantageous. Adopting an encrusting habit may confer greater permanence (the phalanx strategy of Harper & Bell). The most economical shape is circular, minimizing circumference to area, and ensuring that the front of the phalanx directly faces all potential competitors (see Rubin 1982). There are, we suggest, selective pressures favouring a larger lophophore, balancing energetic advantage against competitive success. It is interesting, in this context, to recall observations by Ryland (1975) on encrusting bryozoans in a rich habitat in New Zealand. The range of tentacle numbers over 43 species was 8–25. The mean was 13.9 and the coefficient of variation high (129%); for abundance, however, the mean tentacle number remained similar (14.6) but the coefficient of variation was dramatically reduced to 51%. The conclusion is that, in a competitive situation, species with either very small or very large lophophores are uncommon.

Competition directly involving the feeding apparatus does not occur in the major nonbryozoan encrusters (sponges and modular ascidians), which instead rely on chemical defences (Jackson & Buss 1975: Buss 1976; Dyrynda 1985, this symposium). Allelochemicals, which are also found in bryozoans (Dyrynda 1986), are not confined to encrusting species since they are also used against predators and epizoites which equally attack erect species. Moreover, many other factors influence the success or failure of species in communities of bryozoans, ascidians and sponges. We have insufficient data to establish whether lophophores are smaller in well integrated erect bryozoans than in those of encrusting habit.

### The disposition of zooids in colonies of encrusting Bryozoa

Of significance comparable to the size of modules is their disposition or arrangement in the colony. Zooids in suspension feeders may be organized with their filtering surfaces internal or external to the colony. Internal filterers, such as sponges and compound ascidians, may have their modules not evenly spaced but grouped around large exhalant openings, which facilitate enhanced filtration by passive means (Vogel 1974) and ensure that exhalant water is projected well clear of the inhalant apertures. External filterers may either be passive, as hydroids, Antipatharia and Octocorallia, or active, as Bryozoa and Calyssozoa. Passive filtration demands a colony organization, usually arborescent, that permits water to flow through it (and the integrational sequence we described above for hydroids clearly conforms to that). Active external filterers seem better adapted than passive filterers to exploit the encrusting form of growth, although, except when colonies are very small, excurrent pathways have to be provided (that is, the 'chimneys' of Bryozoa, see below). Other factors being equal it would seem necessary to arrange the zooids in such way as to maximize filtration area.

Assume, in marine Bryozoa, that lophophore funnel tops are circular in outline and essentially non-overlapping. If the zooids were square, of dimensions equal to the lophophore diameter, and in regular transverse rows, then the lophophores would be square-packed and each have an area 78.54% that of its zooid. If, on the other hand, the zooids were regular hexagons, zooids (and lophophores) would be in alternating rows, and each lophophore would cover 90.69% of its zooid (figure 179 in Thompson (1942)). Zooids with a regularly hexagonal frontal area are uncommon among living Bryozoa, and improbable among encrusting forms in which, to accommodate the retracted polypide, length typically exceeds width. Hexagonal (close) packing of lophophores can still be achieved with alternating (quincuncial), rectangular zooids, but only a limited number of shapes is possible. It can be shown that (J. P. Thorpe, personal communication) for a lophophore of top diameter $D$, the rectangle length $l$, and the width $w$, are given by

$$l = D[(n+0.5)^2 + (\cos 30°)^2]^{0.5}$$

$$w = (1.5 \tan 30°)/l,$$

where $n = 0, 1, 2, \ldots, \infty$.

Thus for a lophophore of unit top diameter the only probable zooid sizes are $1 \times 0.866$, $1.732 \times 0.5$ and $2.165 \times 0.4$ units. Observation suggests that zooids of these dimensions (figure 4, centre, left and right, respectively) are characteristic of *Schizoporella unicornis* and most cheilostomes, *Membranipora membranacea* and several flustrines, and various cyclostomes, respectively. Elongate hexagons (figure 4), diamonds or linguiform shapes that pack contiguously can be regarded as rectangles having the same length and same mean width.

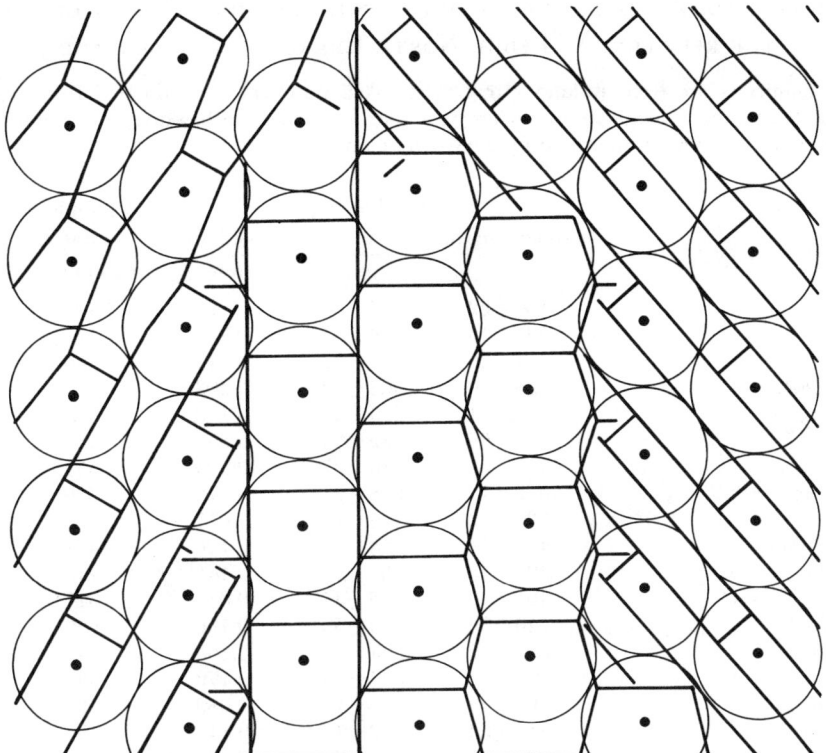

FIGURE 4. Possible zooid shapes in Bryozoa assuming close packing of lophophores. Circles represent the funnel top diameter of close-packed lophophores assuming no overlap. Solid dots indicate the mouth and also the centre of the orifice of the bearing zooid. Three basic zooid shapes (left, centre and right) have been fitted to the single size of lophophore: all shapes necessarily have the same area, 1.10 times greater than that of the lophophore, and are quincuncially arranged. The zooid outlines can be rectangles, hexagons or other shapes that pack contiguously and are reducible to a rectangle of the same length and same mean width. For a lophophore of unit diameter, the centre zooids have dimensions $1 \times 0.87$, the zooids to the left $1.72 \times 0.5$, and the zooids to the right $2.17 \times 0.4$.

Only one data set suitable for testing these predictions has been published (Winston 1979). For 16 encrusting species approximating the first relationship, zooid width averages 0.6 times the length, rather than the predicted 0.87; and lophophore area averages 1.45 times the zooid area (1.32 times if *Celleporaria brunnea*, in which the factor is 3.58, is excluded). To provide a second data set, zooid dimensions have been measured in 21 encrusting cheilostome species from the reef flat at Leigh, New Zealand, source of the lophophore measurements earlier published (table 1 in Ryland 1975). Lophophore area averages 1.10 times zooid area if the whole sample is used, 0.97 times if restricted to the 13 species for which a full set of measurements was obtained (table 1). These ratios are moderately close to those derived from Winston's (1979) data, if the anomalous *C. brunnea* is excluded, and seem consistent with the degree of lophophore interdigitation which observation reveals. Zooid width averages about 0.6 times length, as in Winston, leading to the conclusion that lophophores must be imperfectly close-packed. Perhaps accommodation of the retracted lophophore requires zooids to be about 1.6 times long as wide; or, the average proportions are a function of the maximum zooid width which, in circular colonies, precedes bifurcation of a zooid row. In addition, the outline of lophophores in some species departs from circularity (Ryland 1975), in conformity with the slender proportions of zooids. Further investigation is required to elucidate the explanation.

TABLE 1. DIMENSIONS OF ZOOIDS AND LOPHOPHORES IN ENCRUSTING CHEILOSTOME BRYOZOA FROM THE REEF FLAT AT LEIGH, NORTH ISLAND OF NEW ZEALAND

(Lophophore diameters are from Ryland (1975); asterisked numbers are estimated from tentacle number (correlation in Ryland 1975). Zooid measurements ± standard deviation are based on ten counts for each of five colonies unless a smaller number is indicated by the species name.)

|  | zooid length μm | zooid average width μm | zooid area mm² | lophophore diameter μm | lophophore area mm² |
|---|---|---|---|---|---|
| *Calloporina angustipora* | 460 ± 62 | 264 ± 39 | 0.123 | 406 | 0.129 |
| *Chaperia acanthina* | 564 ± 46 | 376 ± 39 | 0.212 | 514 | 0.207 |
| *C. cervicornis* (1) | 474 ± 25 | 304 ± 32 | 0.144 | 586 | 0.270 |
| *Crassimarginatella papulifera* | 558 ± 36 | 363 ± 29 | 0.203 | 437 | 0.150 |
| *Crepidacantha crenispina* | 418 ± 34 | 305 ± 26 | 0.128 | 390 | 0.119 |
| *Escharoides angela* | 778 ± 69 | 382 ± 34 | 0.297 | 488 | 0.187 |
| *E. excavata* | 773 ± 81 | 380 ± 27 | 0.293 | 554 | 0.241 |
| *Eurystomella foraminigera* (2) | 572 ± 72 | 268 ± 22 | 0.142 | 530 | 0.221 |
| *Exochella tricuspis* | 390 ± 42 | 196 ± 17 | 0.077 | 350 | 0.096 |
| *Fenestrulina thyreophora* | 418 ± 28 | 298 ± 22 | 0.126 | 428 | 0.144 |
| *Hippopodinella adpressa* | 520 ± 73 | 269 ± 26 | 0.141 | 442 | 0.153 |
| *Hippothoa tongima* (= sp. C) | 404 ± 40 | 206 ± 35 | 0.084 | 347* | 0.094* |
| *Macropora grandis* (3) | 1140 ± 79 | 673 ± 58 | 0.766 | 980 | 0.754 |
| *Micropora mortenseni* | 510 ± 72 | 294 ± 30 | 0.150 | 443 | 0.154 |
| *Odontionella cyclops* (1) | 592 ± 65 | 302 ± 18 | 0.179 | 595* | 0.278* |
| *Osthimosia bicornis* (2) | 392 ± 50 | 228 ± 12 | 0.089 | 431 | 0.146 |
| *Retevirgula acuta* (3) | 514 ± 33 | 282 ± 40 | 0.146 | 455 | 0.163 |
| *Schizomavella immersa* | 488 ± 50 | 274 ± 32 | 0.134 | 421 | 0.139 |
| *Smittina torques* | 512 ± 60 | 230 ± 34 | 0.120 | 393 | 0.121 |
| *Steginoporella neozelanica* | 857 ± 78 | 548 ± 42 | 0.471 | 842 | 0.557 |
| *Umbonula bicuspis* (2) | 666 ± 88 | 293 ± 52 | 0.196 | 631 | 0.313 |
| overall mean (all species: $n = 21$) | 569 | 321 | 0.201 | 470 | 0.221 |
| (full data sets only: $n = 13$) | 557 | 321 | 0.190 | 508 | 0.184 |

Much of the apparent variability in the appearance of the zooids of encrusting bryozoans, therefore, is related to size and is inseparable from the dimensions of the lophophore, though combined with variations on a small number of clearly defined basic rectangles. Square packing rarely occurs in cheilostomes: the presumed phenotype *Electra pilosa* f. *verticillata* (Ryland & Hayward 1977, figure 22), on filamentous algae, is one well-categorized exception. Among ctenostomes *Metalcyonidium gautieri* (Hayward 1985, figure 17), a minute erect form, has its zooids similarly in whorls. The Cyclostomata as a group, on the other hand, have definitely evolved away from hexagonally packed zooids. Why?

Harmelin (1976) described a number of evolutionary trends recognizable in three families of Tubuliporina (a suborder of Cyclostomata). In so far as these are colonial characters, he noted (i) a progression from uniserial (*Stomatopora*-like) to multiserial growth: (ii) a tendency toward upright habit; and (iii) trends toward the grouping of zooids. In circular or lobate multiserial colonies the most primitive arrangement of zooid erect portions (peristomes) seems to be quincuncial, such that the lophophores are close-packed. Especially in Tubuliporidae, however, this arrangement is superseded by the local fusion of peristomes to form short transverse rows, usually alternating to the left and right of the growth axis (Harmelin 1976, figure 1 and plate 1, figure 4; Hayward & Ryland 1985, figures 24, 26 and 31). In species that produce the largest colonies such connate rows may become multiseriate fascicles. The colonies in Diastoporidae develop a circular, rather than elongate, outline, and the connate rows, when present, are radial. Significantly, this development is seen also in Lichenoporidae, a quite unrelated cyclostome family with circular colonies. Here, too, the final state, as in *Coronopora*, is with radial fascicles (Hayward & Ryland 1985, figure 47). Extensive colonies, as sometimes found in *Disporella hispida*, may be polycentric, with several mammillae bearing radiating lines of peristomes. When feeding, as shown by Cook (1977) and Winston (1978), excurrent water flows centripetally between the rows, and exits from the summit, which is devoid of zooids. The analogous circulation in tubuliporids would produce medial venting, though in taxa in which peristome height increases from the midline laterally (for example, *Nevianopora*), the circulation could well be reversed: observations here are needed.

Excurrent chimneys in cheilostomes may be associated with morphological features of the colony, as in *Celleporaria* (Winston 1978) and *Hippoporidra* (Cook 1977; Ryland 1979), in which chimneys coincide with the summits of mammillae. This is also likely to be true in *Discoporella*, although unambiguous evidence is lacking (Winston 1978; Chimonides & Cook 1981). Extensive encrusting colonies of many other cheilostomes and carnose ctenostomes display chimneys that are less obviously associated with skeletal morphology (Banta *et al.* 1974; Winston 1978; Cook 1979; Cook & Chimonides 1980; Lidgard 1981). Although some kind of heterozooid may here be involved, the formation of chimneys of any kind always seems to involve distinctive asymmetry (whether ontogenetic or behavioural) of the lophophore and a precise arrangement of zooids showing this asymmetry, implying some kind of coordination (Cook 1979). The extent of colonial feeding patterns in cheilostomes is positively correlated with the degree of integration inferred from their colonial morphology (McKinney 1984). Some at least of the species (or ones closely related to them) are known to have well developed colonial nervous systems (Lutaud 1977), although it is not known whether these are involved in coordination of this kind. The structure of interzooidal communication pores in cyclostomes (Nielsen & Pedersen 1979) seems poorly developed in comparison with cheilostomes (Banta 1969; Lutaud 1979a), and colonial nerves have not convincingly been demonstrated (Lutaud 1979b). It is therefore possible that, either on account of their incomplete protrusion (see below) or for reasons associated with poor coordination, cyclostome lophophores are unable to posture in the manner required to produce chimneys. Certainly we are unaware of reports of excurrent chimneys in cyclostomes having quincuncially arranged peristomes. An alternative possibility is that, since cyclostome lophophores do not project fully from the peristomes during feeding (Ryland 1975), there are direct hydrodynamic advantages to be gained from the replacement of quincuncial peristomes by connate or fasciculated rows.

Further investigation is again required.

### FORM AND FUNCTION IN ERECT COLONIES

Jackson (1979) regarded the erect 'tree-like' growth form in colonial animals as a high-risk strategy owing to its dependence on a small attachment area. However, an erect growth form has two clear advantages: it permits partial escape from the competition among encrusters (see, for example, Sebens 1982b), and it allows greater exploitation of the water column for suspension feeding. The rate of flow in the water column is reduced by friction with the substratum; encrusters inhabit a boundary layer sheltered from mainstream flow, but the taller an erect colony becomes, the more it projects into mainstream flow and the greater the range of flow rates that it experiences (Okamura 1984, 1985).

The ambient water flow has important influences on the form and function of erect colonies. There are two main effects of flow. Of primary importance to suspension-feeding organisms is the delivery of food by flowing water. Second, flow imposes drag forces on erect colonies. These two effects are closely connected because the mechanisms of food particle deposition on to the feeding surfaces depend on the pattern of water flow (Rubenstein & Koehl 1977; LaBarbera 1984) which, in turn, depends on the orientation and other properties of the colony. In this section we describe the influences of water flow on orientation, growth forms and feeding mechanisms, concentrating mainly on non-coralline, passive suspension feeding coelenterates such as hydroids, octocorals and antipatharians (those octocorals and scleractinians which derive nutrition from symbiotic zooxanthellae are excluded because the requirements of photosynthesis impose conflicting influences on colony and module form and function). The discussion leads us to an environmental interpretation of the inverse correlation described above (figures 2 and 3) between the degree of integration and module size.

## Feeding efficiency, drag and planar orientation

The efficiency of particle capture in biological systems is best seen in relation to metabolic cost rather than just to clearance of a filtered volume. If metabolic cost is constant over time, $t$, and directly proportional to mass of living tissue, $m$, the efficiency can be expressed as $Pt^{-1}m^{-1}$, where $P$ is the number of trapped particles of diameter $d$. Clearly the efficiency will vary with the characteristics of the filter, being more efficient at some particle diameters than at others (LaBarbera 1984; Okamura 1985), and with the rate at which particles are delivered to the filter. In passive suspension feeders the particle delivery rate is proportional to, but not the same as, the flow rate in the environment since the filter resists the flow (Vogel 1981).

Erect growth forms contend with both environmental and structural constraints on their efficiency. A measure of the extent to which the surface area is exposed to water flow is the drag, in this case pressure drag (Vogel 1981). In directional currents such as those produced by wave surge or tidal flow, growth forms are often planar, pinnate or fan-shaped, with the plane normal to the flow (see Riedl (1971) for review). This orientation maximizes drag by exposing the maximum surface area to the maximum volume of flowing water. Experimental evidence that this orientation also maximizes particle capture was provided by Levesee (1976). Factors that increase drag in normally oriented planar colonies include increasing flow rate and increasing surface (that is, number and length of branches). To retain a perpendicular orientation with increasing drag, skeletal strengthening is required.

Flat planar colonies are unusual among erect Bryozoa, perhaps because active filtration reduces the dependence on orientation normal to the water flow or because they are more adapted to situations subject to turbulence or multidirectional flow. In *Petralia*, however, rigid fenestrate fans are supported by clustered rhizoids arising from the lower zooids. The frond may be flat or concave, with the zooids opening only on the concave face (cf. discussion below). *Petralia* is found in conditions of moderate current. *Adeona* also has rigid, fenestrate colonies, but the fans are bilaminar and supported by a stout, transversely corrugated, flexible stem. Non-fenestrate bilaminate pedunculate fans occur in *Celleporaria* and *Parmularia*. Observations on the ecology and responses to flow of these bryozoans would be useful.

## Colony flexibility

Most erect non-coralline colonies are flexible. Increasing drag therefore causes bending, the extent depending on the balance of drag and skeletal properties. Bending moderates the increased drag since it reduces the surface exposed to mainstream flow. Reduction of drag by bending has evident importance in extreme environments since it is better to bend than to break. The stems of some shallow-water gorgonians are shaped to facilitate bending in response to the risk of detachment by storm waves (Wainwright & Koehl 1976). In other gorgonians (Muzik & Wainwright 1977) and in some antipatharians from deeper water (Warner 1981) the stems are shaped to resist bending, probably in response to a reduced risk of excessive water movement.

Avoidance of breakage is not, however, the only function of flexibility. In the pinnate hydroid *Abietenaria* the detachment strength of normal flexible colonies is sufficient to resist the drag imposed by impossibly high flow rates (up to $150$ m s$^{-1}$) (Harvell & LaBarbera 1985). Even artificially stiffened colonies were not detached by flows of up to $5$ m s$^{-1}$. Harvell & LaBarbera (1985) measured flow rates adjacent to the polyps of *Abietenaria* over a range of experimental mainstream velocities. They found that the range of actual velocities experienced in normal flexible colonies was less than in artificially stiffened colonies, because the normal colonies bent out of the faster mainstream flows. They suggested that the function of the flexibility of this hydroid was the maintenance of a particular flow régime around the polyps which was optimal for feeding.

## Mechanisms of passive filtration

As well as bending the entire colony, increased flow may bend the filtering elements, for example, the polyps and their tentacles (Riedl & Forstner 1968; Patterson 1984). According to LaBarbera (1984), the sizes of the filtering elements of most passive suspension feeders are such that under environmental flow rates they operate at a low Reynolds number (Re) in conditions of laminar flow; the drag that they experience is therefore mostly skin friction (Vogel 1981). Biological filtration at the small-scale level of the filtering elements has been discussed by Rubenstein & Koehl (1977) and LaBarbera (1984). Their analyses show that 'sieving' (capturing only particles that are larger than the spaces between the filtering elements) is less important than 'aerosol filtration' (capturing smaller particles in addition), and that the commonest aerosol mechanism of particle capture by erect passive suspension feeders is the direct interception of particles by filtering fibres (for example, tentacles) oriented perpendicular to the flow. Direct observation and close-up underwater photography have confirmed that the tentacles of planar hydroids are usually oriented normal to the flow (Warner 1977). In conditions of laminar flow, far less water is intercepted by a tentacle bent parallel to the current than by one normal to it. Unless the particles are independently motile (motile particle deposition: Rubenstein & Koehl (1977)), the number intercepted and therefore captured is similarly reduced. Erect passive suspension feeders which feed in fast currents therefore benefit from any modification that limits the bending of their filtering elements. Reduction in tentacle length is a simple and evident example, and is also a normal consequence of a reduction in zooid size.

*Adhesion and downstream capture*

An essential assumption of aerosol filtration theory is that particles adhere to the filter on contact (Rubenstein & Koehl 1977; LaBarbera 1984). In coelenterates, capture and adhesion is achieved through cnidae or mucous secretions. Forces that oppose adhesion include intrinsic mobility of the captured particle (for example, live zooplankton) and skin frictional drag where the particle projects from the surface of the filtering element. Drag increases with particle size and increasing water velocity, leading to loss of adhesion and reducing particle capture efficiency at increased current speeds, especially for larger particles. Riedl & Forstner (1968) pointed out that gorgonian polyps projecting laterally are bent by faster currents into an environment behind the supporting skeleton in which current speed is reduced and turbulent eddies are present. They suggested that adhesion of particles to tentacles (that is, particle capture) is more effective in slowly moving eddies than in faster mainstream flow. Thus bending of polyps into downstream eddies compensates for the loss of a perpendicular orientation and increases the range of environmental current speeds over which feeding is possible. The effect of the bending of these polyps is therefore analogous to that of the bending of the whole colony of the hydroid *Abietenaria* mentioned above.

Eddies are present on the downstream sides of colonies because, although the filtering elements operate at low Re in largely laminar flow (LaBarbera 1984), the entire colony operates at high Re and is subject to pressure drag (Wainwright & Koehl 1976; Vogel 1981). The presence of downstream eddies in antipatharians at environmental current speeds has been confirmed by direct observation (Warner 1977, 1981). Lasker (1981) observed greater capture rates by downstream polyps of the gorgonian *Bryarium* in a flow tank; the food particles (*Artemia* cysts) were detached from upstream polyps by the current and were made available to downstream polyps by the eddies. A similar result was obtained in an arborescent bryozoan by Okamura (1984), who found that in slow currents upstream zooids captured more particles, but that in fast currents zooids in the central regions of the bushy colonies achieved greater feeding success. In flow tank experiments with the octocoral *Alcyonium siderum* (short cylindrical growth forms), Patterson (1984) found that in slow smooth flow (2.5 cm s$^{-1}$) upstream polyps caught more food particles, but in fast smooth flow (19 cm s$^{-1}$) downstream polyps caught more. In 'rough' (containing turbulence) flow however, there was no upstream–downstream asymmetry in catch rate over a range of current velocities.

*Downstream orientation in planar colonies*

Since in faster flow downstream zooids often achieve greater feeding success than upstream zooids, it is advantageous to colonies inhabiting environments with fast ambient flow to place most or all of their zooids on the downstream sides of their skeletal supports.

However, the directional currents to which planar suspension feeders are exposed are often oscillating, that is, produced by waves or tides. Since these modular organisms are mainly sessile on hard substrata, downstream placement is not always feasible. Apart from the bending of polyps described by Riedl & Forstner (1968), two other possibilities exist: living in unidirectional currents, or reorientation when the current changes.

Mainly unidirectional flow can occur at sites exposed to oceanic currents, or where local conditions cause a preponderance of tidal or wave-driven water to go in one direction. In coral ecosystems water flow may be unidirectional and incursive across the reef, with abundant hydroids appropriately oriented in the flow channels; outflow takes place through passes between the reefs (Penn 1983; Dilly & Ryland 1985; Gibbons & Ryland 1986). Unidirectional currents are also experienced by species which grow epizoically on flexible substrata (Warner 1977; Hughes 1980). Observations of planar organisms in mainly unidirectional currents have revealed examples with downstream placement of modules. Bilaterally symmetrical pinnate hydroids such as *Aglaophenia*, *Plumularia* and *Thecocarpus* are shallowly V-shaped in cross-section and the more streamlined dorsal side of each pinnate fan faces the current. The hydranths arise on the ventral side and face downstream (Svoboda 1976; Warner 1977; Hughes 1980). Svoboda (1976) placed colonies of *Aglaophenia picardi* in a flow tank oriented so that the pinnate fans were parallel to the current; he found that these fans were resorbed and new fans grown which reinstated the normal orientation with the dorsal sides facing the current. In gorgonians, downstream placement of polyps was observed by Muzik & Wainwright (1977) in fan-shaped colonies of *Melithaea* growing in channels through a coral reef. Most species of fan-shaped antipatharians studied by Warner (1977, 1981) showed downstream placement of polyps in mainly unidirectional flow resulting from an oceanic current.

Passive reorientation occurs when colonies can twist or bend so that the same aspect of the filtering surface is presented to the current, whatever its direction. Hydroids such as *Hydrallmania*, which support several bilaterally symmetrical pinnate fans projecting at right angles from a central flexible hydrocaulus, passively reorient by bending. The current always strikes the dorsal aspects of the fans and the hydranths face ventrally downstream (Warner 1977). Single bilaterally symmetrical pinnate hydroid fans also reorient to some extent by twisting. In two species of *Aglaophenia*, *A. harpago* (Svoboda 1973, personal communication) and *A. pluma* (N. G. Cartwright, personal communication), a hinge formed by an oblique constriction at the base of the fan appears to facilitate twisting. Hinge joints of this type are a characteristic feature at the base of the unbranched stems of various sertularian genera (for example, *Dynamena*, *Sertularia*: Millard (1975)). Bilaterally symmetrical pennatulids also orient with their polyps facing downstream; they reorient, probably actively, when the current changes (Magnus 1966; Svoboda 1976).

### Dish-shaped planar colonies

A further refinement of colony morphology is possible in unidirectional currents. In a flat planar colony subject to pressure drag, approaching streamlines diverge and spill past the edges at angles of less than 90°. Recurving the edges of the colony into a dish-shape, to achieve an orientation normal to the streamlines, increases particle capture, as was demonstrated empirically by Warner (1977). The effect is due to the greater amount of water intercepted, and is necessarily associated with an increase in pressure drag. It can be predicted that the recurvature to achieve a normal orientation to divergent streamlines is proportional to the resistance to flow of the colony (that is, the branching density) and to the rate of flow. Antipatharians and gorgonians growing in unidirectional currents form parabolic dishes with the concave side facing the current (Muzik & Wainwright 1977; Warner 1977, 1981), and in bilaterally symmetrical pinnate hydroids the hydrocaulus and hydrocladia are often recurved dorsally, into the current (Warner 1977). Dish-shaped orientations also occur in unitary passive suspension feeders such as stalked crinoids and basket stars (Meyer 1982; Warner 1982).

### Non-planar colonies

In contrast to planar morphology is the bushy or arborescent growth form, found in representatives of colonial coelenterates, sponges and bryozoans. Within this growth form are various grades of integration from the irregular branching of many campanularian hydroids to the ordered spirals of some species of *Sertularia* and *Bugula*. These bushy forms are adapted to environments in which water movement is multidirectional or turbulent (Riedl 1971; Warner 1977). In such environments normally planar species have been found to adopt bushy growth forms (Velimirov 1973). Planar colonies in multidirectional flow are not optimally oriented for maximum feeding success, and in addition they experience twisting stresses (Wainwright & Dylan 1969) which increases the risk of breakage. A bushy morphology avoids twisting and is equally well oriented to the flow whatever the current direction. (In bushy gorgonians and branching scleractinians, which contain zooxanthellae, the growth form is adapted to harvest light as well as to capture suspended particles.)

Colony form in the active suspension-feeding Bryozoa involves great hydrodynamic complexity, in part because planar branch morphologies may become organized during growth into spirals and other non-planar shapes (McKinney 1981; Cheetham & Hayek 1983). Branches are generally narrow, whether joined by bridges or not, to permit water to flow between them. Such forms may be rigidly calcified (Cheetham & Thomsen 1981) and dependent for strength on thickening towards the base of the branches and colony; others achieve flexibility either by the lightness of the calcification (*Bugula, Caberea, Euthyrisella, Flustra*) or by having non-calcified joints separating rigid internodes (*Cellaria, Crisia, Margaretta, Scrupocellaria*). Both rigid and flexible frondose colonies may be bilaminar but in those with unilaminar, non-cylindrical branches, the lophophores are, with few exceptions, directed inwards, as in the spiralling *Bugula* and most scroll-like reteporids. Cheetham & Hayek (1983) have shown that adherence to a restricted range of branching angles and link lengths (distances between successive bifurcations) minimizes branch interference in late growth stages, providing space for protruded lophophores to function. Such non-planar colony morphologies are adapted not only to multidirectional flow situations but to variable velocities, which will favour lophophores situated in different regions of the colony, over a range of ambient flow régimes (Okamura 1984).

### Module size and colony integration

Two points in the foregoing discussion bear on the question of module size in passive suspension feeders. First, smaller polyps with shorter tentacles are less easily bent into a parallel (unfavourable) orientation to the current. Second, smaller food particles are less easily detached from the tentacles by the current. Thus in faster currents it is expected that polyps should be small and should feed on small particles. This conclusion extends those of Sebens (1979) by including the effects of flow on the filter structure and on adhesion. A further extension can be made by considering the nature of food particles. Live zooplankton forms an important part of the diet of all passive suspension-feeding coelenterates (Lewis 1978, 1982; Warner 1981; Sebens & Koehl 1984) and live zooplankters are motile particles. The incidence of motile particle deposition as a mechanism of aerosol filtration increases with decreasing current speed at the expense of direct interception (Rubenstein & Koehl 1977). With decreasing flow the delivery rate of all particles decreases, but the delivery rate of larger zooplankters decreases least because of their motility; it therefore becomes advantageous at reduced current speed to concentrate on the capture of larger zooplankters, using larger polyps.

A final extension of this argument involves colony form. If motile particle deposition is important at lower current speeds, a planar colony form is less important since particles can move between streamlines; colonies therefore become more bushy as current speed decreases. It is also the case that multidirectional flow, the condition associated with bushy growth forms, is rarely encountered in exposed locations (except in the special conditions directly beneath breaking waves); it is more usually observed as turbulence behind some sheltering topographical feature (Warner & Woodley 1975; Meyer 1979; see discussion by Patterson 1984). Thus, combining the arguments presented here, it is expected that in slow, multidirectional-flow colonies should be bushy and polyps large. In faster, directional-flow colonies should be planar and polyps small. The two ends of the spectrum of integration of colony form in hydroids (figure 2) therefore partly represent, on the one hand a disordered response to the disordered nature of large motile particles and slow turbulent flow, and on the other hand an ordered response to the ordered nature of fast directional flow. It is noteworthy that Meyer (1979) came to the same conclusion with respect to a series of crinoid species inhabiting a coral reef. In these unitary passive suspension feeders the filtering modules are the tube feet, and the arms and pinnules may be held in either bushy or planar array (Meyer 1982).

Our thanks are due to M. A. Best, P. L. Cook, P. F. S. Cornelius, P. J. Hayward, F. K. McKinney and A. R. D. Stebbing for their input to this paper, and especially to J. P. Thorpe for deriving the formula relating possible zooid shapes to lophophore diameter and packing.

## References

Banta, W. C. 1969 The body wall of cheilostome Bryozoa. II. Interzoidal communication organs. *J. Morph.* **129**, 149–170.

Banta, W. C., McKinney, F. K. & Zimmer, R. L. 1974 Bryozoan monticules: excurrent water outlets? *Science, Wash.* **185**, 783–784.

Berrill, N. J. 1950 *The Tunicata, with an account of the British species.* London: Ray Society.

Best, M. J. 1985 Some aspects of the ecology and physiology of feeding in marine Bryozoa. Unpublished thesis, University of Liverpool.

Beklemishev, W. N. 1969 *Principles of comparative anatomy of invertebrates: 1. Promorphology* (translated from the 3rd Russian edn by J. M. MacLennan, ed. Z. Kabata) (490 pages.) Edinburgh: Oliver & Boyd.

Bertalanffy, L. von 1957 Quantitative laws in metabolism and growth. *Q. Rev. Biol.* **32**, 217–231.

Boardman, R. S. & Cheetham, A. H. 1973 Degrees of colony dominance in stenolaemate and gymnolaemate Bryozoa. In *Animal colonies: development and function through time* (ed. R. S. Boardman, A. H. Cheetham & W. A. Oliver), pp. 121–220. Stroudsburg: Dowden, Hutchinson & Ross.

Boardman, R. S., Cheetham, A. H. & Oliver, W. A. (eds) 1973 *Animal colonies: development and function through time.* Stroudsburg: Dowden, Hutchinson & Ross.

Bullivant, J. S. 1968 The rate of feeding of the bryozoan *Zoobotryon verticillatum*. *N.Z. Jl. mar. freshw. Res.* **2**, 111–134.

Buss, L. 1976 Better living through chemistry: the relationship between allelochemical interactions and competitive networks. In *Aspects of sponge biology* (ed. F. W. Harrison & R. R. Cowden), pp. 315–327. New York: Academic Press.

Buss, L. 1979a Bryozoan overgrowth interaction: the interdependence of competition for food and space. *Nature, Lond.* **281**, 475–477.

Buss, L. W. 1979b Habitat selection, directional growth and spatial refuges: why colonial animals have more hiding places. In *Biology and systematics of colonial organisms* (ed. G. Larwood & B. R. Rosen), pp. 459–497. London: Academic Press.

Buss, L. & Jackson, J. B. C. 1979 Competitive networks: nontransitive competitive relationships in cryptic coral reef environments. *Am. Nat.* **113**, 223–234.

Chapman, G. 1981 Individuality and modular organisms. *Biol. J. Linn. Soc.* **15**, 177–183.

Chapman, G. & Stebbing, A. R. D. 1980 The modular habit – a recurring strategy. In *Developmental and cellular biology of coelenterates* (ed. P. Tardent & R. Tardent), pp. 157–162. Amsterdam: Elsevier-North-Holland.

Cheetham, A. H. & Hayer, L. C. 1983 Geometric consequences of branching growth in adeoniform Bryozoa. *Paleobiology* **9**, 240–260.
Cheetham, A. H. & Thomsen, E. 1981 Functional morphology of arborescent animals: strength and design of cheilostome bryozoan skeletons. *Paleobiology* **7**, 355–383.
Chimonides, P. J. & Cook, P. L. 1981 Observations on living colonies of *Selenaria* (Bryozoa, Cheilostomata). II. *Cah. Biol. mar.* **22**, 207–219.
Cook, P. L. 1977 Colony-wide water currents in living Bryozoa. *Cah. Biol. mar.* **18**, 31–47.
Cook, P. L. 1979 Some problems in interpretation of heteromorphy and colony integration in Bryozoa. In *Biology and systematics of colonial organisms* (ed. G. Larwood & B. R. Rosen), pp. 193–210. London: Academic Press.
Cook, P. L. & Chimonides, P. J. 1980 Further observations on water current patterns in living Bryozoa. *Cah. Biol. mar.* **21**, 393–402.
Dilly, P. N. & Ryland, J. S. 1985 An intertidal *Rhabdopleura* (Hemichordata, Pterobranchia) from Fiji. *J. Zool., Lond.* **205**, 611–623.
Dyrynda, P. E. J. 1985a Chemical defences and the structure of subtidal epibenthic communites. In *Proc. 19th mar. Biol. Symp.* (ed. P. E. Gibbs), pp. 411–421. Cambridge University Press.
Dyrynda, P. E. J. 1986 Functional allelochemistry in temperate waters: chemical defences of bryozoans. In *Bryozoa: Ordovician to Recent* (ed. C. Nielsen & G. P. Larwood). Fredensborg: Olsen & Olsen. (In the press.)
Gibbons, J. H. & Ryland, J. S. 1986 Intertidal and shallow water hydroids from Fiji. (In preparation.)
Gordon, D. P. 1972 Biological relationships of an intertidal bryozoan population. *J. nat. Hist.* **6**, 503–514.
Gulland, J. 1983 *Fish stock assessment: a manual of basic methods.* Chichester: John Wiley.,
Haeckel, E. 1866 *Generelle Morphologie der Organismen.* Berlin: Reimer.
Harmelin, J. G. 1976 Evolutionary trends within three Tubuliporina families (Bryozoa, Cyclostomata). *Docums Lab. Geol. Fac. Sci. Lyon*, H.S. **3**, 607–616.
Harper, J. L. 1977 *Population biology of plants.* London: Academic Press.
Harper, J. L. & Bell, A. D. 1979 The population dynamics of growth form in organisms with modular construction. In *Population dynamics* (ed. R. M. Anderson, H. D. Turner & L. R. Taylor), pp. 29–52. Oxford: Blackwell.
Harvell, C. D. & La Barbera, M. 1985 Flexibility: a mechanism for control of local velocities in hydroid colonies. *Biol. Bull. mar. biol. Lab. Woods Hole* **168**, 312–320.
Hayward, P. J. 1985 Ctenostome bryozoans. *Syn. Br. Fauna*, N.S. **33**, 1–177.
Herdman, W. A. 1882 Report on the Tunicata. *Scient. Results Challenger Exped., Zool.* **6**, 1–296.
Hughes, R. G. 1980 Current induced variations in the growth and morphology of hydroids. In *Development and cellular biology of coelenterates* (ed. P. Tardent & R. Tardent), pp. 179–184. Amsterdam: Elsevier–North-Holland.
Huxley, T. H. 1851 Observations upon the anatomy and physiology of *Salpa* and *Pyrosoma*. *Phil. Trans. R. Soc. Lond.* **141**, 567–593.
Jackson, J. B. C. 1977 Competition on marine hard substrata: the adaptive significance of solitary and colonial strategies. *Am. Nat.* **111**, 743–767.
Jackson, J. B. C. 1979 Morphological strategies of sessile animals. In *Biology and systematics of colonial organisms* (ed. G. Larwood & B. R. Rosen), pp. 499–555. London: Academic Press.
Jackson, J. B. C. & Buss, L. 1975 Allelopathy and spatial competition among coral reef invertebrates. *Proc. natn. Acad. Sci. U.S.A.* **72**, 5160–5163.
Jackson, J. B. C., Buss, L. W. & Cook, R. E. (eds) 1985 *Population biology and evolution of clonal organisms.* New Haven: Yale University Press.
Jackson, J. B. C. & Winston, J. E. 1981 Modular growth and longevity in bryozoans. In *Recent and fossil Bryozoa* (ed. G. P. Larwood & C. Nielsen), pp. 121–126. Fredensborg: Olsen & Olsen.
Jones, R. 1976 Growth of fishes. In *The ecology of the seas* (ed. D. H. Cushing & J. J. Walsh), pp. 251–279. Oxford: Blackwell.
Jørgensen, C. B. 1975 Comparative physiology of suspension feeding. *A. Rev. Physiol.* **37**, 57–79.
Kaufmann, K. W. 1981 Fitting and using growth curves. *Oecologia (Berl.)* **49**, 293–299.
Kott, P. 1982 Replication in the Ascidiacea: an adaptive strategy in the coral reef environment. In *Proc. 4th int. Coral Reef. Symp.* (ed. E. D. Gomez), vol. 2, pp. 725–733. Manila: University of the Philippines.
LaBarbera, M. 1984 Feeding currents and particle capture mechanisms in suspension feeding animals. *Am. Zool.* **24**, 71–84.
Larwood, G. & Rosen, B. R. (eds) 1979 *Biology and systematics of colonial organisms.* London: Academic Press.
Lasker, H. R. 1981 A comparison of the particulate feeding abilities of three species of gorgonian soft coral. *Mar. Ecol. Prog. Ser.* **5**, 61–67.
Leversee, G. J. 1976 Flow and feeding in fan shaped colonies of the gorgonian coral, *Leptogorgia*. *Biol. Bull. mar. biol. Lab., Woods Hole*, **151**, 344–356.
Lewis, J. B. 1978 Feeding mechanisms in black corals (Antipatharia). *J. Zool. Lond.* **186**, 393–396.
Lewis, J. B. 1982 Feeding behaviour and feeding ecology of the Octocorallia (Coelenterata: Anthozoa). *J. Zool., Lond.* **196**, 371–384.
Lidgard, S. 1981 Water flow, feeding, and colony form in an encrusting cheilostome. In *Recent and fossil Bryozoa* (ed. G. P. Larwood & C. Nielsen), pp. 135–142. Fredensborg: Olsen & Olsen.

Lutaud, G. 1977 The bryozoan nervous system. In *Biology of bryozoans* (ed. R. M. Woollacott & R. L. Zimmer), pp. 377–410. New York: Academic Press.

Lutaud, G. 1979*a* Etude ultrastructurale de "plexus colonial" et recherche de connexions nerveuses interzoidiales chez le bryozoaire chilostome *Electra pilosa* (Linné). *Cah. Biol. mar.* **20**, 315–324.

Lutaud, G. 1979*b* The probability of a plexus in the calcified wall of *Crisidia cornuta* (Linné). In *Advances in bryozoology* (ed. G. P. Larwood & M. B. Abbott), pp. 33–45. London: Academic Press.

Mackie, G. O. 1963 Siphonophores, bud colonies and superorganisms. In *The lower Metazoa: comparative biology and physiology* (ed. E. C. Dougherty), pp. 329–337. Berkeley and Los Angeles: University of California Press.

Magnus, D. B. E. 1966 Zur Okologie einer nachtaktiven Flachwasser–Seefeder (Octocorallia, Pennatularia) im Roten Meer. *Veroff. Inst. Meeresforsch. Bremerhaven, Sonderband* **2**, 369–380.

McKinney, F. K. 1981 Planar branch systems in colonial suspension feeders. *Paleobiology* **7**, 344–354.

McKinney, F. K. 1984 Feeding currents of gymnoloemate bryozoans: better organization with high colonial intergration. *Bull mar. Sci.* **34**, 315–319.

Meyer, D. L. 1979 Length and spacing of the tube feet in crinoids (Echinodermata) and their role in suspension feeding. *Mar. Biol.* **51**, 361–369.

Meyer, D. L. 1982 Food and feeding mechanisms: Crinozoa. In *Echinoderm nutrition* (ed. M. Jangoux & J. M. Lawrence), pp. 25–42. Rotterdam: A. A. Balkema.

Millar, R. H. 1966 Evolution in ascidians. In *Some contemporary studies in marine science* (ed. H. Barnes), pp. 519–534. London: Allen & Unwin.

Millar, R. H. 1970 British Ascidians. *Syn. Br. Fauna, N.S.* **1**, 1–92.

Millard, N. A. H. 1975 Monograph on the Hydroida of southern Africa. *Ann. S. Afr. Mus.* **68**, 1–513.

Muzik, K. & Wainwright, S. 1977 Morphology and habitat of five Fijian sea fans. *Bull. mar. Sci.* **27**, 308–337.

Naumov, D. V. 1969 *Hydroids and hydromedusae of the U.S.S.R.* Jerusalem: Israel Program for Scientific Translations.

Nicholson, H. A. 1872 *A manual of palaeontology*. Edinburgh: Blackwood.

Nielsen, C. & Pedersen, K. J. 1979 Cystid structure and protrusion of the polypide in *Crisia* (Bryozoa, Cyclostomata). *Acta Zool. (Stockh.)* **60**, 65–88.

Oka, H. & Usui, M. 1944 On the growth and propagation of the colonies in *Polycitor mutabilis* (Ascidiae compositae). *Sci. Rep. Tokyo Bunrika Daigaku, ser. B* **7**, 23–53.

Okamura, B. 1984 The effects of ambient flow velocity, colony size, and upstream colonies on the feeding success of Bryozoa. I. *Bugula stolonifera* Ryland, an arborescent species. *J. exp. mar. Biol. Ecol.* **83**, 179–193.

Okamura, B. 1985 The effects of ambient flow velocity, colony size, and upstream colonies on the feeding success of Bryozoa. II. *Conopeum reticulum* (Linnaeus), an encrusting species. *J. exp. mar. Biol. Ecol.* **89**, 69–80.

Patterson, M. R. 1984 Patterns of whole colony prey capture in the octocoral, *Alcyonium siderum*. *Biol. Bull. mar. biol. Lab., Woods Hole*, **167**, 613–629.

Penn, N. 1983 The environmental consequences and management of coral sand dredging in the Suva region, Fiji. Unpublished thesis, University of Wales.

Pütter, A. 1920 Studien über physiologische Ahnlichkeit. VI. Wachstumsahnlichkeiten. *Pflugers Arch. ges. Physiol.* **180**, 298–340.

Riedl, R. 1971 Water movement: introduction (pp. 1085–1088) and animals (pp. 1123–1156). In *Marine ecology* (ed. O. Kinne), vol. 1, part 2. New York: Wiley.

Riedl, R. & Forstner, H. 1968 Wasserbewegung im Mikrobereich des Benthos. *Sarsia* **34**, 103–188.

Rosen, B. R. 1979 Modules, members and communes: a postscript introduction to social organisms. In *Biology and systematics of colonial organisms* (ed. G. Larwood & B. R. Rosen), pp. xiii–xxxv. London: Academic Press.

Rubenstein, D. I. & Koehl, M. A. R. 1977 The mechanisms of filter feeding: some theoretical considerations. *Am. Nat.* **111**, 981–994.

Rubin, J. A. 1982 The degree of intransitivity and its measurement in an assemblage of encrusting cheilostome Bryozoa. *J. exp. mar. Biol. Ecol.* **60**, 119–128.

Rudwick, M. J. S. 1970 *Living and fossil brachiopods*. London: Hutchinson.

Russ, G. R. 1982 Overgrowth in a marine epifaunal community: competitive hierarchies and competitive networks. *Oecol., Berl.* **53**, 12–19.

Ryland, J. S. 1975 Parameters of the lophophore in relation to population structure in a bryozoan community. In *Proc. 9th Eur. mar. biol. Symp.* (ed. H. Barnes), pp. 363–393. Aberdeen University Press.

Ryland, J. S. 1979 Structural and physiological aspects of coloniality in Bryozoa. In *Biology and systematics of colonial organisms* (ed. G. Larwood & B. R. Rosen), pp. 211–242. London: Academic Press.

Ryland, J. S. & Hayward, P. J. 1977 British anascan bryozoans. *Syn. Br. Fauna, N.S.* **10**, 1–188.

Ryland, J. S., Wigley, R. A. & Muirhead, A. 1984 Ecology and colonial dynamics of some Pacific reef flat Didemnidae (Ascidiacea). *Zool. J. Linn. Soc.* **80**, 261–282.

Sebens, K. P. 1979 The energetics of asexual reproduction and colony formation in benthic marine invertebrates. *Am. Zool.* **19**, 683–697.

Sebens, K. P. 1982*a* The limits to indeterminate growth: an optimal size model applied to passive suspension feeders. *Ecology* **63**, 209–222.

Sebens, K. P. 1982b Competition for space: growth rate, reproductive output, and escape in size. *Am. Nat.* **120**, 189–197.
Sebens, K. P. & Koehl, M. A. R. 1984 Predation on zooplankton by the benthic anthozoans *Alcyonium siderium* (Alcyonacea) and *Metridium senile* (Actiniaria) in the New England subtidal. *Mar. Biol.* **81**, 255–271.
Shick, J. V., Brown, W. I., Dolliver, E. G. & Kayer, S. R. 1978 Oxygen uptake in sea anemones: effects of expansion, contraction, and exposure to air and the limitations of diffusion. *Physiol. Zool.* **52**, 50–61.
Spencer, H. 1862 *First principles*. London: Williams and Norgate.
Stebbing, A. R. D. 1973 Competition for space between the epiphytes of *Fucus serratus* L. *J. mar. biol. Ass. U.K.* **53**, 247–261.
Svoboda, A. 1973 Beitrag zur Okologie, Biometrie und Systematik der mediterranen *Aglaophenia* Arten (Hydroidea). Ph.D. thesis, Vienna.
Svoboda, A. 1976 The orientation of *Aglaophenia* fans to current in laboratory conditions (Hydrozoa, Coelenterata). In *Coelenterate ecology and behaviour* (ed. G. O. Mackie), pp. 41–48. New York: Plenum.
Thompson, d'A. W. 1942 *On growth and form*, 2nd edn. Cambridge University Press.
Van Name, W. G. 1945 The North and South American ascidians. *Bull. Am. Mus. nat. Hist.* **84**, 1–476.
Velimirov, B. 1973 Orientation in the sea fan *Eunicella cavolinii* related to water movement. *Helgolander wiss. Meeresunters.* **24**, 163–173.
Vogel, S. 1974 Current induced flow through the sponge *Halichondria*. *Biol. Bull. mar. biol. Lab., Woods Hole* **147**, 443–456.
Vogel, S. 1981 *Life in moving fluids*. Boston: Willard Grant Press (1983, New Jersey: Princeton University Press).
Wainwright, S. A. & Dillon, J. R. 1969 On the orientation of sea fans (genus *Gorgonia*). *Biol. Bull. mar. biol. Lab., Woods Hole* **136**, 130–139.
Wainwright, S. A. & Koehl, M. A. R. 1976 The nature of flow and the reaction of benthic Cnidaria to it. In *Coelenterate ecology and behaviour* (ed. G. O. Mackie), pp. 5–21. New York: Plenum.
Warner, G. F. 1977 On the shapes of passive suspension feeders. In *Biology of benthic organisms* (ed. B. F. Keegan, P. O'Ceidigh & P. J. S. Boaden), pp. 567–576. Oxford and New York: Pergamon Press.
Warner, G. F. 1981 Species descriptions and ecological observations of black corals (Antipatharia) from Trinidad. *Bull. mar. Sci.* **31**, 147–163.
Warner, G. F. & Woodley, J. D. 1975 Suspension feeding in the brittle-star *Ophiothrix fragilis*. *J. mar. biol. Ass. U.K.* **55**, 199–210.
Warren, C. E. & Davis, G. E. 1966 Laboratory studies on the feeding, bioenergetics, and growth of fish. In *The biological basis of freshwater fish production* (ed. S. D. Gerking), pp. 175–214. Oxford: Blackwell.
Winston, J. E. 1977 Feeding in marine bryozoans. In *Biology of bryozoans* (ed. R. M. Woollacott & R. L. Zimmer), pp. 233–271. New York: Academic Press.
Winston, J. E. 1978 Polypide morphology and feeding behavior in marine ectoprocts. *Bull. mar. Sci.* **28**, 1–31.
Winston, J. E. 1979 Current-related morphology and behaviour in some Pacific coast bryozoans. In *Advances in bryozoology* (ed. G. P. Larwood & M. B. Abbott), pp. 247–268. London: Academic Press.
Zeuthen, E. 1953 Oxygen uptake as related to body size in organisms. *Q. Rev. Biol.* **28**, 1–12.

# Modular growth in seed plants

By F. Hallé

*Institut Botanique, Université des Sciences et Techniques du Languedoc, 163 rue Auguste Broussonet, 34000 Montpellier, France*

Modular growth in seed plants may be analysed in terms of three architectural elements: the unit of morphogenesis, the module and the architectural model. Some of the salient features of these structures are reviewed, compared and contrasted. A variety of plant shapes and sizes may be derived schematically from them by two sorts of transformation, gigantism and repetition. The former is uncommon in seed plants, but repetition produces a wide array of constructions. Repetition of the architectural model, a process known as reiteration, leads to a colonial structure characteristic of the crowns of many mature trees. This is often an expression of the plant's opportunistic response to environmental variations in resource availability. The reiterated complexes formed as a result may show some characteristic ontogenetic and phylogenetic sequences to give an architectural continuum of construction.

## 1. Introduction

One of the salient architectural expressions of modular growth in seed plants is the serial repetition of plant organs by apical meristems. If derived from a single apical meristem these organs may form a monopodial axis, and if several meristems are involved the plant may be composed of a branched, sympodial series of axes. Although the idea of modular construction in plants is ancient (White 1979; Cusset 1982), its revival was cast in rather precise morphological terms, centred on the definition and recognition of a basic unit of construction, *l'article* (Prévost 1967; Hallé & Oldeman 1970), subsequently translated as module (Harper & White 1974).

In this paper the expression *modular growth* will be restricted to two different aspects of growth and form in seed plants; one refers to a group of architectural models, called *modular models* (Hallé *et al.* 1978); another refers to the process of repetition of the architectural model which occurs in most ageing trees, but less commonly in herbs, and has been termed *reiteration* (Oldeman 1977). Both involve modular growth and both lead to modular form, but there are significant differences between them, as I shall indicate.

## 2. Units of modular construction

The growth and form of the aerial part of seed plants may be analysed by considering three architectural elements; the unit of morphogenesis, the module, and the architectural model. A unit of morphogenesis is a length of vegetative shoot whose chronological limits are determined by a single period of continuous activity of the apical meristem (figure 1). This may be, but is not necessarily, the same as the unit of extension which is usually detectable on the shoots of trees and shrubs, either in tropical or extratropical climates, as the length of vegetative shoot between two successive groups of bud scale scars (figure 1) (Hallé & Martin 1968).

The module is the leafy axis in which the entire sequence of aerial differentiation is carried out, from the initiation of the meristem that builds up the axis to the sexual differentiation of its apex (Hallé *et al.* 1978). At the base of the module, roots sometimes develop; more frequently they are not expressed but the module retains a rooting ability which can often be used for vegetative propagation (figure 2). Although some seed plants consist of only a single module according to this definition, most have an integrated complex of modules linked together sympodially.

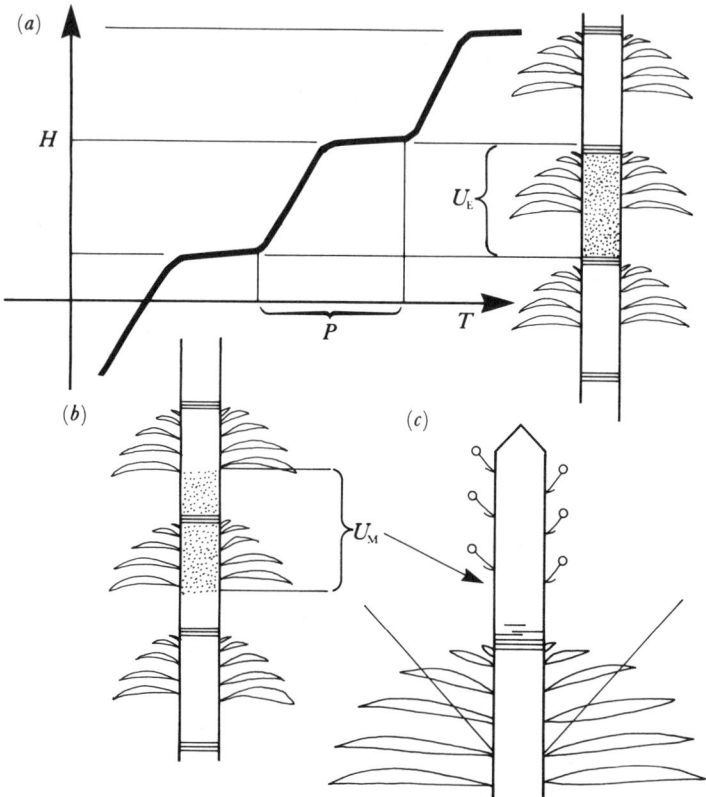

FIGURE 1. (*a*) The growth curve of a shoot with rhythmic growth in *Hevea brasiliensis* (Euphorbiaceae) (rubber tree). $H$, shoot height; $T$, time; $P$, period of the rhythm; $U_E$, unit of extension, whose limits are two successive bud-scale scars. (*b*) The unit of morphogenesis ($U_M$) is a length of vegetative shoot whose chronological limits are determined by a single period of continuous activity of the apical meristem. (*c*) One isolated unit of morphogenesis, with leaves, axillary branching and lateral sexual structures.

The architectural model is the visible expression of the genetical programme of development of the plant, and represents the fully developed, complex plan of assembly of modules into a coherent construction (figure 2). In all but a relatively small number of seed plants it is multimodular. Of approximately 24 architectural models of trees so far described (Hallé *et al.* 1978), six are entirely modular, in the sense that the apical meristem of every module in the model completes the sequence of differentiation and eventually becomes sexual. These are architectural models known as Leeuwenberg, Chamberlain, Tomlinson, Koriba and Prévost (figure 3) in addition to the model of Holttum which consists, by definition, of a single module (Hallé *et al.* 1978).

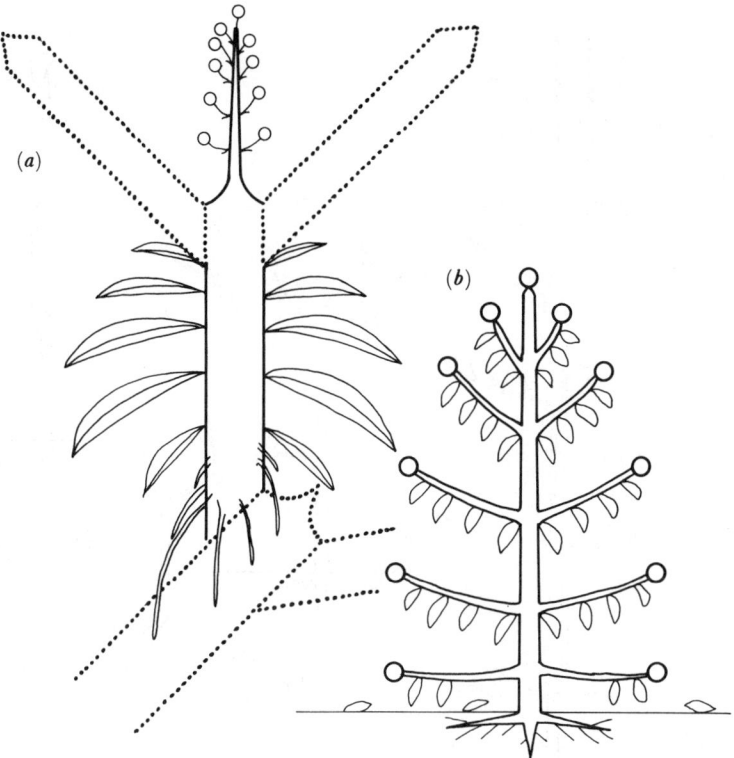

FIGURE 2. (a) A module. (b) an architectural model.

These three architectural levels constitute a hierarchy of elementary structures which have been found to be valuable for interpreting the processes of growth in any kind of seed plant, from small herbs to large trees. The three-dimensional arrangement of leafy axes within architectural models is regular, predictable and probably genetically controlled to a large extent. Models that are entirely modular (figure 3) are easily drawn by computer graphics, since their constructional rules are readily quantifiable (de Reffye 1979, 1983). But not all models are fully modular, insofar as the apical meristems of some shoots do not complete their differentiation and become sexual; in fact most architectural models show this phenomenon (Hallé et al. 1978). Some shoots grow upwards and overtop others to form a trunk, while others remain subordinate and produce leaves and flowers. This physiological and structural differentiation may also be seen in models that are entirely modular (for example, models of Koriba and Prévost: figure 3).

The three structural elements show some obvious differences, but they also show important similarities. As levels of a constructional hierarchy, they differ in that the highest level envelops the lower ones, as outlined above. This may be indicated schematically (figure 4).

On the other hand, they resemble each other: they represent whole sequences of differentiation, growth and organogenesis, photosynthesis, vascular construction, and are terminated by sexuality. They also have a capacity for individual enlargement (gigantism) and for repetition. These processes, especially the latter, are of great significance for the construction of large-scale organisms such as trees.

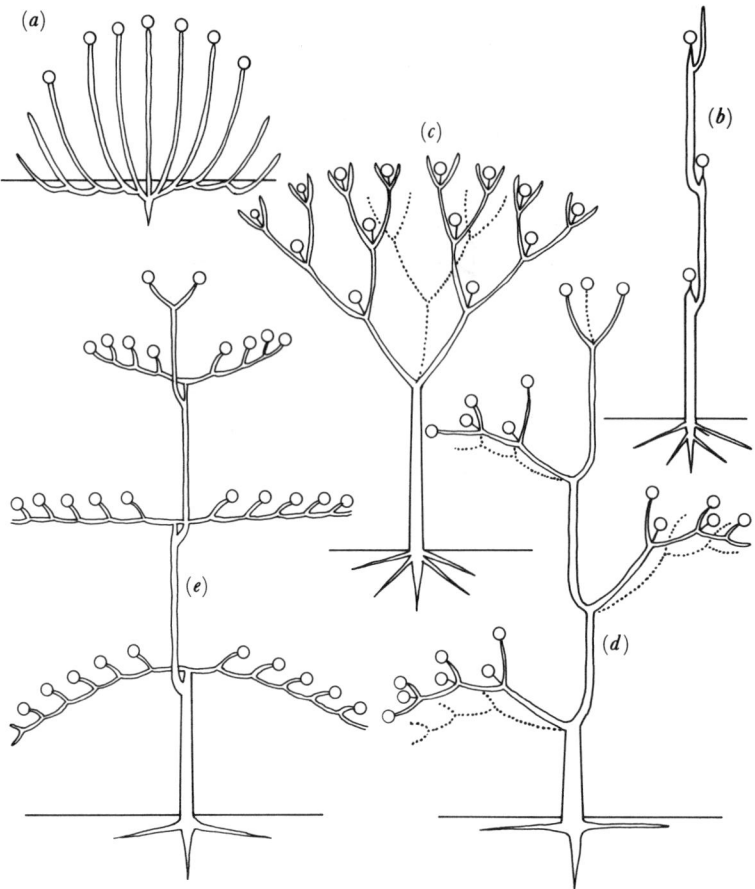

FIGURE 3. Modular architectural models, each drawn schematically. Some typical examples are listed. (a) Tomlinson's model: *Raphia* spp. (Arecaceae), *Musa* spp. (Musaceae), many grasses (Poaceae), a few dicotyledons, such as *Helleborus* sp., *Euphorbia* sp. and *Lobelia* sp. (b) Chamberlain's model: *Cycas* spp. (male plants) (Cycadaceae), *Philodendron* spp. (Arecaceae) and *Talisia* spp. (Sapindaceae). (c) Leeuwenberg's model: *Dracaena* spp. (Dracaenaceae), and a large number of dicotyledons, such as *Rauwolfia* sp., *Senecio* sp., *Croton* sp., *Anthocleista* sp. and *Solanum* sp. (d) Koriba's model: *Ochrosia* spp. (Apocynaceae), *Ochroma* spp. (Bombacaceae), *Homalanthus* and *Hura* spp. (Euphorbiaceae). (e) Prévost's model: *Excoecaria* sp. (Euphorbiaceae), *Cordia* spp. (Boraginaceae) and *Alstonia* spp. (Apocynaceae).

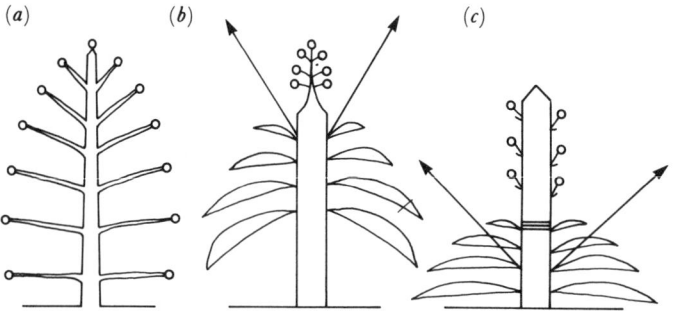

FIGURE 4. The architectural model (a), the module (b) and the unit of morphogenesis (c).

## 3. GIGANTISM AND REPETITION

From the three architectural elements outlined in the previous section, a wide variety of shapes and sizes may be derived schematically by two sorts of transformation which will be referred to as gigantism and repetition. It is suggested that such transformations may have a phylogenetic significance, repetition being a more advanced transformation than gigantism (see figures 5 and 7 for comparison). Schematically one may consider three general developmental patterns in seed plants in which each of these three structural elements may:

(i) remain solitary; their dimensions remain small or may even become 'miniaturized' as herbs (Hallé *et al.* 1978);

(ii) remain solitary but grow increasingly large (gigantism), as in some trees;

(iii) retain their size within some narrow range, but increase in number (repetition) to give large organisms, such as most trees.

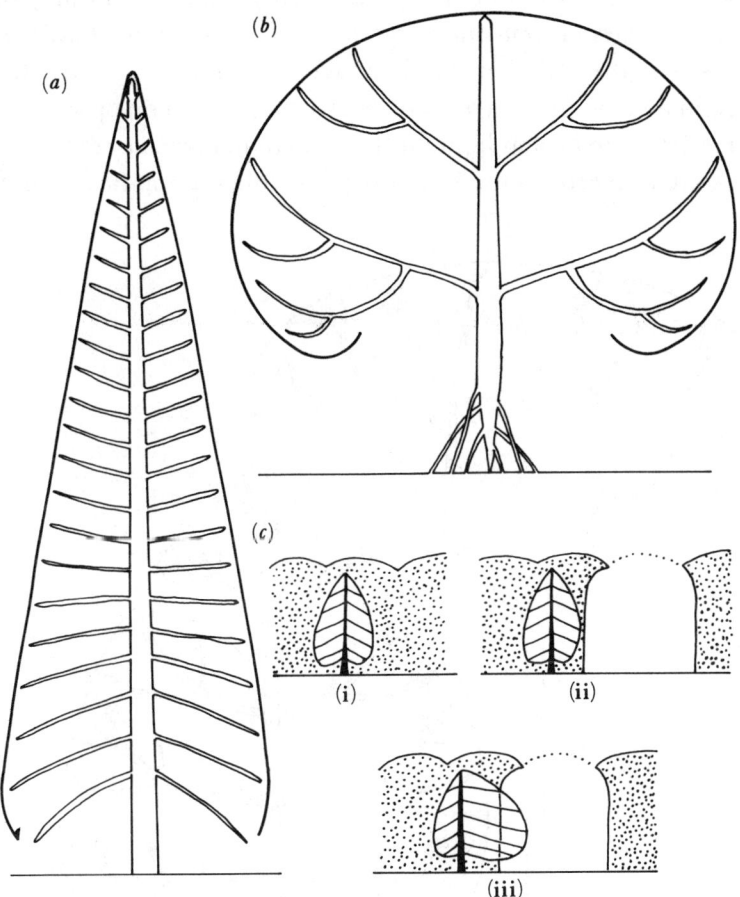

FIGURE 5. Gigantism of a single architectural model in *Araucaria columnaris* (*a*) and *Musanga cecropioides* (*b*). (*c*) The weak and incomplete opportunistic reaction of a mono-model tree to a variable environment. (i) The tree in a closed forest; (ii) the fall of a neighbouring tree creates a gap, giving sudden lateral illumination; (iii) the mono-model tree reacts by a deformation of its crown, but is unable to fill the gap and make full use of the available light.

## (a) Gigantism

When applied to a solitary unit of morphogenesis or to a solitary module, gigantism alone may give rise to large herbs or trees which are monocarpic (for example, *Agave*, *Puya*, *Metroxylon*), but this is relatively rare among seed plants. Gigantism of an architectural model involves the enlargement of the specific plan of construction characteristic of the model (figure 5). In such cases the process of repetition of modules or of units of morphogenesis occurs *within* the model, but the model itself remains solitary. Despite its apparent simplicity, this process is also rare among seed plants. It involves only a few old families, such as Myristicaceae, and several pioneer trees of the tropics: *Anthocephalus*, *Macaranga*, *Solanum*, *Cecropia*, *Musanga* or *Anthocleista*. At least two reasons may be suggested for this. Since only the younger parts of shoots carry leaves and the older parts support them structurally, the progressive enlargement of preexisting shoots without the addition of new (leafy) shoots (by repetition) would give rise to a well-illuminated crown, but one with inadequate photosynthate to support an increasing respiratory burden. On trees with long-lived leaves, such enlargement may be possible, as the architecture of some species of *Araucaria* suggests. Leaf longevity in *Araucaria* spp. may be up to 15 years (Molisch 1938). Another disadvantage of forming large structures by gigantism without repetition in a mono-model tree is the restriction placed on opportunistic growth from dormant apical meristems to exploit light, nutrients and other resources that the organism may encounter in a variable environment. This is depicted diagrammatically in figure 5.

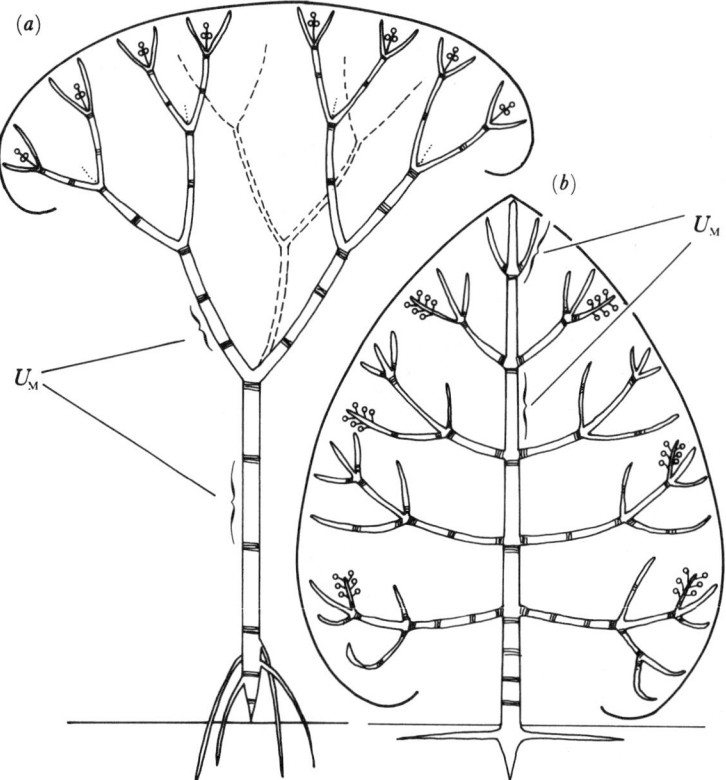

FIGURE 6. Repetition of the unit of morphogenesis ($U_M$) in *Schuurmansia heningsii* (Ochnaceae) (a) and *Pinus silvestris* (Pinaceae) (b).

## (b) Repetition

Repetition of the unit of morphogenesis gives rise to a shoot with rhythmic growth, and usually with laterally borne reproductive structures. Such shoots are easy to observe in many tree species (figure 6).

Repetition of the module gives a modular architectural model (figure 3): the precise and regular arrangement of these modules is usually characteristic of a species in early life. When applied to the architectural model itself, repetition leads to a colonial structure characteristic of the crowns of many mature trees (figure 7). This particular repetition to give a multi-model tree is referred to as *reiteration* and usually starts with the activation of dormant meristems (Oldeman 1977).

The reiteration process is also a manifestation of modular growth, but of a different kind from that previously outlined. Whereas the architectural model appears to be a standard growth response to a narrow range of conditions (for example, trees in the forest understory), the reiterated complex of architectural models represents a more opportunistic response to a greater diversity of conditions. This is shown diagrammatically in figure 7.

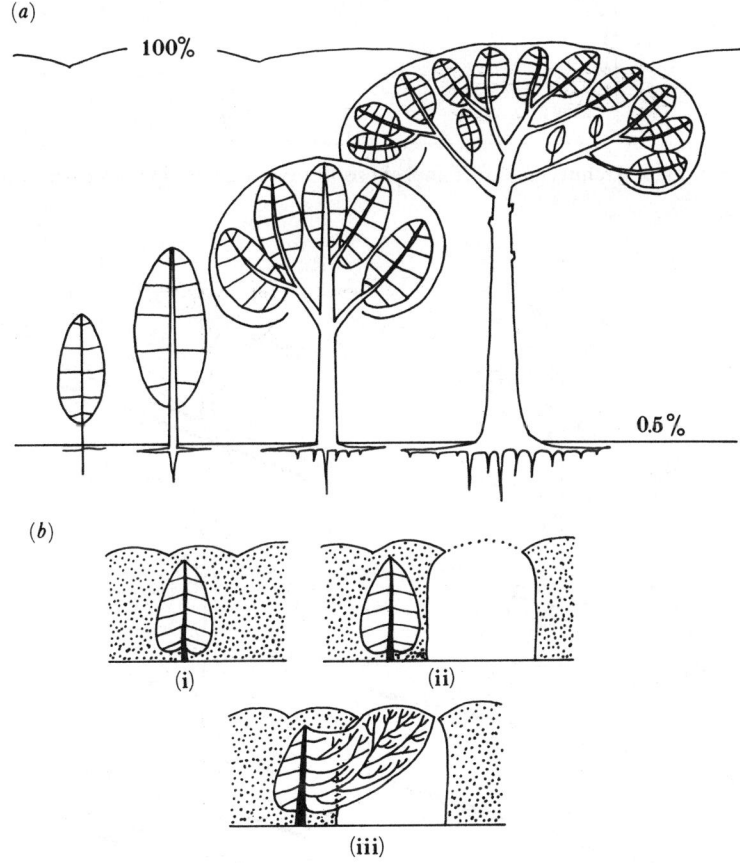

FIGURE 7. (a) The process of repetition applied to the architectural model: reiteration. Relative levels of illumination inside and outside the forest canopy are indicated. (b) The efficient opportunistic reaction of a multi-model tree to a variable environment. (i) The tree in a closed forest; (ii) the fall of a neighbouring tree creates a gap, giving sudden lateral illumination; (iii) the colonial tree reacts by reiteration to fill the gap completely and use all available light.

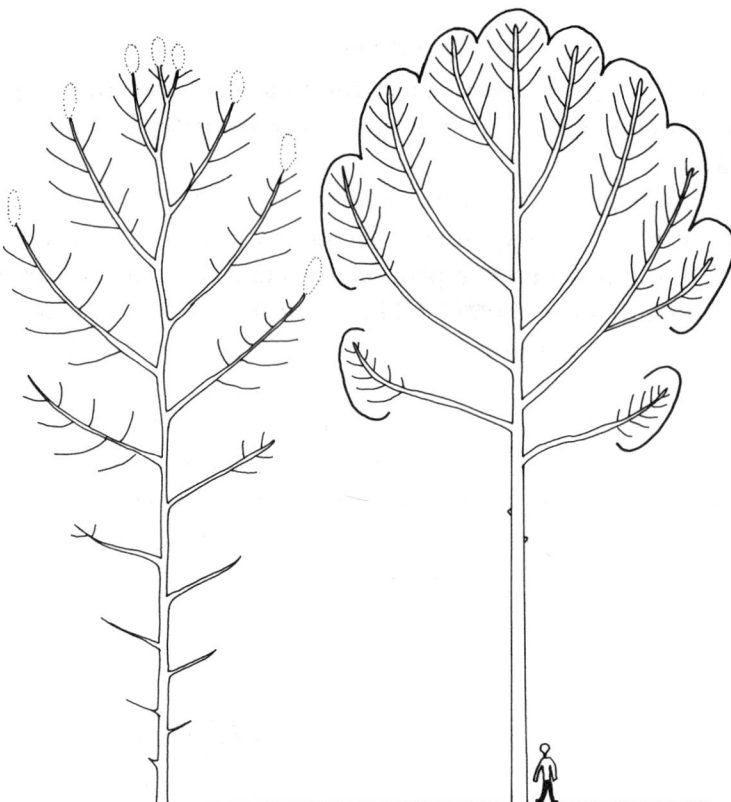

Figure 8. Architectural metamorphosis in two species of Dipterocarpaceae.

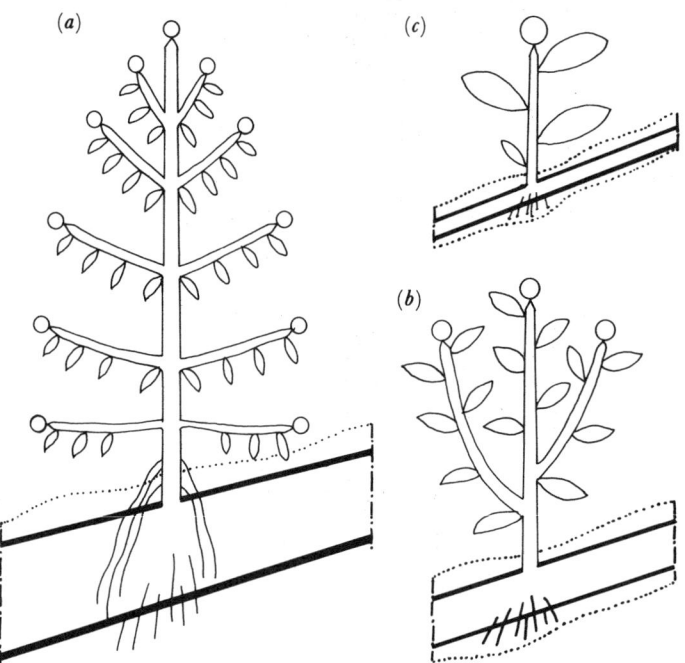

Figure 9. Diminished vegetative growth of successive reiterated complexes in the tree crown during ontogenesis. Canopy roots (Nadkarni 1981) are often easy to observe when the mantle of epiphytic plants wrapping the supporting branch (dotted lines) are removed. (a) Reiterated complexes borne on the trunk or on thickened branches develop like small trees. The architectural model, including its root system, is fully expressed. (b) After several episodes of reiteration, the complexes are smaller and frutescent (shrub-like), and the architectural model is only partly expressed. (c) The smaller axes of the crown support herbaceous reiterations which are neotenic.

In contrast to the rather precise arrangement of modules within architectural models, the spatial distribution of reiterated complexes is neither regular nor predictable. They may appear early in life on an isolated tree, but may be inhibited by the dark forest undergrowth until a tree reaches a height of 20–30 m. In the absence of precise rules for their three-dimensional arrangement, it has so far been difficult to model reiterated complexes with computer graphics, although some promising attempts have been made recently (P. de Reffye, personal communication).

The spatial distribution of reiterated complexes to form a crown shows, none the less, a gradient of predictability. The positions of large reiterated complexes, whose formation and expansion are governed by illumination of the exposed canopy, are impossible to predict morphologically. Their precise extent depends on the levels of interference among themselves. Such interference may be very evident in some trees (for example, *Pinus pinea*, *Nothofagus* sp., *Leptospermum* sp. and many members of the Dipterocarpaceae and Vochysiaceae): the term 'crown shyness' (French *timidité*) has been used to describe this phenomenon (Ng 1977; Hallé 1979; Hallé & Ng 1981).

Reiterated complexes may become organized in a somewhat more predictable manner. This is evident in some dipterocarps whose architectural dynamics have been described by Edelin (1984). Metamorphosis (Edelin 1984) is the process by which the plagiotropic branches characteristic of the sapling (with their limited ability to develop secondary thickening) are abruptly replaced by orthotropic branches imitating the trunk. By acquiring the same orientation of growth and the same thickening ability as the trunk itself, the branches undergoing metamorphosis become adventitious trunks, that is, reiterated complexes (figure 8). Architectural metamorphosis is a fundamental process in the growth and development of most tree species (Edelin 1984).

During the course of evolution as the reiterated complexes become smaller and simpler, their spatial location can be more accurately predicted. During ontogenesis, the successive complexes may become miniaturized and morphologically indistinguishable from a module (figure 9).

4. ARCHITECTURAL CONTINUUM

Modules and reiterated complexes both result from the repetition of a genetic programme of ontogenesis and differentiation. But there appears to be a gradient of predictability in their levels of organization. The architectures of mature trees may be interpreted, with such distinctions in mind, as stages of an architectural continuum. This may be illustrated by some general of tropical woody plants such as *Alstonia* (Prévost 1967), *Solanum* (Prévost 1978), *Cordia* (Edelin & Hallé 1985) and *Tabebuia* (Borchert & Tomlinson 1984), where a complete range of intermediary states between reiterated complexes and modules can be observed (shown schematically in figure 10). The last stage in this continuum is represented by the model of Prévost.

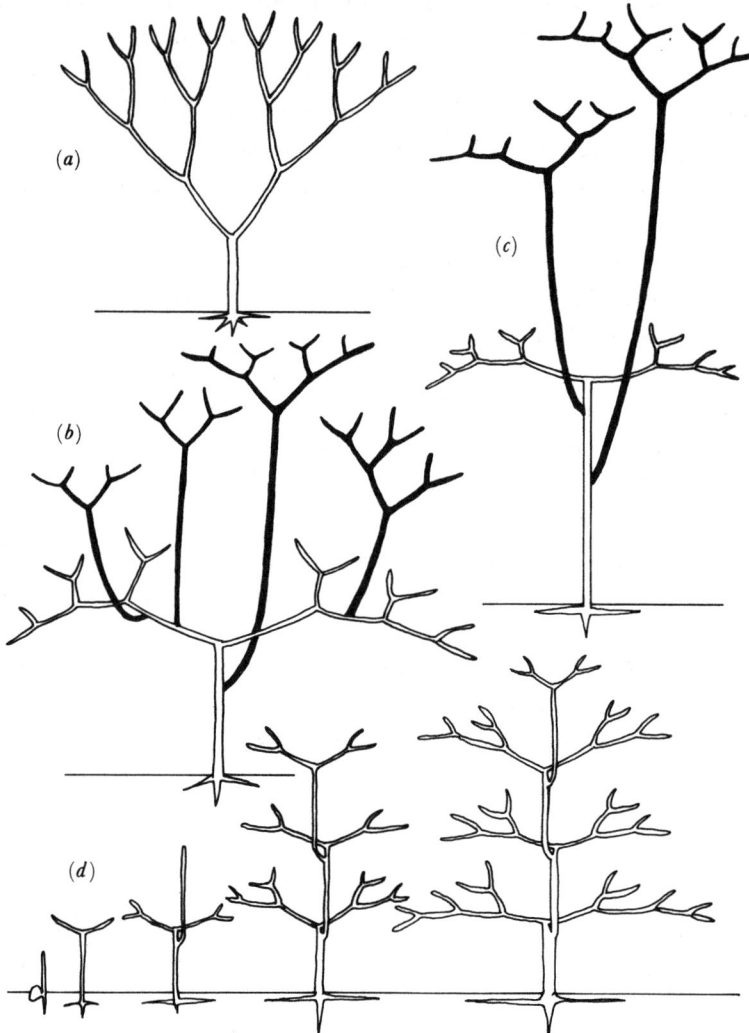

FIGURE 10. Architectural continuum of models and reiterated complexes. (a) Leeuwenberg's model, without reiteration. (b) Diffuse reiteration in Leeuwenberg's model. (c) Concentration of reiterated complexes in the apical part of the trunk module. (d) Ontogenesis of Prévost's model. This architectural continuum occurs in genera such as *Solanum* (Solanaceae), *Cordia* (Boraginaceae), *Tabebuia* (Bignoniaceae) and *Alstonia* (Apocynaceae).

I express my sincere gratitude to James White for his invaluable help in giving my French ideas their English form, which is much more than a mere translation.

## REFERENCES

Borchert, R. & Tomlinson, P. B. 1984 Architecture and crown geometry in *Tabebuia rosea* (Bignoniaceae). *Am. J. Bot.* **71**, 958–969.

Cusset, G. 1982 The conceptual bases of plant morphology. In *Axioms and principles of plant construction* (ed. R. Sattler), pp. 8–86. The Hague: Martinus Nijhoff/Dr W. Junk.

Edelin, C. 1984 L'architecture monopodiale. L'exemple de quelques arbres d'Asie tropicale. Thesis, Université des Sciences et Techniques du Languedoc, Montpellier.

Edelin, C. & Hallé, F. 1985 Architecture et évolution chez deux genres d'arbres tropicaux: *Diospyros* et *Cordia*. *Proc. 110ᵉ Congr. Soc. Savantes, Montpellier*, 1–5 April 1985.

Hallé, F. 1979 Premieres données architecturales sur les Dipterocarpaceae. *Mém. Mus. Nat. Hist. natn. Paris, Ser. B, Bot.* **26**, 20–36.

Hallé, F. & Martin, R. 1968 Étude de la croissance rythmique chez l'Hévéa (*Hevea brasiliensis* Mull. Arg., Euphorbiacées-Crotonoidées). *Adansonia* **8**, 475–503.

Hallé, F. & Ng, F. S. P. 1981 Crown construction in mature Dipterocarp trees. *Malay. For.* **44**, 222–233.

Hallé, F. & Oldeman, R. A. A. 1970 *Essai sur l'architecture et la dynamique de croissance des arbres tropicaux.* Paris: Masson. English translation by B. C. Stone, 1975. *An essay on the architecture and dynamics of growth of tropical trees.* Kuala Lumpur: Penerbit University Malaya.

Hallé, F., Oldeman, R. A. A. & Tomlinson, P. B. 1978 *Tropical trees and forests: an architectural analysis.* Berlin: Springer-Verlag.

Harper, J. L. & White, J. 1974 The demography of plants. *A. Rev. Ecol. Syst.* **5**, 419–463.

Molisch, H. 1938 *The longevity of plants.* New York: E. H. Fulling.

Nadkarni, N. M. 1981 Canopy roots: convergent evolution in rain forest nutrient cycles. *Science, Wash.* **214**, 1023–1024.

Ng, F. S. P. 1977 Shyness in trees. *Nature Malaysiana* **2**, 34–37.

Oldeman, R. A. A. 1977 L'architecture de la forêt Guyanaise. *Mémoires O.R.S.T.O.M.* **73**. Paris: O.R.S.T.O.M.

Prévost, M. F. 1967 Architecture de quelques Apocynacées ligneuses. *Mém. Soc. bot. Fr.* **114**, 23–36.

Prévost, M. F. 1978 Modular construction and its distribution in tropical woody plants. In *Tropical trees as living systems* (ed. P. B. Tomlinson & M. W. Zimmermann), pp. 223–231. Cambridge University Press.

Reffye, Ph. de 1979 Modélisation de l'architecture des arbres par des processus stochastiques. Simulation spatiale des modèles tropicaux sous l'effet de la pesanteur. Application au *Coffea robusta.* Thesis, Université Paris Sud, Centre d'Orsay.

Reffye, Ph. de 1983 Modèle mathématique aléatoire et simulation de la croissance et de l'architecture du caféier robusta. 4. Programmation sur micro-ordinateur du tracé en trois dimensions de l'architecture d'un arbre. Application au caféier. *Café-Cacao-Thé* **27**, 3–19.

White, J. 1979 The plant as a metapopulation. *Ann. Rev. Ecol. Syst.* **10**, 109–145.

# Establishment of modular growth in a tropical tree: *Isertia coccinea* Vahl. (Rubiaceae)

By D. Barthelemy

*Laboratoire de Botanique, Institut Botanique, 163 rue A. Broussonnet, 34000 Montpellier, France*

The developmental architecture of *Isertia coccinea* Vahl. (Rubiaceae) is described in terms of modular construction. In this paper a module is defined as a determinate shoot terminating in an inflorescence.

*Isertia coccinea* Vahl. is a small tree that conforms to Scarrone's architectural model. Before flowering, this tree is not modular and consists of an orthotropic trunk bearing rhythmic tiers of orthotropic branches. In a later stage, the lateral complexes branch sympodially as a result of terminal flowering, the whole tree displaying a partly modular stage. Later still, the trunk can terminate in an inflorescence, the tree's growth thus becoming entirely modular. The partly modular stage can be of long or short duration depending on the environmental conditions. In shaded conditions, this stage is long or indeterminate; in open conditions it can be very short, the entirely modular stage becoming dominant early in the life of the tree.

## Introduction

*Isertia coccinea* Vahl. (Rubiaceae) is a small tree, characteristic of secondary forests. This pioneer has been studied in French Guiana, where it is quite common and abundant. Flowering is precocious, the orange-red tubular flowers being pollinated by hummingbirds. In the first section, the flowering process and architecture of individuals growing in sunny conditions are described. In the second, the influence of environmental conditions (especially light intensity) on the growth form and flowering process of this species is discussed.

## Establishment of modular growth

*Isertia coccinea* Vahl. conforms to Scarrone's architectural model; 'the architecture is determined by an orthotropic rhythmically active terminal meristem which produces an indeterminate trunk bearing tiers of branches, each branch-complex orthotropic and sympodially branched as a result of terminal flowering' (Hallé *et al.* 1978) (figure 1).

In this section we refer only to the normal growth of this tree growing in open conditions in the secondary forest or at the margin of primary forest. In the following descriptions we shall use the terms 'modular growth' or 'module' in the sense of *article* (in French) defined by Prévost (1966) and Hallé (this symposium): a module consists of a determinate shoot terminating in an inflorescence (or otherwise aborting in some way). Thus a tree having a modular construction is made up of a series of equivalent morphological units repeated indefinitely. Most commonly, modules form sympodia.

Before flowering, the young tree is not modular, and conforms to Rauh's architectural model (Hallé & Oldeman 1975; Hallé *et al.* 1978) (see figure 1). The orthotropic trunk grows

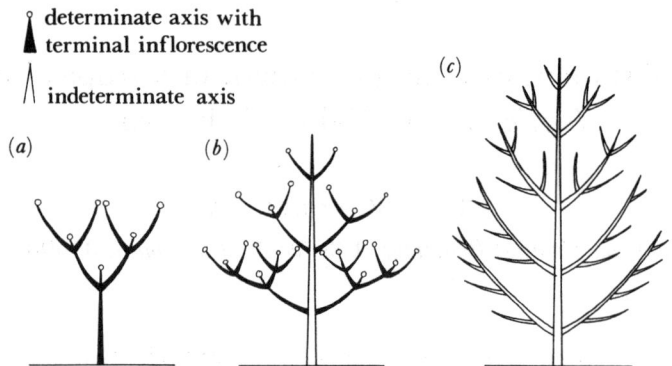

FIGURE 1. (a) Leeuwenberg's model; (b) Scarrone's model; (c) Rauh's model.

rhythmically and produces tiers of orthotropic branches. When the tree becomes older, the first lateral complexes produced can be shed without having flowered.

After a few years, the central axis of each lateral complex flowers terminally, the growth of the tree becoming partly modular (figure 3d).

In all the individuals observed in the above-mentioned environmental conditions, the trunk also bears a terminal inflorescence at an older stage, the whole tree thus displaying an entirely modular growth. The number of internodes produced by the trunk below the terminal inflorescence ranged from 65 to 80 in the individuals observed.

Concerning the proximal modules of each lateral complex, precise analysis of many individuals shows that, in general, the number of internodes produced below the terminal inflorescence of each axis is smaller for those inserted the higher on the trunk, than for the basal ones (figure 2a). Also their structure is simplified, the monopodial growth becoming shorter and shorter from the basal ones to those inserted at the top of the trunk. Correlated with this sequence, the proximal axis of the lower lateral complexes can carry branches on their monopodial growth as well as immediately below the terminal inflorescence, whereas on the proximal axes of the upper complexes, branching occurs only immediately below the terminal inflorescence, that is, in a subterminal position.

Similar sequences occur in the lateral complexes themselves. On large branch complexes, inserted on the lower part of the trunk, the distal units are simpler than the proximal ones, exhibiting a shorter monopodial growth and a smaller number of constitutive internodes (figure 2d).

The last lateral complexes produced by the trunk are entirely modular and branch essentially as in Leeuwenberg's model (figure 1).

In the peripheral part of quite an old tree, all the modules are thus reduced in size. Indeed, at first glance they all seem to be of exactly the same structure. However, the analysis of such a crown shows that the number of internodes of these peripheral modules varies from 6 to 13 (most frequently from 6 to 10), and that these modules are of two kinds: Modules with 6, 7 or 8 internodes, are quite vigorous and composed of a single growth unit; modules of 9 to 13 internodes are slender, much less vigorous, and consist of short internodes. They show two flushes of growth and generally branch very little. Generally these modules, and the few axes they give rise to, soon die.

During the course of their growth, individuals of *Isertia coccinea* become more and more

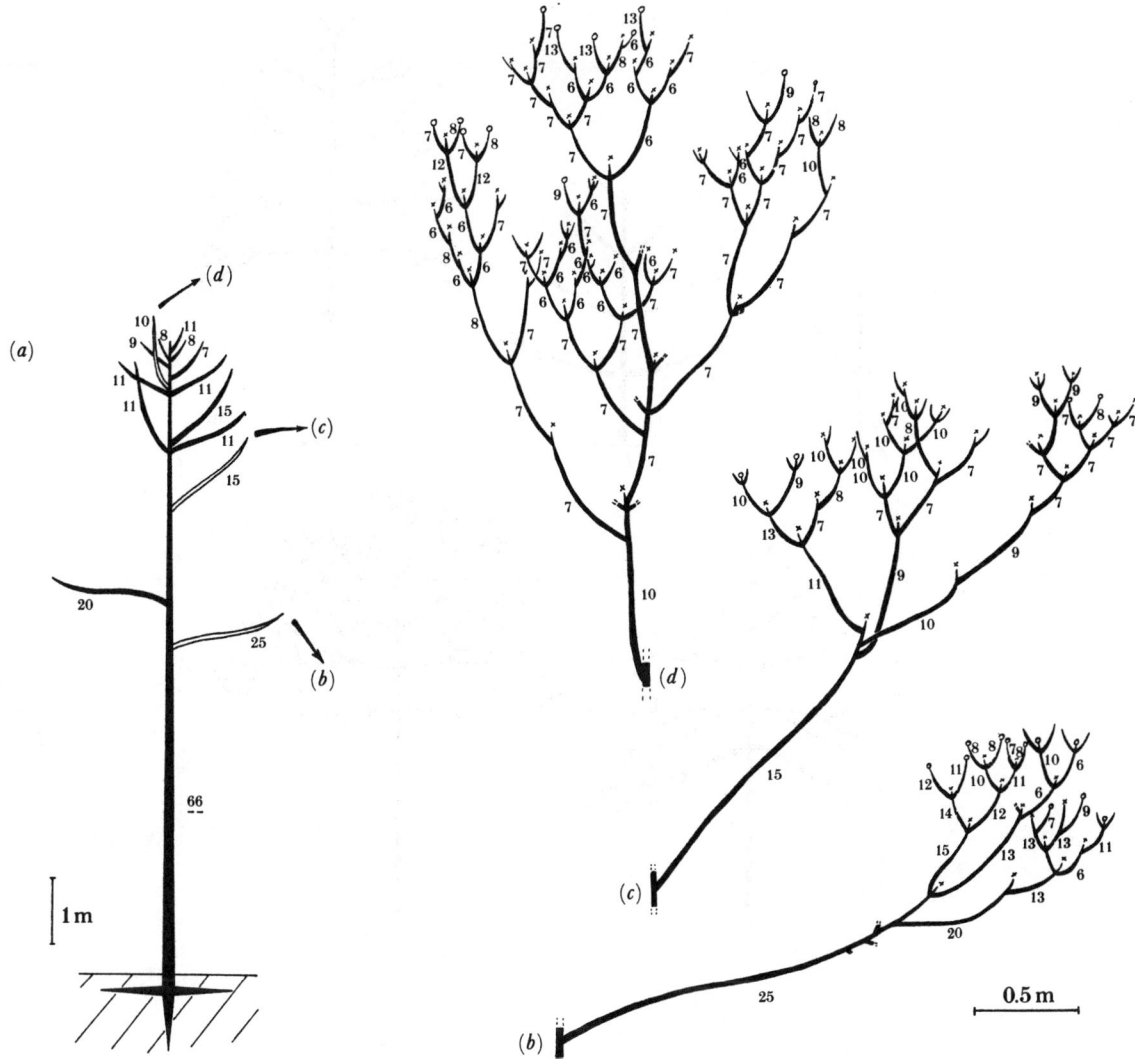

FIGURE 2. Natural individual of *Isertia coccinea* showing the development of modular growth in three lateral complexes (*b, c, d*) inserted at three different levels.

The number of vegetative internodes produced below the terminal inflorescence of each axis shows that the higher the axis is inserted on the trunk, the more modular the lateral branch complexes are, and the more precocious is the flowering process.

(*a*) Trunk and proximal axis of the lateral complexes. The number of internodes for each axis is noted.

(*b*) Basal lateral complex showing the structural development of the modules from the proximal to the distal part.

(*c*) Intermediate construction between the lateral complexes of (*b*) and (*d*).

(*d*) One of the highest lateral complexes displaying entirely modular growth.

modular. The peripheral part of an old tree is entirely modular and composed of modules which are quite similar in structure and which branch immediately below each terminal inflorescence, that is, in a subterminal position. The flowering process, closely related to this growth pattern, is more and more precocious. In the crown of an old tree, all the modules bear a terminal inflorescence, and the whole tree grows as in Leeuwenberg's model, exhibiting prolific flowering.

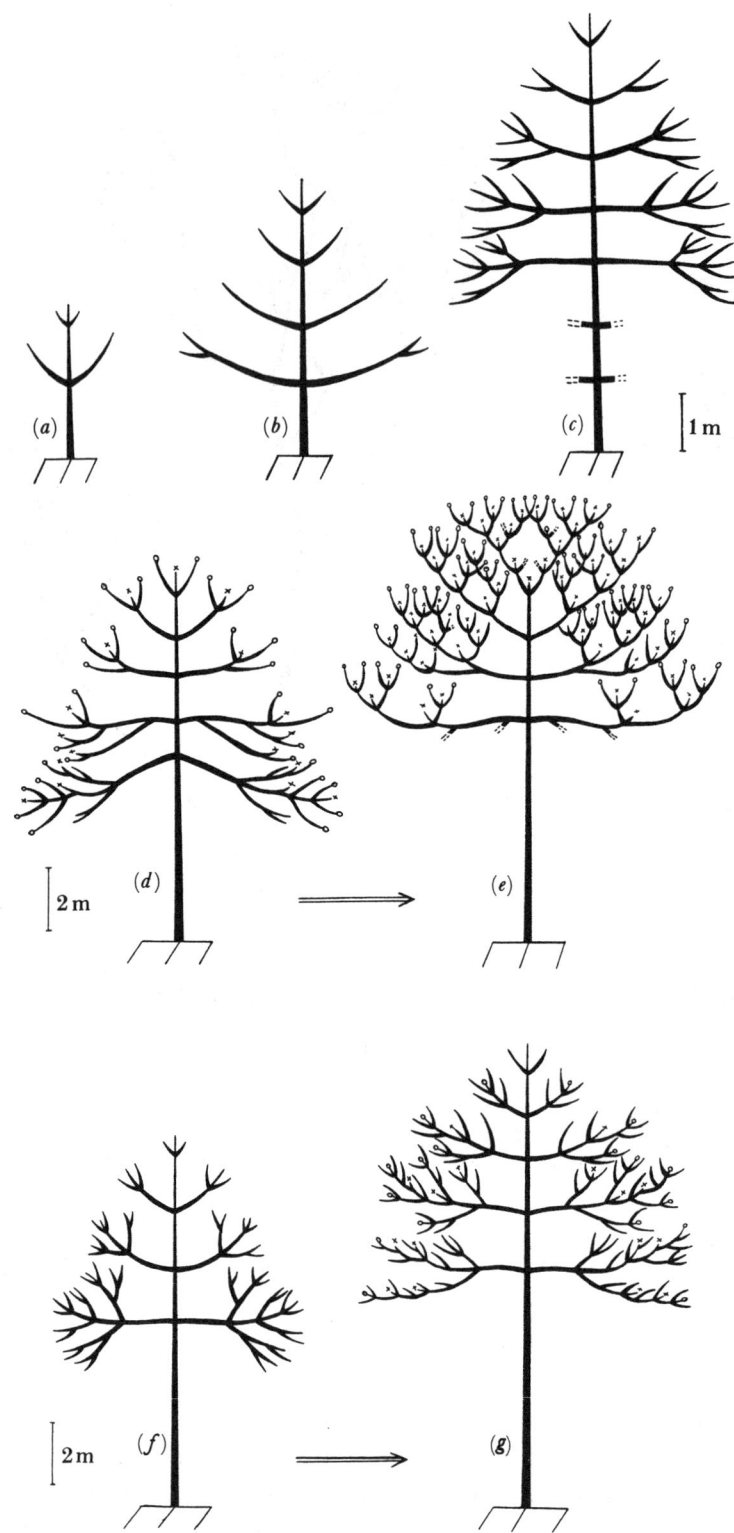

FIGURE 3. Schematic development of Scarrone's model in *Isertia coccinea* Vahl. in open conditions (*d, e*) and in shaded conditions (*f, g*). Before flowering this tree is not modular and consists of an orthotropic trunk rhythmically bearing tiers of orthotropic branches (*a, b, c, f*). In the following stage, the lateral complexes branch sympodially as a result of terminal flowering, the whole tree displaying a partly modular stage (*d, g*).

At an older stage the terminal meristem of the trunk can also flower terminally, the growth of the tree becoming entirely modular. The partly modular stage can be long or short, depending on the environmental conditions: in shaded conditions (*f, g*) this stage is long or indeterminate; in open conditions (*d, e*) this stage can be very short, the entirely modular stage becoming dominant early in the life of the tree.

## The influence of environmental conditions

The vegetative architecture is a constant and stable characteristic of a plant. However, different environmental conditions can involve changes of form. These changes, whatever their importance, are quantitative and do not affect the basic architecture of the species (Hallé 1978).

The most important environmental factor is probably light intensity. The influence of incident light on the branching and flowering behaviour has already been reported for several trees and treelets (Nanda 1962; Kahn 1975; Temple 1975, 1977; Hallé 1978) and further investigations will no doubt provide many other examples. *Isertia coccinea* displays during its growth successively, non-modular, partly modular, and entirely modular architecture. In open conditions, the partly modular stage is short, as the trunk can bear a terminal inflorescence early in the course of its life, sometimes before it is ten years old. Thus in sunny conditions, the whole tree (when it is isolated) exhibits prolific flowering and branching in the course of a few years. Individuals of *Isertia coccinea*, however, can be found in shaded condition under the canopy of the primary forest or shaded by other pioneers such as *Cecropia* sp. (Moraceae). In this particular condition, the partly modular stage can be long or indeterminate, the growth of the trunk remaining monopodial for a long time and the tree displaying a pure Scarrone's model (figure 1).

In individuals shaded by dominant trees, a longer period of monopodial growth is also found throughout the lateral complexes which are sparsely branched compared with individuals freely exposed to bright sunlight. Flowering behaviour is also modified considerably in trees growing under the shade of other trees, and flowering is sparse throughout the whole tree; it is also much less precocious than in open conditions.

The developmental sequence of growth pattern for *Isertia coccinea* in open and shaded conditions is shown in figure 3 in diagrammatic form.

The influence of light intensity on the growth and flowering behaviour of plants is unlikely to be restricted to *Isertia coccinea*, and the current studies of many other species (unpublished data) indicates the same behaviour in many plants. Previous studies (Nanda 1962; Kahn 1975; Temple 1977) also show similar processes in various species, and the reduction of the vegetative stage in individuals of a species growing in open conditions compared with individuals of the same species growing in shaded conditions seems to be a general rule.

## Conclusion

*Isertia coccinea* Vahl. shows clearly the establishment of modular growth in a tree exhibiting orthotropic axes and terminal inflorescences. This establishment of modular growth in Scarrone's model seems to be a frequent process for a number of species ranging from small treelets (unpublished data) to large trees (Nanda 1962).

This example also shows the close relationship that can exist between different architectural models such as Leeuwenberg's model, Scarrone's model and Rauh's model.

In *Isertia coccinea* Vahl. the modular growth is closely related to flowering and in the entirely modular crown of an old individual exposed to bright sunlight flowering and branching is profuse. By comparison, when the tree is smaller and younger, flowering is sparse and in the lower part of the tree some lateral branch complexes can be shed before they have flowered. This is more obvious in individuals growing shaded by other trees; in this situation the monopodial vegetative growth phase is much longer and flowering remains sparse, sometimes for most, if not all, of the life of the tree.

Flowering in this species is probably related to the intensity of light and is much more precocious and profuse for the individuals exposed to bright sunlight than for individuals shaded by other trees. This has very important consequences for the form of the tree as the inflorescences are terminal. The more precocious the flowering is, the earlier modular growth will occur.

## REFERENCES

Hallé, F. 1978 Architectural variation at the specific level in tropical trees. In *Tropical trees as living systems* (ed. P. B. Tomlinson & M. H. Zimmermann), pp. 209–221. Cambridge University Press.

Hallé, F. & Oldeman, R. A. A. 1975 *An essay on the architecture and dynamics of growth of tropical trees.* (156 pages.) Kuala Lumpur: Penerbit University of Malaya.

Hallé, F., Oldeman, R. A. A. & Tomlinson, P. B. 1978 *Tropical trees and forests: an architectural analysis.* (441 pages.) Berlin: Springer-Verlag.

Kahn, S. 1975 Remarques sur l'architecture végétative dans ses rapports avec la systématique et la biogéographie. Thesis, Université des Sciences et Techniques du Languedoc, Montpellier.

Nanda, K. K. 1962 Some observations on growth, branching behaviour and flowering of teak (*Tectona grandis*) in relation to light. *Ind. For.* **88**, 207–218.

Prevost, M. F. 1966 Architecture de quelques Apocynacées ligneuses. *Bull. Soc. Bot. Fr.* **114**, 23–36.

Temple, A. 1975 Ericeae–Etude architecturale de quelques espèces. Thesis, Université des Sciences et Techniques du Languedoc, Montpellier.

Temple, A. 1977 Ericeae–Polymorphisme architectural d'une famille des régions tempérées et tropicales d'altitude. *C. r. hebd. Séanc. Acad. Sci., Paris* **284**, 163–166.

# Growth and form in lower plants and the occurrence of meristems

By A. P. J. Trinci and E. G. Cutter

*Department of Botany, University of Manchester, M13 9PL, U.K.*

Fungi and streptomycetes have a similar morphology and in both groups branching appears to be regulated in a similar manner. Both types of hyphae grow by tip extension but streptomycete hyphae never attain the extension rates commonly observed for fungi. Fungal hyphae are able to attain high rates of extension because a very large volume of protoplasm contributes to tip growth and because a vesicular growth system facilitates the rapid assembly of the tip wall. Growth of fungal and streptomycete mycelia involves the duplication of a physiological unit of growth which consists of a tip and a portion of hypha whose average length remains constant. However, it is not clear that growth of such mycelia is truly modular. Although hyphal fusions within a mycelium are common in higher fungi their significance in the organism's life style is not known. Growth in lower green plants, especially algae, is considered and the question of whether coenocytic algae are modular or not is discussed.

## Introduction

Although hyphae of fungi (3–10 µm in diameter, but occasionally 100 µm) and streptomycetes (0.5–1.5 µm in diameter) differ in diameter, these eukaryotic and prokaryotic organisms produce mycelia which have a very similar morphology (Oliver & Trinci 1985). In both groups, hyphae grow radially outwards from the centre of the colony, branch regularly and execute 'avoidance' reactions. Mycelial growth appears to be an adaptation to life in heterogeneous environments, particularly soil and, as mentioned by Pfenning (1984), this view is supported by the example of convergent evolution in the life styles of the prokaryotic streptomycetes and the eukaryotic fungi. This contribution will describe the growth of hyphae and the formation of mycelia and, where possible, comparisons will be made between fungi and streptomycetes, and algae. Some aspects of growth in the algae and other lower green plants will also be compared.

## Apical wall growth of fungal and streptomycete hyphae

It is well established that fungal hyphae only increase in length by apical extension. The length of the zone involved in extension may be determined either by depositing markers onto the tip and following their subsequent displacement (Castle 1959), or by measuring the distance from the tip to the point at which the hypha first attains its maximum diameter (Trinci & Halford 1975). The latter method was used to show that the extension rate of *Phycomyces* sporangiophores (figure 1) was directly related to extension zone length (Trinci & Halford 1975).

Several attempts have been made to explain tip growth in fungi. Reinhardt (1892) suggested that the tip wall is rigid and that a gradient in the rate of intussusception of new wall material determines the specific growth rate of the surface area of the wall and hence determines tip shape.

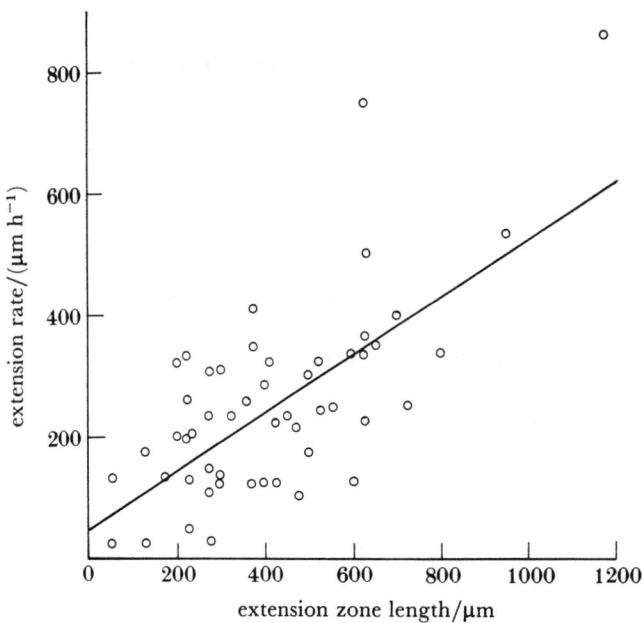

FIGURE 1. Relation between the extension rate of stage I sporangiophores of *Phycomyces blakesleeanus* and the length of their extension zones. The correlation coefficient ($+0.70$) between the rate of extension and extension zone length is highly significant ($p < 0.001$). (Redrawn from Trinci & Halford 1975.)

The 'unitary' growth hypothesis of Bartnicki-Garcia (1973) provided a mechanism to explain how new wall material might be inserted into the rigid tip wall. He proposed that tip extension involves the combined action of enzymes that hydrolyse wall polymers, for example, chitinases and glucanases, and enzymes that synthesize wall polymers, for example, chitin synthase and glucan synthases. However, unlike Reinhardt's model, Bartnicki-Garcia's hypothesis suggests that turgor pressure plays a role in hyphal extension by forcing apart the ends of microfibrils severed by lytic enzymes. Evidence that lytic enzymes may be involved in wall growth is provided by Rosenberger (1979) and Humphreys & Gooday (1984). The latter workers isolated a microsomal chitinase from *Mucor mucedo* that was apparently membrane-bound and which, like chitin synthase in the same organism, was activated by proteases. Humphreys & Gooday suggest that chitinase and chitin synthase may be regulated in concert during hyphal growth. However, as with most previous reports of the possible involvement of lytic enzymes in tip extension, no evidence is presented that allows a discrimination to be made between the action of chitinase in tip growth and its action in branch initiation. This evidence is provided by Kritzman *et al.* (1978) who used an immunofluorescent technique to demonstrate the presence of $\beta$-(1,3)-glucanase at the tips of hyphae of *Sclerotium rolfsii*. In contrast to the models of Reinhardt and Bartnicki-Garcia, Robertson (1959) proposed that tip extension involves a balance between wall synthesis and wall rigidification. The involvement of lytic enzymes in tip growth is not implicit in Robertson's model.

Radioautography can be used to determine the sites at which hyphae form wall polymers. $N$-[$^3$H]acetyl D-glucosamine is a precursor which is rapidly incorporated into chitin in fungal hyphae and into peptidoglycan in streptomycete hyphae; hyphae need only be exposed to this compound for 1 min before being processed for radioautography. The highly polarized nature of the synthesis of these structural polymers is shown in figure 2. Extraction of hyphae (with water at 100 °C or 1 M KOH at 60 °C) of *Schizophyllum commune*, which had been exposed to N-[³H]acetyl D-glucosamine for 10 min before preparing radioautographs, demonstrated that chitin formed at the tip was present in an insoluble form (Sonnenberg 1984). It had previously been shown that the tips of fungal hyphae were always covered with chitin microfibrils (Hunsley & Burnett 1970). By contrast, when growing hyphae of *S. commune* were exposed to [³H]glucose for 10 min, most of the label at the hyphal tip was present as water- or alkali-soluble glucan and very little was present as alkali-insoluble glucan (Sonnenberg 1984). Sonnenberg (1984) used pulse-chase experiments to demonstrate that glucans synthesized at the hyphal apex of *S. commune* moved subapically at a rate (7 µm h$^{-1}$) that was approximately the same as the rate of hyphal extension.

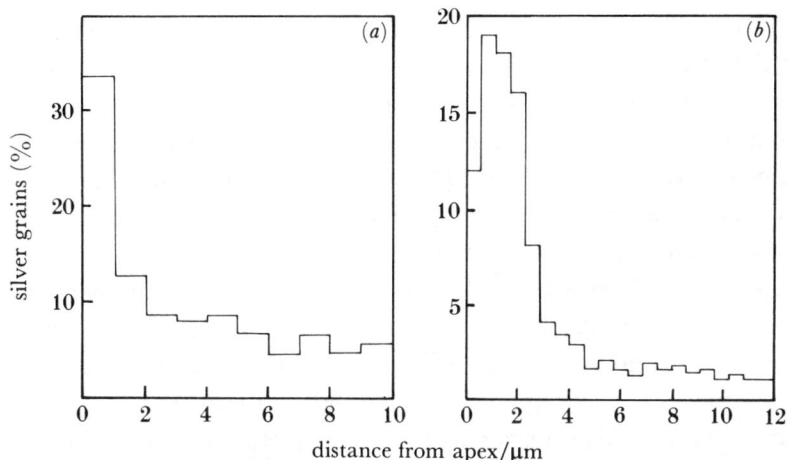

FIGURE 2. Wall synthesis in hyphae of (a) *Streptomyces antibioticus* (redrawn from Braña et al. 1982) and (b) *Schizophyllum commune* (redrawn from Sonnenberg 1984). The *Streptomyces antibioticus* hypha was incubated in N-[³H]acetyl-glucosamine for 1 min. The *S. commune* hypha was incubated in N-[³H]acetylglucosamine for 10 min and the radioactivity was then chased with non-radioactive medium.

Braña et al. (1982) used electron microscopy to determine the distribution of label along *Streptomyces antibioticus* hyphae and found that after exposure to N-[³H]acetyl D-glucosamine for 1 min the highest frequency (30% of the total silver grains counted) occurred within the first 1 µm of the hypha (figure 2). The pattern of apical wall growth found in *Streptomyces antibioticus* was very similar to that observed in fungi (figure 2 and Gooday 1971) and this indicates that primary wall growth in both groups occurs preferentially in apical regions of hyphae.

## WALL RIGIDIFICATION IN FUNGAL HYPHAE

The shape and length of the extension zone of a hypha growing at a linear rate remain constant with time, as does the radius of the hypha at the base of the extension zone. Steele & Trinci (1975) proposed that the period (called the extension zone expansion time) required for a hypha to expand from its minimum to its maximum radius is an inverse measure of the rate at which newly formed wall becomes rigidified. They found that although the extension zones of *Neurospora crassa* hyphae varied from 2.2 to 29.2 µm in length, all hyphae had approximately the same extension zone expansion time (42–54 s). They concluded that wall material inserted into the tip remains plastic for only a relatively short period of time. Although

Robertson (1968) suggested that wall rigidification might involve wall thickening, Trinci & Collinge (1975) failed to detect any difference between the thickness of the wall within and just below the extension zone.

Wessels et al. (1983) and Sonnenberg (1984) have advanced a model for tip growth in S. commune hyphae which is based upon the hypothesis of Robertson (1959). An important feature of this hypothesis is that synthesis and assembly of wall polymers are separate processes. In particular, it is suggested that (1,3)-β-glucan and chitin are inserted into the plastic extension zone wall as separate components which rapidly become cross-linked, and that wall rigidification is associated with this cross-linking process. The enzyme or enzymes that effect the coupling of the β-glucan to chitin must operate outside the protoplasmic membrane because one of the substrates (chitin) is not formed within the cytoplasm. Cross-linking probably occurs concurrently with crystallization of the chitin and glucan chains to produce the β-glucan–chitin complex that is found in the 'rigid' subapical parts of the wall (Sietsma & Wessels 1977, 1979); both components tend to form hydrogen bonds among themselves. Chitin is present at the tip in an insoluble form. However, since nascent chitin is particularly sensitive to chitinase and acid (Gooday & Trinci 1980; Lopez-Romero et al. 1978; Schneider & Seaman 1982; Vermeulen & Wessels 1984), at least some of the chitin in the extension zone may be present in an amorphous condition and thus may be available for cross-linking with other wall polymers.

The conclusion that wall growth and wall rigidification are independent processes is supported by Robertson's (1958) observation that when extension growth of a hypha of *Fusarium oxysporum* was stopped by an osmotic shock, the entire extension zone became rigidified within about 40 s. Similarly, Sonnenberg (1984) found that cross-linking between chitin and β-glucan occurred at the apex of non-growing hyphae of *S. commune* so that eventually the whole apex became rigidified. The results of Robertson, Sonnenberg and Steele & Trinci suggest that wall rigidification is a time-dependent rather than a growth-dependent process.

### Extension rates of fungal and streptomycete hyphae

Streptomycete hyphae never attain the high rates of extension attained by fungal hyphae. For example, hyphae of *Streptomyces coelicolor* extend at 30 °C at a maximal rate of *ca*. 30 μm h$^{-1}$ (E. J. Allan & J. I. Prosser, personal communication) while hyphae of *N. crassa* SY7A extend at 25 °C at a maximal rate of *ca*. 2300 μm h$^{-1}$ (Steele & Trinci 1975). Thus, although convergent evolution has resulted in the production by streptomycetes and fungi of mycelia that are morphologically similar, the constituent hyphae of these mycelia extend at quite different rates. This difference cannot be accounted for in terms of a difference in specific growth rate, since *S. coelicolor* has a specific growth rate of 0.26 h$^{-1}$ (Allan & Prosser 1983) and *N. crassa* SY7A has a specific growth rate of 0.27 h$^{-1}$ (Trinci 1973a). The difference in the extension rates of streptomycete and fungal hyphae can be attributed to two aspects of their growth. First, the volume of protoplasm contributing to hyphal extension in fungi is much greater than that contributing to the extension of streptomycete hyphae; the volume of the peripheral growth zone (the region of a hypha contributing to tip extension (Trinci 1971)) of a leading hypha of *N. crassa* SYR-17-3A extending at a linear rate is about $260 \times 10^3$ μm$^3$ (calculated from data in Trinci (1973a)) while the equivalent zone in *S. coelicolor* is only about 80 μm$^3$. The terminal 6.8 mm (Trinci 1973a) of an *N. crassa* hypha can contribute to tip extension because protoplasm formed in intercalary compartments can migrate towards the hyphal apex via pores present in the septa. Such migration is not possible in streptomycete hyphae because their septa lack pores; in streptomycetes the maximum length of hypha contributing to tip extension is less than *ca.* 100 μm. Some fungi (Fiddy & Trinci 1976*a*; Valla 1984) resemble streptomycetes in that tip extension is largely or wholly supported by the apical compartment. In other fungi the volume of protoplasm in a growing hypha remains virtually constant and migrates forward at the same rate as tip extension (Robinow 1963; Heath & Heath 1978; Gow & Gooday 1984). The second important difference is that extension of a fungal hypha, unlike that of a streptomycete hypha, involves the fusion of vesicles with the tip wall. Thus, extension rates of up to 100 μm min$^{-1}$ can be achieved in fungi because a large volume of protoplasm synthesizes membrane and cell wall precursors that are transported to the hyphal tip where they are rapidly integrated with the existing protoplasmic membrane and wall. Vesicular systems of tip growth may be a way of ensuring that expansion of membrane and wall are closely integrated and similar systems are involved in apical growth of root hairs (Bonnett & Newcomb 1966), pollen tubes (Rosen 1964) and algal filaments (Sievers 1965; Otto & Brown 1974).

In algae such as *Vaucheria*, in which typical tip growth occurs (Kataoka 1975*a*), new regions of wall growth can be induced by unilateral light, leading to bending (Kataoka 1975*b*), or by illumination with blue light just below the apex, leading to branching (Kataoka 1975*b*; Aberg 1978). The lobes of desmids such as *Micrasterias* also exemplify tip growth (Kallio & Lehtonen 1981). By using a fluorescent indicator, it has recently been shown that each area of growth that will give rise to a lobe can be identified by accumulation of membrane-bound calcium (Meindl 1982). These results indicate that the template for the complex pattern of the *Micrasterias* cell lies in the plasma membrane, and support Kiermayer's (1981) view that the incorporation of vesicles into the plasma membrane depends on a membrane-recognition process between vesicle and plasma membranes. Meindl considers that this mechanism may be mediated by membrane-bound calcium.

Microvesicles, called chitosomes, have been isolated from a wide range of fungi (Bracker *et al.* 1976; Bartnicki-Garcia *et al.* 1979). Chitosomes are 40–70 nm in diameter and serve as conveyors of chitin synthase zymogen from its point of synthesis to the region of apical extension. The observation that a chitin microfibril can be formed by a single chitosome indicates that a chitosome must contain sufficient chitin synthase to synthesize the many chains that make up a microfibril. Each chitin chain is presumably made by a separate chitin synthase unit and the chains collectively crystallize into a microfibril as they are synthesized. Various reports (Archer 1977; Braun & Calderone 1978; Wessels & Sietsma 1979) suggest that active chitin synthase is predominately located in the protoplasmic membrane where it would be accessible to substrates and effectors from the cytoplasm and, providing that it spanned the membrane, could 'spin out' the growing chitin chain to the wall. An aggregate of enzyme monomers could thus give rise directly to the microfibrils in the wall. Similarly, $(1,3)$-$\beta$-glucan synthase has been shown to be associated with the protoplasmic membrane of *Saprolegnia monoica* (Girard & Fèvre 1984). However, Bartnicki-Garcia *et al.* (1984) have shown that the protoplasmic membrane fraction isolated from the slime mutant of *N. crassa* contained only a small portion of the chitin synthase of this fungus. This group of workers believe that the bulk of chitin synthase in growing fungi is in chitosomes rather than in the protoplasmic membrane and they think that chitosomes may be released into the wall (Bartnicki-Garcia *et al.* 1979).

## Polarity

Transcellular electrical currents have been shown to occur in many eukaryotes and electrical currents have been demonstrated around the tips of extending hyphae and rhizoids of various fungi (Kropf *et al.* 1983; Armbruster & Weisenseel 1983; Gow 1984; Stump *et al.* 1980; Horwitz *et al.* 1984) and algal filaments (Weisenseel & Kichereer 1981). The direction of flow of the electrical current (*ca.* 0.25–0.30 μA cm$^{-2}$ at the tip) in *Achlya bisexualis* was normally inward for the apical 350 μm of the hypha and outward further back, and the net inward and outward currents were approximately equal (figure 3). These measurements were made approximately

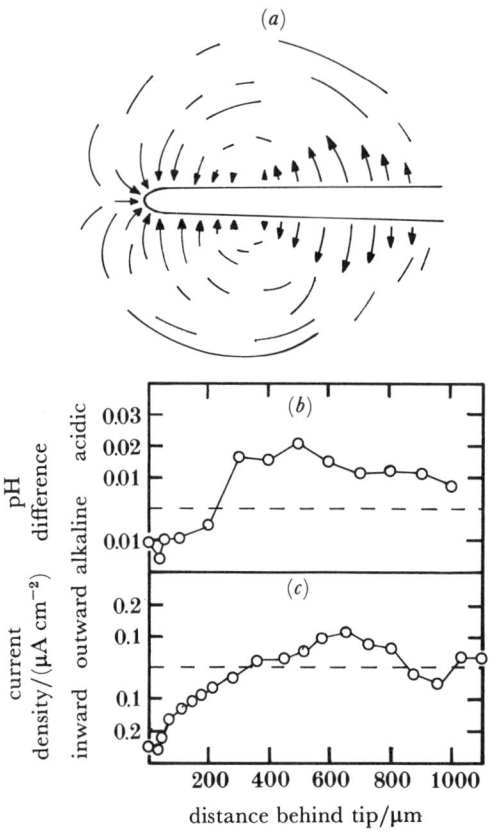

FIGURE 3. Transhyphal electrical current in hyphae of *Achlya bisexualis* and longitudinal pH gradient in the surrounding medium. (*a*) Pattern of transhyphal ion–electric current around the hypha; (*b*) extracellular pH profile in medium adjacent to the hypha; (*c*) electrical current profile of the same hypha. (Redrawn from Gow (1984) and Gow *et al.* (1984).)

30 μm from the hyphae, thus current density at the membrane surface may be considerably greater than the recorded values. The pH of the medium immediately external to the tips of extending hyphae of *A. bisexualis* was always alkaline (on average by about 0.022 pH units) with respect to the bulk, unbuffered culture medium, but the pH of the medium adjacent to segments of hyphae located 250–1000 μm from the tip was on average about 0.028 pH units more acidic than the bulk medium. The average location of maximum alkalinization (*ca.* 38 μm behind the apex) around an extending hypha corresponded approximately to the region (*ca.* 30–60 μm behind the apex) of maximum inward current, and the extent of alkalinization of

the medium around a hyphal tip was quantitatively consistent with an influx of protons at a rate given by the observed inward electrical current (Gow et al. 1984). However, although the zone of alkalinity around the tip matched that of inward current, the distal acidic zone extended beyond the region of outward current. Gow et al. (1984) suggest that some of this distal acidity results from the production of acidic metabolites such as lactic or succinic acids. Hyphae have been seen to drive currents even though they were not extending, and some hyphae have been seen to extend in the absence of electrical current flow (Kropf et al. 1984).

Figure 4 illustrates the hypothesis proposed by Kropf et al. (1984) to account for the profiles of pH and current observed around hyphae of *A. bisexualis*. They suggest that electrical current is driven through the hypha because protons are expelled by distally located proton-translocating ATPase and flow back into the tip by symport (co-transport) with amino acids. Thus, this hypothesis suggests that a fungal hypha is a spatially extended chemiosmotic system, with the nutrient symporter preferentially located in the hyphal tip. The actual distribution of symporters and ATPases is not known: symporters may be concentrated at the tip, or ATPases may be excluded from the tip, or both. Kropf et al. (1983) showed the transhyphal electrical current required the presence of amino acids and was abolished by raising the external pH from 6.5 to 8.5.

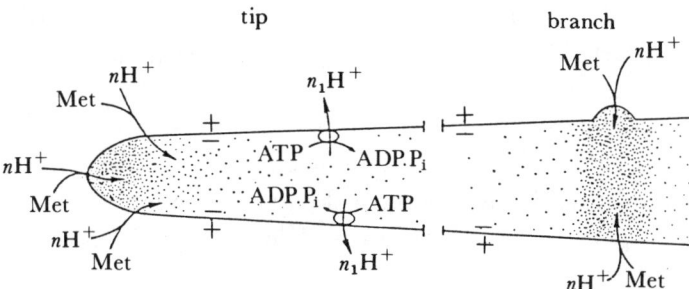

FIGURE 4. Model proposed to account for the generation of transcellular ion currents by *Achlya bisexualis*. The inward current results from symport of $n$ protons with one methionine molecule (Met). The outward current is thought to be generated by an electrogenic ATPase that extrudes $n_1$ protons per ATP hydrolysed (from Kropf et al. 1984).

Support for the hypothesis shown in figure 4 is provided by the work of Slayman and others. Slayman (1965 a, b) inserted microelectrodes into hyphae of *N. crassa* and found that they maintained a membrane potential of approximately $-200$ mV. Most of the potential was indifferent to $K^+$ diffusion and extremely sensitive to inhibition of mitochondrial respiration. The potential was subsequently shown to be generated by an electrogenic, proton-translocating ATPase which was plugged through the protoplasmic membrane (Slayman et al. 1973; Scarborough 1976). In *N. crassa* the proton current generated by the ATPase drives the accumulation of various nutrients such as glucose (Slayman & Slayman 1974), phosphate and ammonium ion (Slayman 1977). The proton-translocating ATPase of the protoplasmic membrane of eukaryotes appears to be distinct from that of mitochondria (Harold 1982).

Other eukaryotic organisms that form tip growing filaments have also been shown to drive transcellular currents. These include germinating algal eggs (Nuccitelli & Jaffe 1974), root hairs (Weisenseel et al. 1979) and pollen tubes (Weisenseel et al. 1975). The transcellular currents generated in these systems are similar to those observed in fungal hyphae, that is, the current is oriented in the same axis of polarity as the cells, with positive current entering the growing apex and leaving further back.

H. MacKinnon & N. A. R. Gow (personal communication) have shown that *N. crassa*, like zygotes of *Fucus* (Peng & Jaffe 1976), spores of *Equisetum* and *Funaria* (Bentrup 1968; Chen & Jaffe 1979) and pollen grains (Marsh & Beams 1945), became oriented in external electrical fields. They found that when grown in electrical fields of 20–40 mV per cell, spores and mycelia of *N. crassa*, respectively, formed germ tubes and branches preferentially on sides facing the anode, and growing hyphae became preferentially oriented towards the anode. MacKinnon & Gow suggest that (since under normal conditions the hyphal tip is electropositive) growth tends to occur preferentially towards the positive poles of both endogenous and applied fields.

Jaffe (1977) has suggested that transcellular currents in filamentous systems may constitute a force that directs cell constituents (vesicles?), either by electrophoresis of charged vesicles within the cytoplasm or of particles within the fluid phase of the membrane, to sites of wall synthesis. Evidence for self-electrophoresis as a mechanism of growth localization is provided by studies of the oocyte-nurse cell syncytium of *Hyalophora cecropia* (Woodruff & Telfer 1980; DeLoofe 1983).

By using eggs of *Pelvetia* and labelled calcium, Robinson & Jaffe (1975) showed that a substantial calcium current passed through the cells while they were being polarized by unilateral light. Fields of $0.1$ V cm$^{-1}$ or more could be produced. During the induction of rhizoids by light, membrane components that are involved in the inward movement of ions are translocated to the dark side of the cell, leading to a localized increase in intracellular $Ca^{2+}$ (Quatrano 1978). One means of controlling the redistribution of macromolecules, etc., might be by electrophoretic segregation of particles (Quatrano 1978).

Local application of a calcium ionophore tends to establish the rhizoid on that side. The drug is thought to act by rendering the membrane leaky to calcium ions. Enhanced localized entry of calcium ions may be part of the positive feedback loop that establishes the rhizoid pole (see Jaffe 1982).

Polarity may be established, however, not just in tip-growing regions, but in every cell of a multicellular filament. By refinement of earlier work on *Griffithsia* and *Cladophora*, Duffield *et al.* (1972) excised and cultured single intercalary cells of the filamentous red alga, *Griffithsia*, and showed that within a few days such cells regenerated a shoot cell at the original apical end, and a rhizoidal cell at the base. These experiments suggest that each cell of a filament is polarized, perhaps because of its position with respect to a longitudinal gradient.

### Growth of mycelia formed by fungi and streptomycetes

When cultured on a solid medium under conditions that support unrestricted growth (Righelato 1975), a spore of *Geotrichum candidum* or a hyphal fragment of *Coprinus cinereus* forms a mycelium whose total hyphal length and number of branches increase exponentially at approximately the same specific rate (Trinci 1974; Butler 1984). The ratio between the total hyphal length of a *G. candidum* mycelium and its number of branches eventually attains a value ($G$, the hyphal growth unit) which remains approximately constant (*ca.* 110 μm) as the mycelium increases in size. Hyphal growth unit length is strain- (figure 5) and species-specific and in fungi values ranging from 35 μm (*Cunninghamella* sp.) to 602 μm (*Fusarium avenaceum*) have been observed (Bull & Trinci 1977).

In *Achlya bisexualis*, a new zone of inward electrical current precedes the emergence of a branch by about 20 min and predicts its location, and during branch initiation the intensity of the

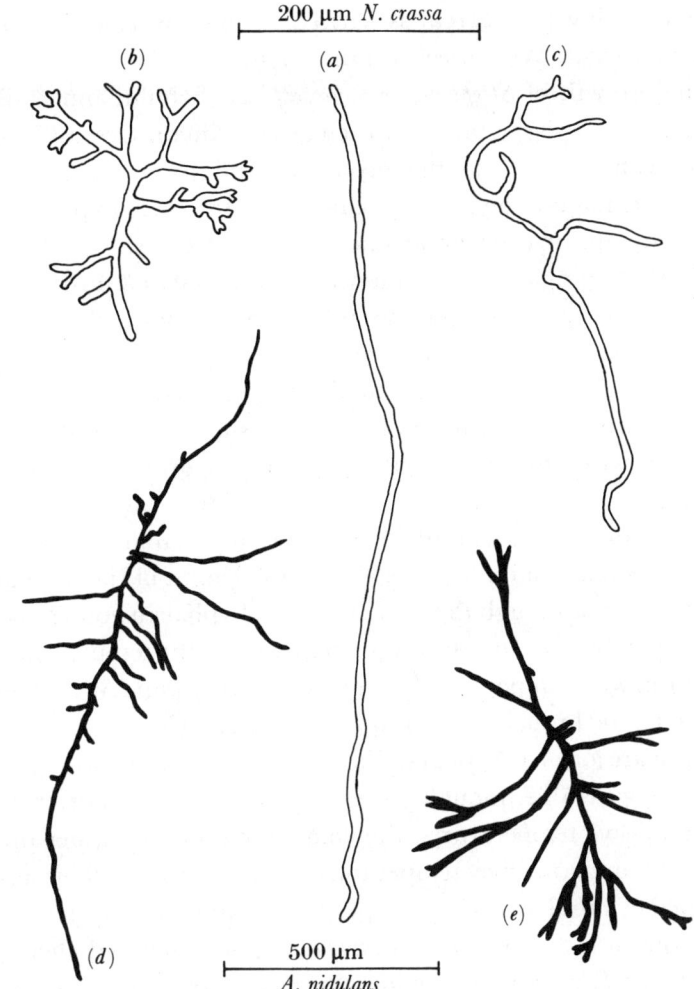

FIGURE 5. Mycelia of *Neurospora crassa* ((a)–(c)) and *Aspergillus nidulans* ((d) and (e)) grown at 25 and 37 °C, respectively. (a) *N. crassa* SYR-17-3A; wild type strain; (b) *N. crassa spco* 12; spreading colonial mutant; (c) *N. crassa spco* 1; spreading colonial mutant; (d) *A. nidulans* parental strain; lateral branches; (e) *A. nidulans sep* A2; lateral and dichotomous branches. (a)–(c) Redrawn from Trinci (1973b); (d) and (e) redrawn from Trinci & Morris (1979).)

inward current at the tip of the parent hypha diminishes and sometimes reverses. Despite these changes in current intensity and current direction before and during branch emergence, the tip of the parent hypha continues to elongate at a constant rate (Kropf *et al.* 1983). By contrast, there is no detectable alteration in the pH profile of medium adjacent to hyphae which are undergoing branching (Gow *et al.* 1984); it is possible that metabolic acid (see above) swamps the small proton influx produced as a result of branch initiation. Kropf *et al.* (1983) suggest that reversal of the electrical current during branching of *A. bisexualis* may be due to the presence of other ion fluxes that mask the inward current carried by protons. Kropf *et al.* (1984) propose that symporters for methionine and other amino acids are concentrated at a new branch point (figure 4). Future sites of out-growth of rhizoids from *Pelvetia* zygotes (Nuccitelli 1978) and pollen tubes from pollen grains (Weisenseel *et al.* 1975) are also preceded and predicted by a localized influx of electrical current, suggesting that the generation of a transcellular current is also a very early event in the establishment of cell polarity in these organisms; in *Pelvetia* and in pollen tubes the inward current is carried in part by calcium and potassium ions, respectively (Nuccitelli 1978; Weisenseel & Jaffe 1976).

Although mycelial growth of *Streptomyces hygroscopicus* (Schuhmann & Bergter 1976) and *S. coelicolor* (Allan & Prosser 1983) is biphasic, their growth kinetics resemble those of *G. candidum*, namely, the total hyphal length of the mycelium and its number of branches increase exponentially at approximately the same specific rate. However, hyphal growth unit lengths recorded for streptomycete mycelia lie at one extreme of the range observed for fungi, for example, *S. coelicolor* has a hyphal growth unit of *ca.* 32 μm (Allan & Prosser 1983). Thus, when all nutrients are present in excess and no inhibitors are formed, growth of fungal and streptomycete mycelia may be considered in terms of the duplication of a hypothetical growth unit which consists of a tip and a certain length of hypha. However, the portion of hyphae actually supporting growth of a tip will vary from a very short length just after branch initiation to a maximum length when the hypha attains the linear growth rate characteristic of the conditions and strain.

If the mean radius of the constituent hypha of a mycelium remains constant as the mycelium increases in size, the volume (and mass?) as well as the length of the hyphal growth unit will remain constant. However, although the mean radius of hyphae in young mycelia of most fungi does remain approximately constant, the radius of hyphae of the temperature-sensitive mutant, *A. nidulans sep* A2 varies with temperature (Trinci & Morris 1979). At permissive temperatures septa are formed and lateral branches are produced by *sep* A2. However, at 37 °C (the restrictive temperature) no septa are formed, hyphal radius is increased and dichotomous as well as lateral branches are formed (figure 5). Although mycelia of *sep* A2 have a shorter hyphal growth unit at 37 °C than at permissive temperatures, the volume of protoplasm per tip is approximately the same at 37 °C as at permissive temperatures (Trinci 1984). This result suggests that branching is regulated by the changes in protoplasmic volume (mass?) that accompany growth. When Robinson & Smith (1979) grew *G. candidum* in glucose-limited chemostat culture, they found that hyphal diameter increased with increase in dilution rate, hyphal growth unit length decreased with increase in dilution rate but hyphal growth unit volume remained approximately constant (figure 6). Thus, under conditions of restricted growth in chemostat culture, growth of *G. candidum* involves the duplication of a growth unit whose volume (and mass?) remains constant but whose length and diameter vary with dilution rate.

Riesenberger & Bergter (1979) grew *Streptomyces hygroscopicus* at different dilution rates in glucose-limited chemostat culture and obtained results that were similar to those reported by Robinson & Smith (1979) for *G. candidum* (figure 6). Thus, branch initiation in fungi and streptomycetes is regulated by a mechanism which involves the organism monitoring its own volume (mass?). A similar 'sizer' mechanism has been invoked by Nurse (1981) to explain the regulation of DNA synthesis and mitosis in *Schizosaccharomyces pombe*, and by Fiddy & Trinci (1976b) to explain the regulation of mitosis in *A. nidulans*.

## Colony growth

Exponential growth of a mycelium will only continue provided that all nutrients (including oxygen) are present in excess, the pH of the medium does not become inhibitory for growth and no inhibitory substances are produced as a result of growth. Growth on a solid substrate will eventually result in conditions at the centre of the colony becoming less favourable for growth than was initially the case. Figure 7 shows that a glucose gradient develops beneath

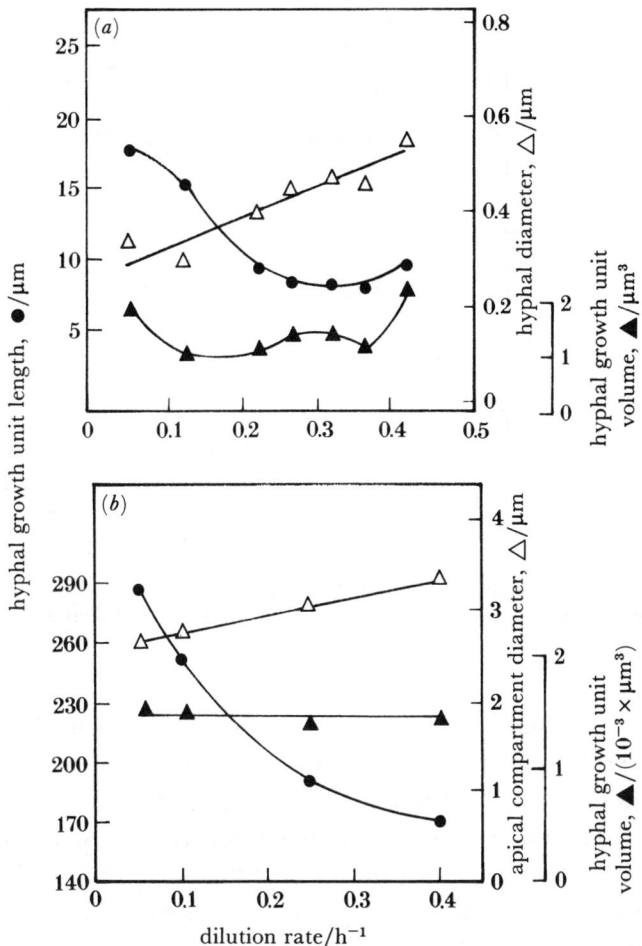

FIGURE 6. Effect of dilution rate on hyphal growth unit length, hyphal growth unit volume and hyphal diameter of mycelia of (a) *Streptomyces hygroscopicus* (redrawn from Riesenberg & Bergter (1979)) and (b) *Geotrichum candidum* (redrawn from Robinson & Smith (1979)).

and around *Rhizoctonia* colonies cultured on medium containing 10 μM glucose. This gradient is established in response to uptake of glucose by the mycelium and diffusion of glucose from uncolonized to colonized parts of the substrate. The development of conditions below the centre of a colony that are unfavourable for growth will inevitably result first in a deceleration in growth rate and eventually in a cessation of growth (Trinci & Thurston 1976). For most fungal colonies, however, the marginal hyphae will continue to extend at a linear rate so that eventually the colony may attain a diameter of several metres, for example, as in 'fairy rings'. The development of unfavourable conditions for growth at the centre of a colony is usually associated with plugging of septal pores (Trinci & Collinge 1973). This event reduces 'communication' within the colony and is associated with the induction of sporulation.

## VEGETATIVE HYPHAL FUSIONS

In fungi, fusions may occur between adjacent hyphae belonging to a single mycelium or, more rarely, between hyphae of two different strains of the same species. Hyphal fusions between two different strains may result in the formation of a heterokaryon, and subsequent operation

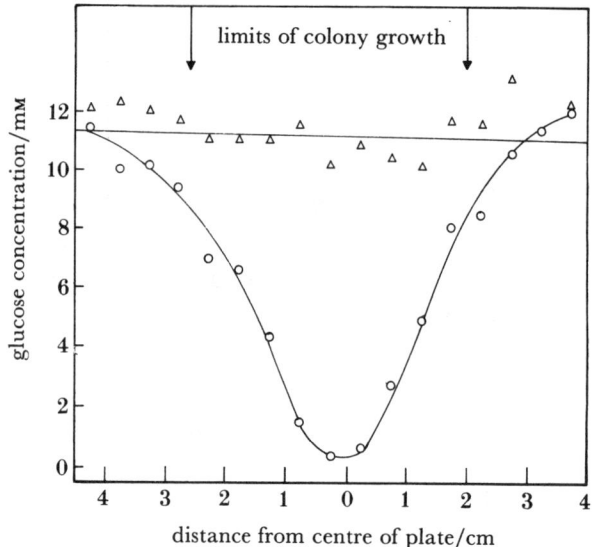

FIGURE 7. Glucose concentration in the medium below a colony of *Rhizoctonia cerealis* (○) and in the medium of an uninoculated plate of the same medium (△). (Previously unpublished data of G. D. Robson and A. P. J. Trinci.)

of the parasexual cycle (Pontecorvo 1956) in this mycelium may lead to genetic recombination. Consequently, heterokaryosis has generally been considered to be an important component of the genetic systems of fungi, particularly for fungi lacking sexual reproduction.

Vegetative hyphal fusions are common in the Ascomycotina, Basidiomycotina and Deuteromycotina, but are completely lacking in 'lower' fungi including the Zygomycotina, although gametangial fusions do occur in the latter group during sexual reproduction. Vegetative hyphal fusions usually develop in the old (slowly growing?) parts of a mycelium, but they have also been observed at the margin of some colonies (Galun *et al.* 1981) where conditions are usually considered to be favourable for vegetative growth (Trinci 1971). Buller (1931, 1953) suggested that vegetative hyphal fusions only occur between growing hyphal tips, but Aylmore & Todd (1984) observed occasional fusions between the tip of one hypha of *Coriolus versicolor* and the side wall of another. When two dikaryotic hyphae of *C. versicolor* fused, the participating hyphal compartments exhibited a donor–recipient relationship with respect to the subsequent behaviour of their nuclei; the two nuclei in the recipient compartment degenerated and were replaced by daughter nuclei formed by division of the pair of nuclei in the donor compartment. After completion of this 'nuclear disintegration–replacement reaction' both compartments were able to initiate branches and undergo mitosis. Fusions of the above type were observed between genetically identical (self-fusions) and between genetically different (non-self-fusions) hyphae of *C. versicolor*.

## CELL FUSION IN ALGAE

If an intercalary cell of the filamentous red alga *Griffithsia* is punctured and killed, the cell above (distal to) it develops a rhizoid, and the subjacent cell gives rise to an atypical (repair) shoot cell, with some rhizoidal characters, such as colour banding. The two new cells grow towards one another within the wall of the punctured cell, and eventually fuse, forming a functional intercalary cell (Waaland & Cleland 1974; Waaland 1984). Later, Waaland

& Watson (1980) showed that decapitated filaments would only form repair shoots when a rhizoid was present within a period of 4–7 h after decapitation. They obtained evidence of the occurrence of an endogenous hormone, rhodomorphin; this was subsequently partly purified, and identified as a glycoprotein (Watson & Waaland 1983). Since the hormone acted as an attractant between rhizoids and repair shoots, it was suggested that a comparable substance might regulate the series of complex cell fusions that occur during sexual reproduction in the red algae (Waaland & Watson 1980).

This hormone is species-specific, since fusion did not occur between cells of different species of *Griffithsia* (Waaland 1975). However, fusions were obtained between cells of male and female filaments of the same species. Most of the hybrid filaments gave rise to tetrasporangial branches typical of the diploid plant, but eventually male or female branches were formed. This suggested that these male–female hybrids functioned as a dikaryon (Waaland 1978).

## Vegetative incompatibility

As mentioned above, heterokaryosis and the parasexual cycle are often considered as important components of the genetic systems of fungi. However, for several fungi it has been shown that genetically determined incompatibility systems restrict the formation of heterokaryons within a species and hence restrict genetic recombination. In *Podospora anserina* the macroscopic manifestation of a vegetative incompatibility reaction is the appearance between neighbouring mycelia of a 'barrage' or a demarcation zone, which is pigmented and consists of hyphal debris (Esser & Kuenen 1967). In *C. versicolor*, vegetative incompatibility reactions prevent the formation of the type of unit mycelium described by Burnett & Partington (1957). Instead a single piece of wood may be colonized by several genetically distinct mycelia of *C. versicolor* which do not fuse (Rayner & Todd 1977; Todd & Rayner 1978). Thus, vegetative incompatibility mechanisms help to maintain the 'individuality' of mycelia.

*Rhizoctonia solani* has been divided into nine anastomosis groups on the basis of hyphal fusions (anastomoses) (Parameter *et al.* 1969). Hyphae of one *R. solani* isolate will fuse with hyphae of all other isolates in the same anastomosis group, but will not fuse with the hyphae of isolates belonging to other anastomosis groups. However, successful fusions usually result in the vacuolation and death of five or six hyphal compartments on either side of the fusion compartments (Flentje *et al.* 1967). Thus, a vegetative incompatibility reaction prevents the formation of a heterokaryon even between isolates belonging to the same anastomosis group. DNA homology studies by Kuninga & Yokosawa (1984*a*, *b*) have shown that an incompatibility reaction can even occur between strains which have very high (97.6–100%) DNA homology values.

Caten (1971) showed that the frequency of intrastrain incompatibility in *Aspergillus versicolor* and *Aspergillus terreus*, two imperfect species, was similar to that observed in four perfect species (*Aspergillus amstelodami*, *Aspergillus glaucus* group, *Aspergillus heterothallicus* and *A. nidulans*). Thus, despite the added genetic significance of heterokaryosis in the absence of sexual reproduction, vegetative incompatibility is apparently no less extensive in fungi lacking sexual reproduction than in fungi with sexual reproduction.

The natural functions of vegetative incompatibility within a fungal species remain uncertain. Two possibilities exist. First, vegetative incompatibility could act as a genetic isolation mechanism so that natural populations consist of a mixture of asexually reproducing clones

(Grindle 1963 a, b). Second, vegetative incompatibility may reduce the spread within a species of harmful cytoplasmic mutations and viral infections. Caten (1972) showed that the cytoplasmically inherited mutant character 'vegetative death' transferred freely between pairs of vegetative compatible isolates of *A. amstelodami* but not at all, or with a very reduced frequency, between pairs of incompatible isolates. Similarly, it has been shown that vegetative incompatibility prevents the transfer of dsRNA between isolates of *Endothia parasitica*; the dsRNA affects the morphology and virulence of this fungus (Anagnostakis & Waggoner 1981). A possible physiological benefit of hyphal fusions is that they may facilitate the transport of food reserves between various parts of a mycelium (provided that septal pores are not blocked (Trinci & Collinge 1973)) and hence they may enable the whole or large parts of the mycelium to cooperate in the formation of fruit bodies.

## Meristems in lower plants

In many of the simpler algae, growth is rather diffuse and is not related to a specialized region of growth such as a meristem. Sometimes growth may be concentrated at the base of the filament, as in *Rivularia*, or at the apex, where there may or may not be a distinctive apical cell. In the filamentous brown alga *Sphacelaria*, which has an evident apical cell, the subapical cell can replace a damaged apical cell within 24 h (Ducreux 1977); this must involve considerable cytoplasmic reorganization. *Sphacelaria* also exhibits a kind of apical dominance; both the angle of the branches, and their determinate growth, are altered by damage to the apical cell. *Chara* and *Nitella* both have distinctive apical cells that give rise to cells of different function, which themselves divide to give rise to the short nodal and long internodal cells so characteristic of these organisms. The apical cell of *Dictyota* and similar algae may occasionally divide equally to give rise to equal branches of the thallus, as well as unequally to give rise to the other cells of the thallus. Larger brown algae also have apical cells that exert apical dominance (Moss 1965, 1967, 1970). Apical cells are characteristic of most bryophytes and ferns.

A type of marginal meristem occurs in the flat plate-like algae *Coleochaete* and *Padina*, and a type of intercalary meristem in the stipes of *Laminaria* and related algae. The meristoderm of these algae contributes to growth in thickness.

Apart from studies of apical meristems, however, relatively little work has been carried out on the mode of growth or its control in these meristems.

## Modular organisms

To what extent do these various modes of growth in lower plants give rise to modular constructions? Mycelial organisms possess some of the properties normally associated with modular organisms. First, during unrestricted growth they are constructed from growth units that consist of a hyphal tip associated with a certain amount of protoplasm, the average volume of which remains approximately constant. Second, except during the initial stages of growth, a mycelium does not behave as an integrated whole, but instead responds to the local environmental conditions to which its constituent parts are exposed. In so doing, 'communication' within a colony becomes progressively reduced. Finally, in most mycelial organisms senescence is a local event. Provided that suitable cultural conditions are maintained, a colony can continue to expand in diameter for an indefinite period of time, as witnessed by the formation of 'fairy rings'. If modules are defined as units of construction (Harper 1977), rather than as the developmental products of a single apex (White 1984), ferns, *Equisetum*, *Selaginella*, *Lycopodium* and many mosses could be regarded as modular, in various ways, in much the same way as higher plants. Some of the colonial green algae, even, might be considered to be truly modular. For example, *Pandorina* consists of similar cells all of which can form daughter colonies. However, Chapman (1981) considers that a module is a multicellular unit. By this definition, the simplest modular photosynthetic organisms might be algae such as *Nitella* or the more complex *Chara*, constructed of units comprising a node and an internode. Growth of such modules would be limited, however. On the other hand, organisms of apparently modular construction occur among the coenocytic algae, for example, *Bryopsis* and *Caulerpa*. In such cases, far from being multicellular, the module could comprise only part of a cell.

In higher plants there are usually correlative effects between modules, but this might be less likely in the case of *Caulerpa*, in which, for example, it has been shown that indoleacetic acid is uniformly distributed rather than forming a gradient (Brennan & Jacobs 1980).

The extent to which lower plants can be equated with higher plants as modular organisms thus remains debatable.

We wish to thank Dr N. A. R. Gow for providing us with previously unpublished information and Dr D. Moore for his helpful comments on the manuscript.

### References

Aberg, H. 1978 Light and branch formation in the alga *Vaucheria dichotoma* (Xanthophyceae). *Physiol. Plant.* **44**, 224–230.

Allan, E. J. & Prosser, J. I. 1983 Mycelial growth and branching of *Streptomyces coelicolor* on solid medium. *J. gen. Microbiol.* **129**, 2029–2036.

Anagnostakis, S. L. & Waggoner, P. E. 1981 Hypovirulence, vegetative incompatibility, and the growth of cankers of chestnut blight. *Phytopathology* **71**, 1198–1202.

Archer, D. B. 1977 Chitin biosynthesis in protoplasts and subcellular fractions of *Aspergillus fumigatus*. *Biochem. J.* **164**, 653–658.

Armbruster, B. L. & Weisenseel, M. H. 1983 Ionic currents traverse growing hyphae and sporangia of the mycelial water mould *Achlya debaryana*. *Protoplasma* **115**, 65–69.

Aylemore, R. C. & Todd, N. K. 1984 Hyphal fusion in *Coriolus versicolor*. In *The ecology and physiology of the fungal mycelium* (ed. D. H. Jennings & A. D. M. Rayner), pp. 103–126. Cambridge University Press.

Bartnicki-Garcia, S. 1973 Fundamental aspects of hyphal morphogenesis. In *Microbial differentiation* (ed. J. O. Ashworth & J. E. Smith), pp. 245–267. Cambridge University Press.

Bartnicki-Garcia, S., Bracker, C. E., Lippman, E. & Ruiz-Herrera, J. 1984 Chitosomes from the wall-less 'slime' mutant of *Neurospora crassa*. *Arch. Microbiol.* **139**, 105–112.

Bartnicki-Garcia, S., Ruiz-Herrera, J. & Bracker, C. E. 1979 Chitosomes and chitin synthesis. In *Fungal walls and hyphal growth* (ed. J. H. Burnett & A. P. J. Trinci), pp. 149–168. Cambridge University Press.

Bentrup, F. W. 1968 Die Morphogenese Pflanzlucher Zellen im elektrischen Feld. *Z. PflPhysiol.* **59**, 309–339.

Bonnett, H. T. & Newcomb, E. H. 1966 Coated vesicles and other cytoplasmic components of growing root hairs of radish. *Protoplasma* **1**, 59–75.

Bracker, C. E., Ruiz-Herrera, J. & Bartnicki-Garcia, S. 1976 Structure and transformation of chitin synthase particles (chitosomes) during microfibril synthesis *in vitro*. *Proc. natn. Acad. Sci. U.S.A.* **73**, 4573–4574.

Braña, A. F., Manzanal, M.-B., & Hardisson, C. 1982 Mode of cell wall growth of *Streptomyces antibioticus*. *FEMS Microbiol. Lett.* **13**, 231–235.

Braun, P. C. & Calderone, R. A. 1978 Chitin synthesis in *Candida albicans*. Comparison of yeast and hyphal forms. *J. Bact.* **135**, 1472–1477.

Brennan, T. & Jacobs, W. P. 1980 Polarity and the movement of [$^{14}$C]indol-3-ylacetic acid in the coenocyte, *Caulerpa prolifera*. *Ann. Bot.* **46**, 129–131.

Bull, A. T. & Trinci, A. P. J. 1977 The physiology and metabolic control of fungal growth. *Adv. microb. Physiol.* **15**, 1–84.

Buller, A. H. R. 1931 *Researches on fungi*, vol. 4. London: Longmans Green.
Buller, A. H. R. 1953 *Researches on fungi*, vol. 5. London: Longmans Green.
Burnett, J. H. & Partington, M. 1957 Spatial distribution of fungal mating-type factors. *Proc. R. phys. Soc. Edinb.* **26**, 61–68.
Butler, G. M. 1984 Colony ontogeny in basidiomycetes. In *The ecology and physiology of the fungal mycelium* (ed. D. H. Jennings & A. D. M. Rayner), pp. 53–71. Cambridge University Press.
Castle, E. S. 1959 Growth distribution in the light-growth response of *Phycomyces*. *J. gen. Physiol.* **42**, 697.
Caten, C. E. 1971 Heterokaryon incompatibility in imperfect species of *Aspergillus*. *Heredity, Lond.* **26**, 299–312.
Caten, C. E. 1972 Vegetative incompatibility and cytoplasmic infection in fungi. *J. gen. Microbiol.* **72**, 221–229.
Chapman, G. 1981 Individuality and modular organisms. *Biol. J. Linn. Soc.* **15**, 177–183.
Chen, T. H. & Jaffe, L. F. 1979 Forced calcium entry and polarized growth of *Funaria* spores. *Planta* **144**, 401–406.
Ducreux, G. 1977 Étude expérimentale des corrélations et des possibilités de régénération au niveau de l'apex de *Sphacelaria cirrhosa* Agardh. *Ann. Sci. nat. Bot. ser.* 12, **18**, 163–179.
Duffield, E. C. S., Waaland, S. D. & Cleland, R. 1972 Morphogenesis in the red alga, *Griffithsia pacifica*: regeneration from single cells. *Planta* **105**, 185–195.
DeLoofe, A. 1983 The meroistic insect ovary as a miniature electrophoresis chamber. *Comp. Biochem. Physiol.* **74A**, 3–9.
Esser, K. & Kuenen, R. 1967 *Genetics of fungi*. New York: Springer Verlag.
Fiddy, C. & Trinci, A. P. J. 1976a Nuclei, septation, branching and growth of *Geotrichum candidum*. *J. gen. Microbiol.* **97**, 185–192.
Fiddy, C. & Trinci, A. P. J. 1976b Mitosis, septation and the duplication cycle in *Aspergillus nidulans*. *J. gen. Microbiol.* **97**, 169–184.
Flentje, N. T., Stretton, H. M. & McKenzie, A. R. 1967 Mutation in *Thanetephorus cucumeris*. *Aust. J. biol. Sci.* **20**, 1173–80.
Galun, M., Malki, D. & Galun, E. 1981 Visualization of chitin-wall formation in hyphal tips and anastomoses of *Diploidia natalensis* by fluorescein-conjugated wheat agglutinin and [$^3$H]$N$-acetyl-D-glucosamine. *Arch. Microbiol.* **130**, 105–110.
Girard, V. & Fèvre, M. 1984 β-1-4- and β-1-3-glucan synthesis are associated with the plasma membrane of the fungus *Saprolegnia*. *Planta* **160**, 400–406.
Gooday, G. W. 1971 An autoradiographic study of hyphal growth of some fungi. *J. gen. Microbiol.* **67**, 125–137.
Gooday, G. W. & Trinci, A. P. J. 1980 Wall structure and biosynthesis in fungi. In *The eukaryotic microbial cell* (ed. G. W. Gooday, D. Lloyd & A. P. J. Trinci), pp. 207–251. Cambridge University Press.
Gow, N. A. R. 1984 Transhyphal electrical currents in fungi. *J. gen. Microbiol.* **130**, 3313–3318.
Gow, N. A. R. & Gooday, G. W. 1984 A model for the germ tube formation and mycelial growth form of *Candida albicans*. *Sabouraudia* **22**, 137–143.
Gow, N. A. R., Kropf, D. L. & Harold, F. M. 1984 Growing hyphae of *Achlya bisexualis* generate a longitudinal pH gradient in the surrounding medium. *J. gen. Microbiol.* **130**, 2967–2974.
Grindle, M. 1963a Heterokaryon compatibility of unrelated strains in the *Aspergillus nidulans* group. *Heredity, Lond.* **18**, 191–204.
Grindle, M. 1963b Heterokaryon compatibility of closely related wild isolates of *Aspergillus nidulans*. *Heredity, Lond.* **18**, 397–405.
Harold, F. M. 1982 Pumps and currents. A biological perspective. *Curr. Top. Membranes Transp.* **16**, 485–516.
Harper, J. L. 1977 *Population biology of plants*. London: Academic Press.
Heath, I. B. & Heath, M. C. 1978 Microtubules and organelle movements in the rust fungus *Uromyces phaseoli* var. *vignae*. *Cytobiologie* **16**, 393–411.
Horwitz, B. A., Weisenseel, M. H., Dorn, A. & Gressel, J. 1984 Electric currents around growing *Trichoderma* hyphae, before and after autoinduction of conidiation. *Pl. Physiol.* **74**, 912–916.
Humphreys, A. M. & Gooday, G. W. 1984 Properties of chitinase activities from *Mucor mucedo*: Evidence for a membrane-bound zymogenic form. *J. gen. Microbiol.* **130**, 1359–1366.
Hunsley, D. & Burnett, J. H. 1970 The ultrastructural architecture of the walls of some hyphal fungi. *J. gen. Microbiol.* **62**, 203–218.
Jaffe, L. F. 1977 Electrophoresis along cell membranes. *Nature, Lond.* **265**, 600–602.
Jaffe, L. F. 1982 Developmental currents, voltages and gradients. In *Developmental order: its origin and regulation* (ed. S. S. Subtelny & P. B. Green), pp. 183–215. 40th Symp. Soc. Devl Biol. New York: Alan R. Liss, Inc.
Kallio, P. & Lehtonen, J. 1981 Nuclear control of morphogenesis in *Micrasterias*. In *Cytomorphogenesis in plants* (ed. O. Kiermayer), pp. 191–213. Wien, New York: Springer-Verlag.
Kataoka, H. 1975a Phototropism in *Vaucheria geminata* I. The action spectrum. *Pl. Cell Physiol.* **16**, 427–437.
Kataoka, H. 1975b Phototropism in *Vaucheria geminata* II. The mechanism of bending and branching. *Pl. Cell Physiol.* **16**, 439–448.
Kiermayer, O. 1981 Cytoplasmic basis of morphogenesis in *Micrasterias*. In *Cytomorphogenesis in plants* (ed. O. Kiermayer), pp. 147–189. Wien, New York: Springer-Verlag.
Kritzman, G., Chet, I. & Henis, Y. 1978 Localization of β-(1,3)-glucanase in the mycelium of *Sclerotium rolfsii*. *J. Bact.* **134**, 470–475.

Kropf, D. K., Lupa, M. D., Caldwell, J. C. & Harold, F. M. 1983 Cell polarity; endogenous ion currents precede and predict branching in the water mould *Achlya*. *Science, Wash.* **220**, 1385–1387.

Kropf, D. K., Caldwell, J. C., Gow, N. A. R. & Harold, F. M. 1984 Transcellular ion currents in the water mould *Achlya*. Amino acid proton symport as a mechanism of current entry. *J. Cell Biol.* **99**, 486–496.

Kuninaga, S. & Yokosawa, R. 1984a DNA base sequence homology in *Rhizoctonia solani* Kuhn. IV. Genetic relatedness within AG-4. *Ann. Phytopathol. Soc. Japan.* **50**, 322–330.

Kuninaga, S. & Yokosawa, R. 1984b DNA base sequence homology in *Rhizoctonia solani* Kuhn. V. Genetic relatedness with AG-6. *Ann. Phytopathol. Soc. Japan.* **50**, 346–352.

Lopez-Romero, E., Ruiz-Herrera, J. & Bartnicki-Garcia, S. 1978 Purification and properties of an inhibitory protein of chitin synthetase from *Mucor rouxii*. *Biochim. biophys. Acta* **525**, 338–345.

Marsh, G. & Beams, H. W. 1945 The orientation of pollen tubes of *Vinca* in the electric current. *J. cell. comp. Physiol.* **25**, 195–204.

Meindl, U. 1982 Local accumulation of membrane-associated calcium according to cell pattern formation in *Micrasterias denticulata*, visualized by chlorotetracycline fluorescence. *Protoplasma* **110**, 143–146.

Moss, B. 1965 Apical dominance in *Fucus vesiculosus*. *New Phytol.* **64**, 387–392.

Moss, B. 1967 The apical meristem of *Fucus*. *New Phytol.* **66**, 67–74.

Moss, B. 1970 Meristems and growth control in *Ascophyllum nodosum* (L.) Le Jol. *New Phytol.* **69**, 253–260.

Nuccitelli, R. 1978 Ooplasmic segregation and secretion in the *Pelvetia* egg is accompanied by a membrane generated electrical current. *Devl Biol.* **62**, 13–33.

Nuccitelli, R. & Jaffe, I. E. 1974 Spontaneous current pulses through developing fucoid eggs. *Proc. natn. Acad. Sci. U.S.A.* **71**, 4855–4859.

Nurse, P. 1981 Genetic analysis of the cell cycle. In *Genetics as a tool in microbiology* (ed. S. W. Glover & D. A. Hopwood), pp. 291–316. Cambridge University Press.

Oliver, S. G. & Trinci, A. P. J. 1985 Modes of growth of bacteria and fungi. In *Comprehensive biotechnology*, vol. 1, (ed. A. T. Bull & H. Dalton), pp. 153–181. Oxford: Pergamon.

Otto, D. W. & Brown, R. M. 1974 Developmental cytology of the genus *Vaucheria* 1. Organisation of the vegetative filament. *Br. Phycol. J.* **9**, 111–126.

Parameter, J. R., Sherwood, R. T. & Platt, W. D. 1969 Anastomosis grouping among isolates of *Thanatephorus cucumeris*. *Phytopathology* **59**, 1270–1278.

Peng, H. B. & Jaffe, L. F. 1976 Polarization of fucoid eggs by steady electrical fields. *Devl Biol.* **53**, 277–284.

Pfenning, N. 1984 Microbial behaviour in natural environments. In *The microbe 1984* (ed. D. P. Kelly & N. G. Carr), pp. 23–50. Cambridge University Press.

Pontecorvo, G. 1956 The parasexual cycle in fungi. *A. Rev. Microbiol.* **10**, 393–400.

Quatrano, R. S. 1978 Development of cell polarity. *A. Rev. Pl. Physiol.* **29**, 487–510.

Rayner, A. D. M. & Todd, N. K. 1977 Intraspecific antagonism in natural populations of wood-decaying basidiomycetes. *J. gen. Microbiol.* **103**, 85–90.

Reinhardt, M. O. 1892 Das Wachstum der Pilzhyphen. Ein Beitrag zur Kenntniss des Flachenwachstums vegetabilischer Zellmembranen. *Jb. wiss. Bot.* **23**, 479–566.

Riesenberger, D. & Bergter, F. 1979 Dependence of macromolecular composition and morphology of *Streptomyces hygroscopicus* on specific growth rate. *Z. allg. Mikrobiol.* **19**, 415–430.

Righelato, R. C. 1975 Growth kinetics of mycelial fungi. In *The filamentous fungi*, vol. 1, (ed. J. E. Smith & D. R. Berry), pp. 79–102. London: Arnold.

Robertson, N. F. 1958 Observations on the effect of water on the hyphal apices of *Fusarium oxysporum*. *Ann. Bot.* **22**, 159–173.

Robertson, N. F. 1959 Experimental control of hyphal branching and branch form in hyphomycetous fungi. *J. Linn. Soc., Bot.* **56**, 207–211.

Robertson, N. F. 1968 The growth process in fungi. *A. Rev. Phytopath.* **6**, 115–136.

Robinow, C. F. 1963 Observations on cell growth, mitosis and division in the fungus *Basidiobolus ranarum*. *J. Cell Biol.* **17**, 123–152.

Robinson, K. R. & Jaffe, L. F. 1975 Polarizing fucoid eggs drive a calcium current through themselves. *Science, Wash.* **187**, 70–72.

Robinson, P. M. & Smith, J. M. 1979 Development of cells and hyphae of *Geotrichum candidum* in chemostat and batch culture. *Trans. Br. mycol. Soc.* **72**, 39–47.

Rosen, W. G. 1964 Chemotropism and fine structure of pollen tubes. In *Pollen physiology and fertilization* (ed. H. F. Linkens), pp. 159–166. Amsterdam: North Holland.

Rosenberger, R. F. 1979 Endogenous lytic enzymes and wall metabolism. In *Fungal walls and hyphal growth* (ed. J. A. Burnett & A. P. J. Trinci), pp. 265–277. Cambridge University Press.

Scarborough, G. A. 1976 The *Neurospora* plasma membrane ATPase in an electrogenic pump. *Proc. natn. Acad. Sci. U.S.A.* **73**, 1485–1488.

Schneider, E. F. & Seaman, W. L. 1982 Structure of chitin in the cell walls of *Fusarium sulphureum*. *Can. J. Microbiol.* **28**, 531–535.

Schuhmann, E. & Bergter, F. 1976 Mikroskopische untersuchungen zur wachstumskinetik von *Streptomyces hyroscopicus*. *Z. allg. Mikrobiol.* **16**, 201–215.

Sietsma, J. H. & Wessels, J. G. H. 1977 Chemical analysis of the hyphal wall of *Schizophyllum commune*. *Biochim. biophys. Acta* **496**, 225–239.

Sietsma, J. H. & Wessels, J. G. H. 1979 Evidence for covalent linkages between chitin and β-glucan in a fungal wall. *J. gen. Microbiol.* **114**, 99–108.

Sievers, A. 1965 Elektronenmikroskopische Untersuchungen zür geotropischen Reaktion. I. Über Sonderheiten im Feinbau der Rhizoide von *Chara foetida*. *Z. PflPhysiol.* **53**, 193–213.

Slayman, C. L. 1965a Electrical properties of *Neurospora crassa*: effects of external cations on the intracellular potential. *J. gen. Physiol.* **49**, 69–92.

Slayman, C. L. 1965b Electrical properties of *Neurospora crassa*: respiration and the intracellular potential. *J. gen. Physiol.* **49**, 93–116.

Slayman, C. L. 1977 Energetics and control of transport in *Neurospora*. In *Water relations in membrane transport in plants and animals* (ed. A. M. Jungreis, T. Hodges, A. M. Kleinzeller & S. G. Schultz), pp. 69–86. New York: Academic Press.

Slayman, C. L., Long, W. S. & Lu, C. Y. 1973 The relationship between ATP and an electrogenic pump in the plasma membrane of *Neurospora crassa*. *J. Membr. Biol.* **14**, 305–308.

Slayman, C. L. & Slayman, C. W. 1974 Depolarization of the plasma membrane of *Neurospora* during active transport of glucose: evidence for a proton-dependent cotransport system. *Proc. natn. Acad. Sci. U.S.A.* **71**, 1935–1939.

Sonnenberg, A. G. M. 1984 Biosynthesis and assembly of fungal wall polymers. Ph.D. thesis, University of Groningen, Haren.

Steele, G. C. & Trinci, A. P. J. 1975 The extension zone of mycelial hyphae. *New Phytol.* **25**, 583–587.

Stump, R. F., Robinson, K. R., Harold, R. I. & Harold, F. M. 1980 Endogenous electrical currents in the water mould *Blastocladiella emersonii* during growth and sporulation. *Proc. natn. Acad. Sci. U.S.A.* **77**, 6673–6677.

Todd, N. K. & Rayner, A. D. M. 1978 Genetic structure of a natural population of *Coriolus versicolor* (L. ex Fr.) Que'l. *Genet. Res.* **32**, 55–65.

Trinci, A. P. J. 1971 Influence of the peripheral growth zone on the radial growth rate of fungal colonies. *J. gen. Microbiol.* **67**, 325–344.

Trinci, A. P. J. 1973a Growth of wild type and spreading colonial mutants of *Neurospora crassa* in batch culture and on agar medium. *Arch. Mikrobiol.* **91**, 113–116.

Trinci, A. P. J. 1973b The hyphal growth unit of wild type and spreading colonial mutants of *Neurospora crassa*. *Arch. Mikrobiol.* **91**, 127–136.

Trinci, A. P. J. 1974 A study of the kinetics of hyphal extension and branch initiation of fungal hyphae. *J. gen. Microbiol.* **81**, 225–236.

Trinci, A. P. J. 1984 Regulation of hyphal branching and hyphal orientation. In *The ecology and physiology of the fungal mycelium* (ed. J. H. Jennings & A. D. M. Rayner), pp. 23–52. Cambridge University Press.

Trinci, A. P. J. & Collinge, A. J. 1973 Structure and plugging of septa of wild type and spreading colonial mutants of *Neurospora crassa*. *Arch. Mikrobiol.* **91**, 355–364.

Trinci, A. P. J. & Collinge, A. J. 1975 Hyphal wall growth in *Neurospora crassa* and *Geotrichum candidum*. *J. gen. Microbiol.* **91**, 355–361.

Trinci, A. P. J. & Halford, E. A. 1975 The extension zone of stage I sporangiophores of *Phycomyces blakesleeanus*. *New Phytol.* **74**, 81–83.

Trinci, A. P. J. & Morris, N. R. 1979 Morphology and growth of a temperature sensitive mutant of *Aspergillus nidulans* which forms aseptate mycelia at non-permissive temperatures. *J. gen. Microbiol.* **114**, 53–59.

Trinci, A. P. J. & Thurston, C. F. 1976 Transition of the non-growing state in eukaryotic micro-organisms. In *The survival of vegetative microbes* (ed. J. R. Postgate & T. R. G. Grey), pp. 50–80. Cambridge University Press.

Valla, G. 1984 Hyphal extension and branch initiation in *Polyporus arcularius*: a biolaser microsurgical investigation. *Can. J. Bot.* **62**, 2788–2792.

Vermeulen, C. A. & Wessels, J. G. H. 1984 Ultrastructural differences between wall apices of growing and non-growing hyphae of *Schizophyllum commune*. *Protoplasma* **120**, 123–131.

Waaland, S. D. 1975 Evidence for a species-specific cell fusion hormone in red algae. *Protoplasma* **86**, 253–261.

Waaland, S. D. 1978 Parasexually produced hybrids between female and male plants of *Griffithsia tenuis* C. Agardh, a red alga. *Planta* **138**, 65–68.

Waaland, S. D. 1984 Positional control of development in algae. In *Positional controls in plant development* (ed. P. W. Barlow & D. J. Carr), pp. 137–156. Cambridge University Press.

Waaland, S. D. & Cleland, R. 1974 Cell repair through cell fusion in the red alga *Griffithsia pacifica*. *Protoplasma* **79**, 185–196.

Waaland, S. D. & Watson, B. A. 1980 Isolation of a cell-fusion hormone from *Griffithsia pacifica* Kylin, a red alga. *Planta* **149**, 493–497.

Watson, B. A. & Waaland, S. D. 1983 Partial purification and characterization of a glycoprotein cell fusion hormone from *Griffithsia pacifica*, a red alga. *Pl. Physiol.* **71**, 327–332.

Weisenseel, M. H., Dorn, A. & Jaffe, I. F. 1979 Natural $H^+$ currents traverse growing roots of barley (*Hordeum vulgare* L.). *Pl. Physiol.* **64**, 512–518.

Weisenseel, M. H. & Jaffe, I. F. 1976 The major growth current through lily pollen tubes enters as $K^+$ and leaves as $H^+$. *Planta* **133**, 1–7.
Weisenseel, M. H. & Kicherer, R. M. 1981 Ionic currents as control mechanism in cytomorphogenesis. In *Cytomorphogenesis in plants* (ed. O. Kiermayer), pp. 379–399. Wien, New York: Springer-Verlag.
Weisenseel, M. H., Nuccitelli, R. & Jaffe, I. F. 1975 Large electrical currents traverse growing pollen tubes. *J. Cell Biol.* **66**, 556–567.
Wessels, J. G. H. & Sietsma, J. H. 1979 Wall structure and growth in *Schizophyllym commune*. In *Fungal walls and hyphal growth* (ed. J. H. Burnett & A. P. J. Trinci), pp. 27–48. Cambridge University Press.
Wessels, J. G. H., Sietsma, J. H. & Sonnenberg, A. S. M. 1983 Wall synthesis and assembly during hyphal morphogenesis in *Schizophyllum commune*. *J. gen. Microbiol.* **129**, 1607–1616.
White, J. 1984 Plant metamerism. In *Perspectives on plant population ecology* (ed. R. Dirzo & J. Sarukhán), pp. 15–47. Sunderland, Massachusetts: Sinauer Associates.
Woodruff, R. I. & Telfer, W. H. 1980 Electrophoresis of proteins in intercellular bridges. *Nature, Lond.* **286**, 84–86.

# Modular growth and form of corals: a matter of metamers?

By B. R. Rosen

*Department of Palaeontology, British Museum (Natural History),
Cromwell Road, London SW7 5BD, U.K.*

Following a lead from botanists, this paper offers a re-interpretation of the morphology and growth of scleractinian, rugosan and tabulate corals in terms of iterated morphological units (modules), with an emphasis on the shape and organization of colonies. It presents a new theoretical basis for modules, a new descriptive framework for corals above the zooidal level, a review of practical applications and a comparison of coral modules with other kinds of organic iteration.

Consideration of cloning is a prerequisite and leads to a revival of the hypothesis that colonies have arisen by arrested budding (clonoteny), where zooids are paedomorphic to varying degrees with respect to the primitive state of clonally produced, detached (clonoparous) individuals. A second hypothesis follows, that the mouth structure is universally homologous in all corals. R. Riedl's work (*Order in living organisms*. Chichester: John Wiley (1978)) on morphological organization and hierarchy can therefore be applied to corals to obtain a non-arbitrary theoretical basis for recognizing modules and distinguishing them from other kinds of repetition of parts.

There are five criteria for modularity, four topological and one homological: (i) a modular structure is a three-dimensional tesselation; (ii) modules correspond to homogeneous units of hierarchical subdivision; (iii) an organism is modular if subdivision at the highest (or very high) level reveals homogeneous units; (iv) if homogeneous units occur in an unbroken series of subdivisions, they are all modules at their respective levels (hence modules of modules, that is, cormidia), and the finest unit is the fundamental module; (v) homogeneous units are those which are homologous with each other (homonoms). On this basis, and allowing for interzooidal connective structures, the fundamental module of corals is the zooid (not the polyp alone). Differences between zooids, as suggested by other authors, are of degree, not kind.

Colony form can therefore be specified by using a scheme that starts with zooids, and is based on (i) organization (component morphology and modular arrangement), and (ii) shape, for every modular level present. The basis for higher level modules is reiterative, either homomodular, or heteromodular (subdivided into polymorphic or polystatic). Branching organization and branching shape should be distinguished, and either can occur at any modular level, or even not at all.

Modules bring potentially greater precision to growth studies, statistically and topologically, either heuristically or to test hypotheses about fecundity, senescence, determinate growth, variation and inherited architecture. Higher level modularity has previously been largely ignored. Density banding, dye markers and computer modelling, ideally in combination, are likely to produce the most useful results in the near future, and should lead to advances in taxonomy and phylogenetics, in the understanding of the morphological consequences of the coral–zooxanthellae symbiosis, and in the inference of past conditions from fossil corals.

In a wider phyletic context, zooidal (= fundamental) modules are meristematic and broadly metameric, though it follows from R. B. Clark (*Zool. Jb.* **103**, 169–195 (1980)) that this does not automatically signify a close phylogenetic relationship with annelids, or with any other metameric organisms. Coral 'metamers' are probably non-serial and clonal because the coral mode of life is usually vegetative. They exist at a higher organizational level than these other metamers.

*'What is not identically repeated, we do not understand.'* (Riedl 1978)
*'The actinozooid is a living thing which knows no time of youthful vigour, no waxing to a period of adult life, no waning to senility – it knows no age – it practically knows no natural death.'* (Wood-Jones 1907, 1910)

## 1. INTRODUCTION

The purpose of this paper is to adapt and interpret the present body of knowledge of growth and form of corals in terms of modular construction and thereby explore the idea of modules as it applies to corals. This is not a contrived exercise: the modular approach in botany has already proved to be a fertile one, and the same may prove so for corals, too.

There is a long tradition of coral biologists borrowing from botany especially because of the analogous shapes of many plants and corals and their analogous synecology. Crossland (1913) has written of coral gardens, flowers and bushes. Squires (1964) refers to coral thickets and coppices; and stems, buds and branches abound throughout the coral literature. Phytosociology has had a significant influence on coral ecology (see Scheer (1974, 1978) for review of origins of this) and Connell (1978) has made direct comparative observations on the ecology of tropical rain forests and coral reefs. A consequence of the well-known symbiosis between many corals and plants (that is, dinoflagellate algae, or zooxanthellae) is that light is equally important to both plants and zooxanthellate corals, and suggests that coral and plant analogies are not always superficial. In all, the old term zoophytes, which included corals, seems completely appropriate in this context, even though it has long since been rejected nomenclaturally.

The modular idea is not of course botanical in the first place, but its application to organisms has so far been dominated by botanists. As botanists have used the idea in different ways, however (White 1984), I have found it more satisfactory to go back to its more general meaning in design and technology to explore its application to corals. This has proved a better route for making comparisons between coral modules, plant modules and modules of other organisms than trying to apply botanical concepts from the outset (van Valen 1978). For this reason, this paper first considers coral modules in terms of design and topology, then in terms of coral biology, especially homology, function, colony organization and growth, and it concludes with a broader discussion of metamers, meristems, and modules. This exercise is primarily a perceptual one, rather than a practical one based on experiments and new observations. I have tried to provide a balance between detail to satisfy coral specialists, and a broader perspective for other readers.

With the advent of a relatively new concept and term like 'modules', we must ask whether it will help us to advance our understanding of the organisms concerned. Failure to do so will simply ensure that the idea will subside into obscurity as an academic curiosity. For a way forward, however, there is the stimulus of Harper & Bell's (1979) eight-point exposition of the dynamic significance of modularity in organisms in general (see also White 1984). This leads me to suggest five areas for which a modular approach might prove fruitful in coral studies, three of which stem directly from Harper & Bell's points (to which I make cross reference).

(i) Ecology (points 1, 6–8); especially new insights into the basis of 'community' patterns of corals and other associated sessile vegetative organisms, for example, on reefs.

(ii) Evolutionary processes (points 1, 3–5, 7): see Jackson & Coates (this symposium); hence insights into the origins of coloniality. Also, 'being the right size' (Haldane 1927) suggests comparisons between sizes of (a) colonies, (b) solitary corals, and (c) polyps within colonies.

(iii) Growth and form (points 2, 4–7): greater understanding of ecophenotypic and phenotypic form, hence inherited architecture and architectural models as for tropical trees (Hallé et al. 1978).

(iv) Taxonomy and phylogenetics: drawing on (iii), and especially in the scope for more thorough specification of colony morphology.

(v) Symbiosis: in particular, the broad empirical correlation of the colonial habit with the symbiosis with zooxanthellae. Potential insights lie in comparing zooxanthellate and non-zooxanthellate corals in (i)–(iv) above. Form through time (see, for example, Wells 1954; Coates & Oliver 1973) may throw light on the age and evolution of the symbiosis.

There are also broader possibilities in the modular approach than for either corals or plants on their own, essentially by using corals to test the general ideas suggested by Harper & Bell (1979). The first task, however, is to establish how the diverse array of coral forms and iterative morphology can be placed in a rational modular frame of reference. This should provide the necessary practical basis for further observations and experiments at particular modular levels.

## 2. Background

### (a) Corals

'Coral' has no formal taxonomic status. I use it here to refer to the anthozoan order Scleractinia (*sensu* Wells 1956), and the extinct anthozoan subclasses Rugosa and Tabulata (*sensu* Hill 1981). Emphasis is on Scleractinia because the biology of extinct groups is very much more conjectural. There is nevertheless a surprising amount of relevant work on Rugosa and Tabulata considering the assumptions that have to be made, and the labour-intensive methodology that has been used.

The colonial habit occurs in numerous coelenterates, but in corals it is also allied to the characteristic of an enduring, substantial and often very elaborate calcareous skeleton accumulated throughout a lifetime which may be as long as hundreds, perhaps thousands, of years (Potts *et al.* 1985; Hughes & Jackson 1985). Coral polyps are therefore very much attached to their own record of growth history (figure 1), though Sammarco (1982) has noted that polyps can occasionally detach themselves from their skeletons and recommence colony growth elsewhere. Although the skeleton is dead, its volume and mass may far exceed the amount of living soft tissue. Thus in a typical domed colony of *Porites*, say 2 m across, the living tissue is virtually a veneer which is, at most, only as deep as about 0.5% of the colony's radius. In finely branched or sheet-like corals, however, the relative proportion of skeleton is much lower than in massive corals. There are obvious analogies between a coral skeleton and a tree's heartwood.

For appropriate illustrations of the living corals mentioned here, especially of their skeletal morphology, the most comprehensive works are the eastern Australia monographs (Veron & Pichon 1973, 1980, 1982; Veron & Wallace 1984; Veron *et al.* 1977). For larger illustrations of living colony form, see more general works like that by Wood (1983) and Randall & Myers (1983). For fossil forms, see Wells (1956) and Hill (1981).

(b) *Coral colonies*

It is conventional to regard a coral consisting of a single polyp as a solitary coral, and those with numerous polyps as colonies (figure 1). The skeletal counterpart to a polyp is the calice, the cavity occupied by the polyp. Over the time that a colony polyp exists (perhaps several years or more: there are no adequate data), it contributes to colony growth in two possible ways: laterally by budding new polyps, and longitudinally by continually renewing its calicinal

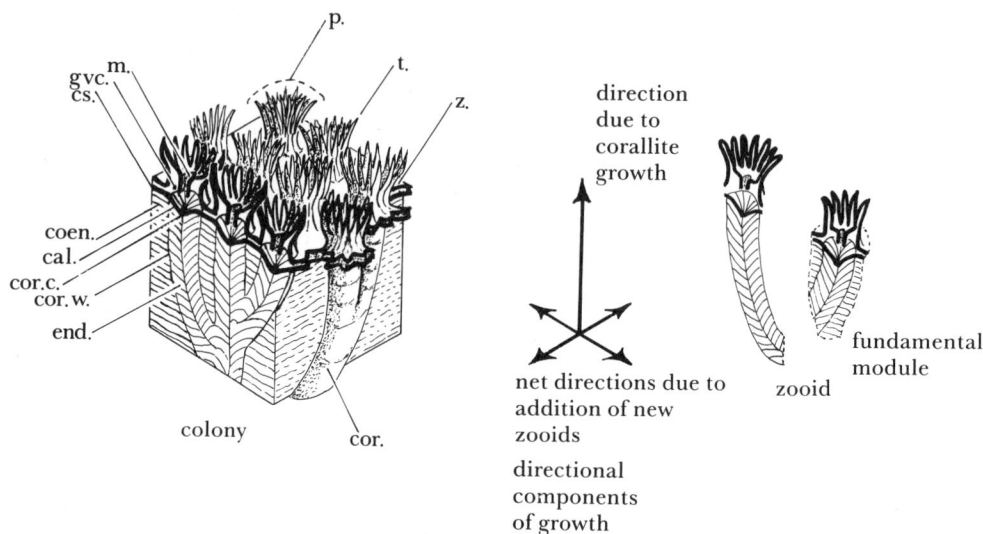

FIGURE 1. General relationships of polyps to skeleton in a plocoid colonial coral (compare with figure 2b). Polyp tissues in bold outline on upper surface and skeletal tissues below. Detailed anatomy is omitted. Representation of skeleton as in figure 2. A zooid (that is, polyp with its calice and corallite) and its share of coenenchymal tissues (if present), represents a fundamental module, and depending on the coral, might be between 1 mm and 20 mm in diameter, typically 2–10 mm. Note continuity of gastrovascular cavity between polyps. In some corals the skeletal components are perforate and the polyp tissues invest the uppermost skeletal layers. Abbreviations: cal., calice; coen., coenosteum; cor., corallite; cor.c., corallite centre; cor.w., corallite wall; cs. (c.e.z.), coenosarc (communal edge zone); end., endotheca; gvc., gastrovascular cavity; m., mouth (=substantive component); p., polyp; t., tentacle; z., zooid.

structure, distally, along its own longitudinal axis, the latter resulting in a tube-like structure, the corallite. It is the combined corallite length and arrangement in a colony that gives a colony its particular form (figures 2 and 3). It is therefore impossible to consider colony form of corals without reference to both polyps and their corallites, even though the bulk of a corallite is dead tissue. In the absence of a suitable term for a polyp with its corallite, and because of its fundamental importance as a unit of colony growth and form, I shall follow bryozoological usage and refer to the two together as a zooid (figure 1).

Unfortunately for the non-specialist or someone seeking to apply a modular approach to corals, this is not the complete picture. Many corals consist of numerous mouths rather than discrete polyps (for example, some of the fungiids), and these corals too are generally regarded as colonies. This may not be rigorous if we follow Wells' (1971) argument that they are 'cerberoid', that is, single-polyp forms with many mouths, but it is convenient to accept that they are colonies at this stage, and return to the problem later (§ 3c). In particular, it is possible to think of these repeated mouths as incomplete polyps or polypoids. Their skeletal counterparts are calicinal or corallite centres without walls defining their limits (figures 2d, h, 3d, h). These centres consist of distinct structures, or points of convergence of the radial plates (septa) which correspond to the mesenteries of the polyp, or both. In some corals, these skeletal centres are cryptic, and only verifiable by reference to the soft tissue anatomy (for example, *Hydnophora*; figure 3*h*, and see Matthai (1926), figure 16).

### (*c*) *Zooidal budding*

The development of new zooids in a coral colony is now invariably referred to as budding in the scleractinian literature, and as increase in the rugosan and tabulate literature. Blastogeny has also been used (Fedorowski & Jull 1976) but not widely adopted. One term previously in wide use was asexual reproduction but this has been rejected by some authors (Rosen 1979; Harper & Bell 1979). Various kinds of budding have been defined, but they all belong to one of two major categories, intratentacular and extratentacular (Wells 1956; Oliver 1968; Hill 1981). Both processes amount to expansion of the gastrovascular cavity of the original polyp, with a concomitant opening of a new mouth into this expanded cavity (figure 1), and it would therefore be easy to overstate the biological significance of this distinction. In faviid corals, Matthai (1926) noted that both modes of budding can occur even in a single colony. Corals that differ in their modes of budding can have a similar outward form (for example, *Favia* and *Montastrea*).

Nevertheless, budding modes do have a considerable influence on colony architecture. Figure 2 shows how the combined characters of budding mode and corallite wall development determine whether zooids exhibit a branching pattern, and therefore also whether it is possible to observe zooidal genealogy from skeletal morphology. In corals that have neither a branching relationship between the corallites, nor any other indication of polyp genealogy, it is probably impossible to determine a true polyp-by-polyp growth history other than by direct observations of their growth.

In the Rugosa and Tabulata, we cannot be sure how the various modes of budding inferred from their corallite architecture relate to the presumed polyps, but Oliver (1968) has pointed out their likely correlation with intra- and extratentacular budding in Scleractinia.

It is common to refer to newly budded zooids as daughters. In fact, no sexuality of zooids is meant to be implied, and in living Scleractinia, the zooids of some taxa are hermaphroditic and in other taxa they are specifically male or female (Fadlallah 1983; Harrison 1985). Even the notion of parent and offspring zooids may also be misleading, or difficult to apply. In addition to the reasons shown in figure 2 (compare *a, b, c, f* with *d, e, g, h*), budding in some corals is pseudodichotomous, sometimes making it difficult to distinguish parent zooid from offspring (for example, *Favia* species). (Matthai (1926) insisted that longitudinal division of the mouth, and hence truly dichotomous budding, does not occur in living corals.) In other corals, parenthood is effectively collective among two or more zooids, as within the meander systems of meandroid corals (figure 3*d*) where a new zooid can appear between two pre-existing ones with mesenterial tissues derived from both of them (Matthai 1926; Stephenson & Stephenson 1933).

### (*d*) *Growth and form of corals*

Growth and form have been studied extensively by coral workers, but only rarely in a way that can be considered relevant to their modularity. The study of growth and form in terms of the repeated units within a colony has been relatively overlooked, whereas the gross outer shapes of corals and the ways in which these grow have always attracted more attention. Closer to modular considerations is the recent increase in understanding of variation in form between corallites within a colony (see, for example, Oliver 1968; Foster 1983, 1985) and its significance within populations of colonies (see, for example, Best *et al.* 1984; Foster 1984). Even so, much of this is concerned with the components within modules rather than the spatial arrangements and life histories of colony units.

The most relevant and organized body of knowledge of this kind concerns modes of budding in coral colonies, but while this tells us how new zooids arise within a colony and how this affects their arrangment, we have yet to learn about their complete subsequent histories. How often, and when, do they themselves produce offspring zooids, and how long do they survive recognizably within a colony? Wood-Jones (1907, 1910; see quotation at the head of this paper) thought polyps were virtually immortal, but Beklemishev (1969) claimed that they are short lived. Such questions are prompted by Harper's view that colony units can be treated demographically (Harper 1977; Harper & Bell 1979); they constitute a metapopulation (White 1979). In this approach, growth is the net result of the population dynamics of colony units, and form is a synoptic chart or census of the colony units at a particular instant. This raises an old coral question in a new guise: how far can we distinguish the genetic component of colony demography from external influences?

It is useful to distinguish internal and external form in corals because the arrangement of zooids and their interconnecting tissues within a colony is often not evident or deducible from its surface appearance (for example, figure 2*b*, *e*, *f*). Internal form, even of living corals, has to be investigated by preparation of dead material. Similar external forms of corals can have very different internal forms (for example, figure 3*b*) though differences in outward form can usually be related to differences in internal organization.

Somewhat confusingly, 'form' in coral workers' usage often refers to external form, sometimes also called growth form or growth habit, with the added implication that it refers only to those subspecific, plastic, morphological features that are due in part to external influences. In practice, however, it has been a familiar but notorious difficulty to distinguish (i) genotypic form, (ii) phenotypic form, (iii) environmental selection of phenotypic form, and (iv) direct response of growth to external influences (see Foster (1979) for a review of this problem). There is an obvious need to distinguish the form of a particular specimen from more abstract or generalized notions of form based on populations or specimen suites. This allows us to eliminate fortuitous features and progress towards inherited colony architecture, by analogy with tree architecture (Hallé *et al.* 1978).

For growth, too, it follows that there are internal and external aspects, particular and generalized, genetically or externally influenced. Growth is a resultant of lateral and longitudinal directional components, the balance between them differing within colonies, between colonies, and between taxa (figure 1). The use of the modular concept should help us to be more precise in our investigations, by providing standard, specific features to count or observe in a group of animals whose growth and form have otherwise proved elusively complex and irregular to quantify and analyse except at the grossest levels.

## 3. Cloning, fusion and fission

Remarks on cloning in corals are not only relevant in their own right as a feature of growth, but are essential to an understanding of the different kinds of zooids seen in corals.

Colony growth by budding implies that a colony is a clone (Oliver 1968). In fact, the ambiguities of the word colonial have tempted authors to qualify it by using 'clonal' (Rosen 1979; Ryland 1981), or to give outright preference to clonal, instead (Jackson & Coates, this symposium). We can therefore distinguish cloning in which polyps remain in contact with the rest of the colony, from that in which they become completely detached. The first might be called *clonoteny*, and the second *clonopary*, and each corresponds to Rosen's (1979) continuous and discontinuous categories, respectively. For population studies, the distinction is important, but both phenomena probably have a common origin in being the same process taken to different lengths (Wood-Jones 1907, 1910). The solitary zooidal form is invariably regarded as primitive (see, for example, Wells 1956), and, assuming that clonopary is comparably primitive, coloniality can most simply be regarded as having evolved from solitary forms by paedomorphosis of bud development. Clonoteny is incomplete clonopary. (This is an old idea (see Clark 1964, p. 22), but I exclude from it here the phylogenetic implications with respect to other phyla (that is, the corm theory).)

Taking this point further, this same paedomorphic trend can account for some of the major features of zooidal form and arrangement in colonies, by still further evolutionary suppression of bud development (Wood-Jones 1907, 1910). Zooids without walls, as in thamnasteroid (figure 2$d$), aphroid and astreoid colonies, are less complete than the free-branching zooids with walls seen in fasciculate colonies (figure 2$a$). Coates & Oliver (1973) regard the fasciculate (phaceloid) condition as the most primitive kind of coral colony.

There may also be more than one level of bud development within a single colony of a species (*polystatic clonogeny*). I interpret some meandroid corals in this way (figure 3$d$ and §5$d$); and *Goniopora stokesi* is both clonotenous and clonoparous (Boschma 1923; Rosen & Taylor 1969; Veron & Pichon 1982). Scheer (1959, 1960) on the other hand interpreted the cloned colonies as an unusual development of larval brooding and presumably therefore as a sexually produced phenomenon. There also appears to be some plasticity of clonogenic level in that Sammarco's (1982) observation of polyp detachment in *Seriatopora* may represent an environmentally induced switch from clonoteny to full clonopary.

Clonopary in solitary corals predictably results in solitary offspring (for example, *Fungia*; see Wells 1966), but in colonial corals, the offspring clones might in theory be solitary or colonial. Stoddart (1983) concluded that *Pocillopora damicornis* can produce larvae asexually, but, apart from this, cloning of colonial offspring is more widely documented (see, for example, Hughes & Jackson 1980, 1985; Highsmith 1982). In *G. stokesi* there appears to be specific genetic control of this, though authors differ on the outward details. Thus Boschma (1923) believed that clonoparous zooids are previously attached to their parent colony by a weak skeletal connection, but Veron & Pichon (1982) state that they are held only be extended soft tissue (coenosarc).

Colonies also result from cloning by fragmentation or partial mortality. The genetic element here is less specific, Highsmith (1982) having argued that this kind of cloning is genetic in a broad way, being a susceptibility 'selected for over evolutionary time and incorporated into the life history of many corals'.

The concepts of ramets and genets (as in Harper 1977) can be applied to corals (Rosen 1979;

Heyward & Collins 1985), but because fusion of corals is also widespread, a single colony cannot be assumed to be either a single genet or a single ramet. Fusion of juvenile colonies of the same species has long been known (see, for example, Duerden 1902; Stephenson 1931; Boschma 1929), but more recently there has been increasing recognition of mature colony fusion (see, for example, Hughes & Jackson 1980, 1985). Buddemeier *et al.* (1974, figure 5) reproduced a fine X-ray photograph of a single colony of *Goniastrea retiformis* over 20 cm across with a concordant outline, but with an internal structure showing 'coalescence of two apparently originally independent colonies'.

Conversely, there is also now a substantial body of knowledge of incompatibility reactions between neighbouring corals starting with Lang's (1971) now classic paper on *Scolymia* (a solitary coral). Since such reactions can occur at any taxonomic level, and even between clonally distinct colonies of the same species (see Lang 1984, p. 24, for review), it is clear that the pattern of genetic control of whether fusion or non-fusion takes place is complex, and at the moment we cannot generalize about a possible limiting genetic distance for fusion.

In general, the smallest unit of clonal growth is the zooid, or larval zooid, but in at least some fungiids, the solitary forms can regenerate from broken segments consisting of as little as one-sixth of the original, this being that portion of the original which lies between an adjacent pair of primary septa (Wells 1966).

## 4. MODULES: THEORY

### (a) *Generalities*

Chapman (1981) has discussed the general history of the use of the word 'module', including its non-biological use. Its biological use has been adapted from architecture and design, this twentieth century use being itself an adaptation of older meanings. Its first botanical use is apparently due to Prévost (1967, in White 1984). Its first zoological use was proposed by Finks (1971), in a very different sense, auspiciously, from the botanical usage. Its rapidly spreading current usage in both botany and zoology, however, is due to Harper (1977), who was the first to suggest that modular organisms included corals.

White (1984) has distinguished the original botanical meaning from Harper's broader application ('any convenient unit'), which according to White corresponds to plant metamerism. Flexibility of choice is the essence of Harper modules, provided that the unit chosen shows (collectively) 'demographic properties' within a plant or colony or genet (Harper & Bell 1979). As White (1984) has pointed out the inconsistency among botanists, which botanical usage is most readily applicable to corals?

The *de facto* notion of a coral module is actually the polyp (Harper 1977; Harper & Bell 1979; Rosen 1979; Ryland 1981; Hughes 1983). Chapman & Stebbing (1980) had evidently also arrived at this independently. Polyps (or zooids) however do not generally correspond to Prévost modules (§7). Few would deny the importance of the polyp as a colony unit, but choice of modules for corals has never been discussed. I have therefore taken an approach derived from Riedl (1978), and first consider the formal analysis of morphological order and repetition, and then the functional and homological significance of the resulting fundamental units.

## (b) Topology

The modular idea is essentially a visual and topological one, but the choice of repeated units in corals is wide, including septa, polyps, branches, tentacles, verrucae, mesenteries, mouths and dissepiments (figures 1–3). It is of great importance that many of these repeated elements are parts of other repeated structures, sometimes combined in collective hierarchies. Riedl (1978) has argued that this kind of morphological hierarchy is not an accident of human subjectivity, but an absolute phenomenon. We can start with the whole organism and make successive morphological subdivisions into progressively smaller units. Zooids contain septa, and septa consist of trabecular elements; branches bear polyps and these bear tentacles. I suggest that the following four criteria can be used to find modular order in coral colonies.

(i) A structure is modular if it can be divided completely into a pattern of repeated similar units, without any omitted features. Boundaries of modules should abut and there should be no morphological 'no-man's land' between them (figure 1). In two dimensions, a tesselation is modular, and in corals or plants, *a modular structure is a three-dimensional tesselation* (though the boundaries may have to be drawn in communal tissue or in free space between the visible units). The zooids of a cerioid coral (figure 2*c, g*) are obviously modules, but their mouths alone are not, because this would exclude other repeated units. (For practical purposes of course, key parts like mouths may be used as indicators of zooids (figure 1), but this alone does not make them modules.)

(ii) A second reason why, in this chosen example, mouths and columellae are not modules, is that they occur at the same level of hierarchical subdivision as various other features like tentacles and septa: (zooids (mouth) (columella) (septa) (tentacles) (etc.)). The zooid is the next higher level unit which embraces all these features (figure 1). We can therefore distinguish morphological subdivision which yields only units that are similar (A (B) (B) (B) (B)) from that which yields a variety of different units (A (B) (C) (D) (E)). These are, respectively, homogeneous and heterogeneous subdivisions. *Modules should correspond to homogeneous units of hierarchical subdivision.* The zooids of most kinds of coral obviously fulfil this condition (figure 2), with respect to, say, the whole colony, or, if there are also branches (figure 3*a, e*), to colony branches: (colony (zooids)) or, (colony (branches (zooids))).

(iii) In practice, morphological analysis through successive levels within most organisms reveals that they consist of units of both homogeneous and heterogeneous subdivision at one level or another. Hence, *an organism is modular if subdivision at the highest (or very high) level reveals homogeneous units.* Trees and some corals have a first subdivision which is heterogeneous, consisting of stem and branches, but modular organization is clear thereafter.

(iv) Analysis of some organisms reveals that not only the first subdivision is homogeneous, but that there may be a sequence of two or more further consecutive homogeneous subdivisions, like (colony (branches (zooidal groups (zooids)))), before the sequence is interrupted by heterogeneous subdivision (figure 3).

*If homogeneous units occur in an unbroken series of subdivisions, they are all modules at their respective level (hence modules of modules, that is, cormidia), and the finest unit is the fundamental module.* From the present point of view, it is of no direct relevance that homogeneous units of some kind may reappear after heterogeneous interruption. In corals, after heterogeneous division of zooids into different parts, homogeneity reappears at the levels of cells, skeletal sclerites, etc., but these are not modules of corals.

With these criteria, a solitary coral (or single zooid: figure 1) is not modular because the first stage of subdivision is heterogeneous, consisting of an array of different zooidal parts. Colonial corals are modular because it is usually easy to show that the above criteria are met, though it is the zooid that is the basis of the fundamental unit, not the polyp on its own (figure 1). A few corals, however, seem to be anomalous. In *Hydnophora*, for example, the finest units of homogeneous subdivision seem to be the monticules (mont. in figure 3h), and in *Acropora* (figure 3a), the polymorphism of the zooids makes subdivision at the zooidal level seem heterogeneous. These are obviously limitations to purely topological analysis, and recourse must be made to biology (below).

Taking the zooid to be the fundamental module of a coral, notwithstanding the apparent anomalies, we can see that the heterogeneous units found at the next lower level of subdivision always include one mouth and its skeletal counterpart, the corallite centre (figure 1). The various parts that make up a zooid cannot be modules, but can be regarded as components, the mouth or centre being the substantive component of the fundamental module. Completeness of zooidal development can be gauged from the suite of other, accessory, components which are also present in the fundamental module. The notion of a fundamental zooid is similar to that of the unit cell in crystallography, except that it is physically and geometrically less rigid.

### (c) *Homology*

Riedl (1978) recognizes seven forms of similarity. Among these the similarity of units found in homogeneous subdivisions constitute homonoms, that is, they are homologous with respect to each other within that particular organism and within their own hierarchical level. Ideally, for present purposes, they should also be homologous from one coral to another, and therefore for the group as a whole. This would be difficult to argue for many higher-level modules, but I do propose this for fundamental modules (zooids).

Wells (1971) has apparently argued the contrary. He states that the repeated units produced by intratentacular budding are not the same as those produced by extratentacular budding, because the first are 'incomplete individuals marked by lack of directive mesenterial couples and often not morphologically individualized', whereas the second are 'complete homomorphic individuals.' Coates & Oliver (1973), in their important contribution to our understanding of coloniality in corals, followed Wells.

Against this, there are three considerations. Firstly the zooids produced by intratentacular budding share an important feature with extratentacular zooids, namely the mouth. This is not a simple structure but a whole functional centre of action within a colony, especially for feeding, reproduction and excretion (Matthai 1926). If the mouth is homologous throughout corals (or Scleractinia, at least), then the zooids must also be homologous, regardless of their budding mode or completeness of development.

Second, if the mouths developed in the zooids of each kind of budding mode are not homologous, we must assume that they have evolved twice, independently, in the two kinds of corals. Moreover, we should expect differences in the anatomy, histology or development of these mouths. It becomes especially difficult to envisage a convincing evolutionary pathway for corals (for example, *Favia* species) which bud both intra- and extratentacularly, even within a single colony (Matthai 1914; Veron *et al.* 1977).

Third, Matthai (1914) also stressed that the symmetry differences referred to by Wells (1971) did not correspond exactly to budding modes.

In all, it is a simpler proposition to regard all mouths as homologous. Thus even if the zooids in some taxa are more complete than in others, they can still be regarded as homologous with each other. Polymorphic zooids within a colony would also be homologous. We therefore have a single conceptual zooid, which is more or less modified, or more or less arrested in its development, and which is sometimes polymorphic, according to the taxon.

Two more general points can be mentioned in support of this homology argument. First, the solitary zooidal state is invariably regarded as primitive with respect to colonial forms (Wells 1956; Coates & Oliver 1973). Second, it is interesting that the mouths within a polystomatous meander (figure 3d) in *Platygyra sinensis* function reproductively as separate zooids, producing their own egg–sperm bundles during spawning (see photograph by J. Oliver in Talbot (ed.) (1984) p. 115, lower right).

Homology overcomes the limitations of the foregoing topological analysis, not only in the case of polymorphism, but also for corals whose interzooidal, coenosteal or wall modifications dominate the visual pattern of repetition (for example, figure 3c, h; and the striking, upward extensions of interzooidal tissue in *Pectinia paeonia*). In such cases, homology directs us to the mouth first, to discover the fundamental module. The interzooidal features are secondary. They presumably have an important, if as yet unknown, colony function, and in *M. verrucosa* (figure 3c) and *H. microconos* (figure 3h), they are the substantive components of higher level modules consisting also of zooids clustered close to, or around these features. In other corals, such interzooidal features are less regular, as in the collines of *Pachyseris* and *Pavona varians*, or the 'carinae' of frondose *Pavona*, and are less obviously modular. In *Pectinia paeonia*, *Stylocoeniella* and *Stylocoenia*, there is a single discrete interzooidal structure for each zooid, so it belongs to the same level of modularity as the zooid.

In conclusion, although it is possible to make useful inferences about corals by using Harper modules, I have deliberately tried to qualify convenience with topology and homology in the belief that this will advance our understanding of the morphological comparison (and evolution) of different taxa, and stabilize terminology. The conclusion that the zooid is the primary coral module is not, perhaps, surprising, but it is new in so far as it (i) unites polyp and corallite; (ii) recognizes no fundamental difference between different kinds of zooid; and (iii) provides a fixed point of reference for relating all other kinds of repeated feature to each other.

## 5. Modular specification of colony form

### (a) *Colony form = zooidal organization + colony shape*

In the previous section, I used morphological analysis to clarify the prime significance of the zooid as a colony unit for corals in general. To specify coral colonies morphologically, we now reverse this approach by starting with the components that make up a zooid and repeating this specification for all the modular levels present in colony. For fossil corals, we have to use the corallite or corallite centre as the fundamental module because polyp tissues are absent.

The need to distinguish internal and external morphology has already been mentioned. Internal morphology concerns the organization of zooids, whereas external morphology includes the expression of this organization at the colony surface together with the colony's outline shape. Colony form can therefore be thought of as the sum of its zooidal organization (internal and external) and its shape. Although shape may be modular too, this is not always so. If shape is modular it generally reflects an internal modular organization, but shape modules should be specified at the appropriate level of organizational module, because, ordinally, their levels may not correspond.

To date, interest in colony morphology has centred on the fundamental zooid, and generalized shape. Little attention has been paid to higher level modular groupings of zooids. For full colony specification, therefore, I suggest that the procedure might be as follows.

(i) Fundamental module form: (a) components; (b) spatial arrangement of modules; (c) shape of module*.

(ii)* Second-level module form: (a) zooidal complement and other specifically second-level components; (b) spatial arrangement of second-level modules; (c) shape of second-level module*.

(iii)* (and so one, as necessary): (a), (b), and (c) as (ii).

Asterisks indicate features which may not be present or specifiable. Components (figures 1 and 2) include the substantive zooidal components and accessory components; the accessory components are zooidal (endothecal) and coenenchymal (exothecal), the latter being applicable to some corals only. Together, the components, their relationship to each other, and their shape (item (c)) describe the module (figure 1). A module at the first level becomes the substantive component at the second level, and so on for each modular level present (figure 3). Particular components either endothecal or exothecal, may be characteristic of a particular level (for example, figure 3g, h). There may be polymorphism of modules at any level, and these are specified consecutively within each level.

Items (a) and (b) together specify organization. Item (c) refers only to outward shape as it affects the outline of the whole colony. The hierarchy of shape modules (if present at all) may differ from that of the organizational modules in that there may not be a shape module for every organizational level. This approach does not make any assumptions about what is genetic and what is not, but information collected and organized by this scheme may be generalized as coral architecture as a step towards inferring inherited form.

### (b) Components of fundamental modules (i a)

The substantive component, the mouth and its skeletal centre, may be difficult to discern on superficial examination, especially in those meandroid corals where in the skeleton, one centre can merge with the next (in contrast to figure 3d). In *Pachyseris*, the mouths are minute and there are no tentacles (Yonge 1930). Exact location of mouths in such corals cannot be inferred from skeleton alone. Accessory zooidal components include septa, tentacles, mesenteries, columella, endothecal dissepiments and various kinds of wall structures (figures 1 and 2). The total complement gives an indication of how complete the zooid is with respect to its notional solitary counterpart. Some wall structures are specialized with parts that project upward from the calicinal surface (*Stylocoeniella*, *Stylocoenia*). It may be necessary to specify more than one kind of fundamental module, as in *Acropora* (figure 3a) and some circumoral corals (figure 2h).

Coenenchymal components are present only in plocoid corals (figures 1, 2b, e, f, 3c), the zooids being spaced from one another and laterally connected by components like costae, exothecal dissepiments and their soft tissue counterpart, communal edge zone. Such interconnections are absent from the Rugosa (Hill 1981). In corals without zooidal walls, however (especially thamnasteroid forms (figure 2d)), the distance between zooidal centres may be so great that the intervening tissue becomes a kind of coenenchyme that merges continuously with zooidal

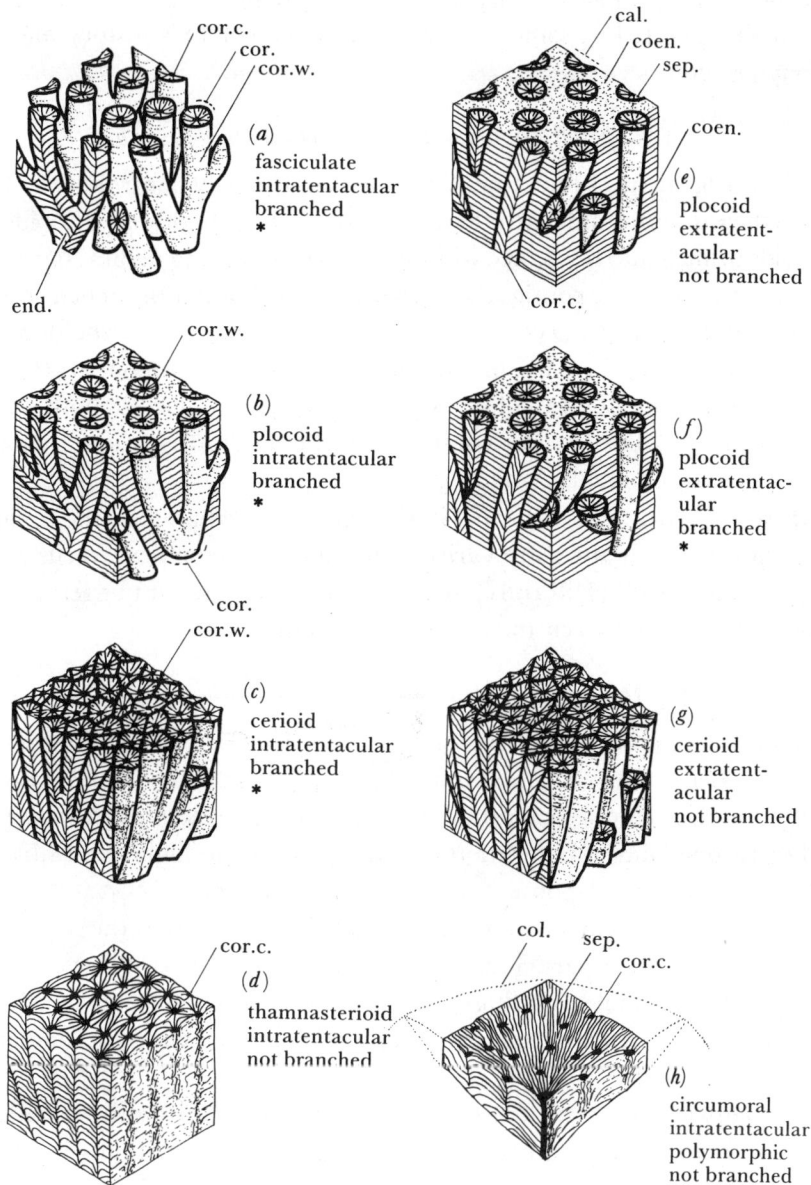

FIGURE 2. Examples of first level (fundamental) modular organization. All examples are schematic, showing only selected components, and are not to scale. Left-hand faces represent cut longitudinal sections, usually sagittal. Right-hand faces represent more or less broken surfaces. Upper faces represent calicinal surfaces without polyp tissues (but compare with figure 1). Polyp tissues are generally continuous between zooids except in (a) where the polyps often become completely separated. Note that left hand faces are idealized and in practice, a cut face would be unlikely to show such consistent alignments and orientation of corallites. In practice, therefore, (b) may appear similar to (e) and (f), and (c) similar to (g). In (b), (c), (e), (f) and (g), walls may be perforate or incomplete and hence transitional in appearance to (d). Abbreviations: col., colony outline; sep., septum; *, genealogy of zooids is directly recorded in the skeleton of these examples, but may be deducible in other examples from other evidence. Other abbreviations as figure 1.

tissues. This tissue may also develop specialized surface structures of its own, especially at higher modular levels.

Topologically, the boundary of the fundamental module takes in the whole zooid down into the skeleton to the point at which corallite growth commenced, and also includes a half-share of any coenenchymal tissue lying between itself and each of its neighbours (figure 1). There will of course be some modules in many colonies that are not seen at the colony surface at all, having died or degenerated at some earlier point in the colony's history and been obscured by later colony growth.

### (c) *Spatial arrangement of fundamental modules* (i b)

The spatial arrangement of fundamental zooids is primarily controlled by budding modes but is also specified in terms of length, location and angle of budding of zooids. In fasciculate corals, the zooids branch into free space without lateral connecting tissues (figure 2a). In cerioid corals (figure 2c, g), zooids are contiguous, either branched or not branched (but see discussion of intermural budding in Oliver (1968)). In plocoid corals (above), the zooids may be branched (figure 2b, f) or independent, resting proximally on coenosteal components (figure 2e), though their lateral arrangement at the colony surface may be similar in both cases (*Favia* and *Montastrea*, respectively). The first is a consequence of either intra- or extratentacular budding and the second of extratentacular budding.

Many of these standard terms for zooidal arrangement are based on the notion of walls as a single feature, but wall structures are varied and sometimes complex (Wells 1956; Mori *et al.* 1977; Mori & Minoura 1980; Hill 1981), suggesting that new descriptive terminology is needed, especially for consistency between the major coral groups.

### (d) *Organization of second level modules* (ii a) *and* (ii b)

Higher level groupings of zooids, or cormidia, if present, are all reiterations in the broader sense of Harper & Bell (1979), but the Montpellier school use *réitération* more restrictedly. Thus Dauget (1985) has described regeneration of whole colony growth as *réitération*, whereas here, this would be just one kind of reiteration which occurs at the highest possible modular level for that particular coral (in response to environmental factors).

The fundamental modules alone may constitute the components at the second modular level, but there is also sometimes a particular feature like a specialized wall or coenosteal structure associated with such zooidal groupings, like the coenosteal verrucae of *Montipora verrucosa* (figure 3c). Some typical arrangements of fundamental modules within second and higher level modules are shown in figure 3.

There are three ways in which these groupings are built up (with or without their own components): (i) homozooidal and (ii) heterozooidal, which are either (ii a) polymorphic, or (ii b) polystatic. These three categories are not mutually exclusive. They are sometimes clearly genetic (*Hydnophora*) and sometimes environmental (many of the poritids), often with the first modified by the second.

In homozooidal groupings, the zooids are all of the same kind, forming a recognizable discrete structure that results from a particular kind of growth pattern that is sometimes centred around particular points or axes (figure 3b, e). X-ray techniques reveal these groupings very well (and their relation to colony shape), and show how they can vary from being relatively weakly defined (Buddemeier *et al.* 1974, figure 2: *Porites lobata*) to very definite structures (Buddemeier *et al.*, figure 4: *Psammocora togianensis*).

In heterozooidal groupings, the zooids are not all identical, but are composed of particular combinations of the different kinds of zooid. This is visually obvious in the case of polymorphism (*Acropora*: figure 3a), but not in the polystomatous groupings of some meandroid corals. When they consist of numerous meanders (figure 3d), one kind of zooid is represented by the centres without complete walls, and the other by the wall that encloses the centres. This may seem nonsense, but is a consequence of growth. The existence of numerous distinct meander systems implies that from time to time during colony growth, a new zooid, or an existing zooid within a meander, develops a complete wall (centre of figure 3*d*). The majority of zooids, however, do not do so, so remaining within the walls of those that do. These zooids-within-zooids are, therefore, an example of the idea of paedomorphic clonogeny (§3), but within a single colony, two different stages of arrest occur simultaneously (*polystasis*).

In *Hydnophora* (figure 3*h*) and *Manicina areolata*, there is normally only a single continuous meander system, so there is no polystasis. In some colonies of *Manicina*, however, there are two or three meanders, but these have been interpreted as the result of fusion of different larvae (Boschma 1929; Yonge 1936). Thus fusion can also give rise to higher level modules in the homozooidal category (see also the *Goniastrea retiformis* of Buddemeier *et al.* (1974), mentioned already). *M. areolata* also shows another kind of homozooidal module: its meander system divides into branches (me.br. in figure 3*d*). *Isophyllia sinuosa* shows a five-rayed variant on this theme (Matthai 1926). In corals like the meandroid faviids and mussids where this meander branching is common, the branched sections constitute second level modules, and the distinct meander systems, if present, represent third level modules (by fusion or polystasis).

Second level modules may just be growth aggregations or superimpositions (figure 3*b*, and Buddemeier & Kinzie (1976), figures 1A (*Porites*) and 1B (*Pavona*)), but they can also be categorized as fasciculate, cerioid, plocoid, etc., as for first level modules (see caption to figure 3*d*).

### (*e*) *Organization of third and higher level modules* (iii*a*), (iii*b*), *etc.*

The zooidal groups of the second level may themselves be grouped (especially figure 3*b*, *d*, *e*, *f*), and the same approach outlined for lower levels can be followed, with suitable modification where necessary. The previous distinction of homozooidal and heterozooidal groups can be generalized for all levels as homomodular and heteromodular. Some examples of three-tier modularity have already been mentioned in connection with meandroid corals whose meanders are branched. Another example of three-tier modularity, based entirely on homomodular groups, is found in verrucose forms of *Pocillopora* whose verrucae represent second level structures (figure 3*g*), and whose flattened, columnar branches represent third level modules. Table *Acropora* consists of polymorphic branches fused or anastomosed into a horizontal plate. Its branches are second level modules, but its plates are sometimes reiterated and therefore represent third level homomodular units. Dauget's (1985) examples of traumatic *réiteration*, already mentioned, also belong here. A number of meandroid corals show a third or higher level organization when the whole colony has the form of a ramose branching or platey system (*Platygyra zelli*, *Merulina ampliata*, *Hydnophora rigida*, and *H. exesa*).

### (*f*) *Shape* (*c*)

Each level of modular organization may also contribute directly to shape, so that a shape module usually has a corresponding organizational module, but the reverse is not necessarily true. Similar outward shapes may conceal different modular organizations (Krasnov & Preobrazhenskiy 1972).

The simplest colony shapes are massive (more or less hemispherical), mound-like or encrusting sheets, and are not therefore modular in their shape. Their organization is very varied

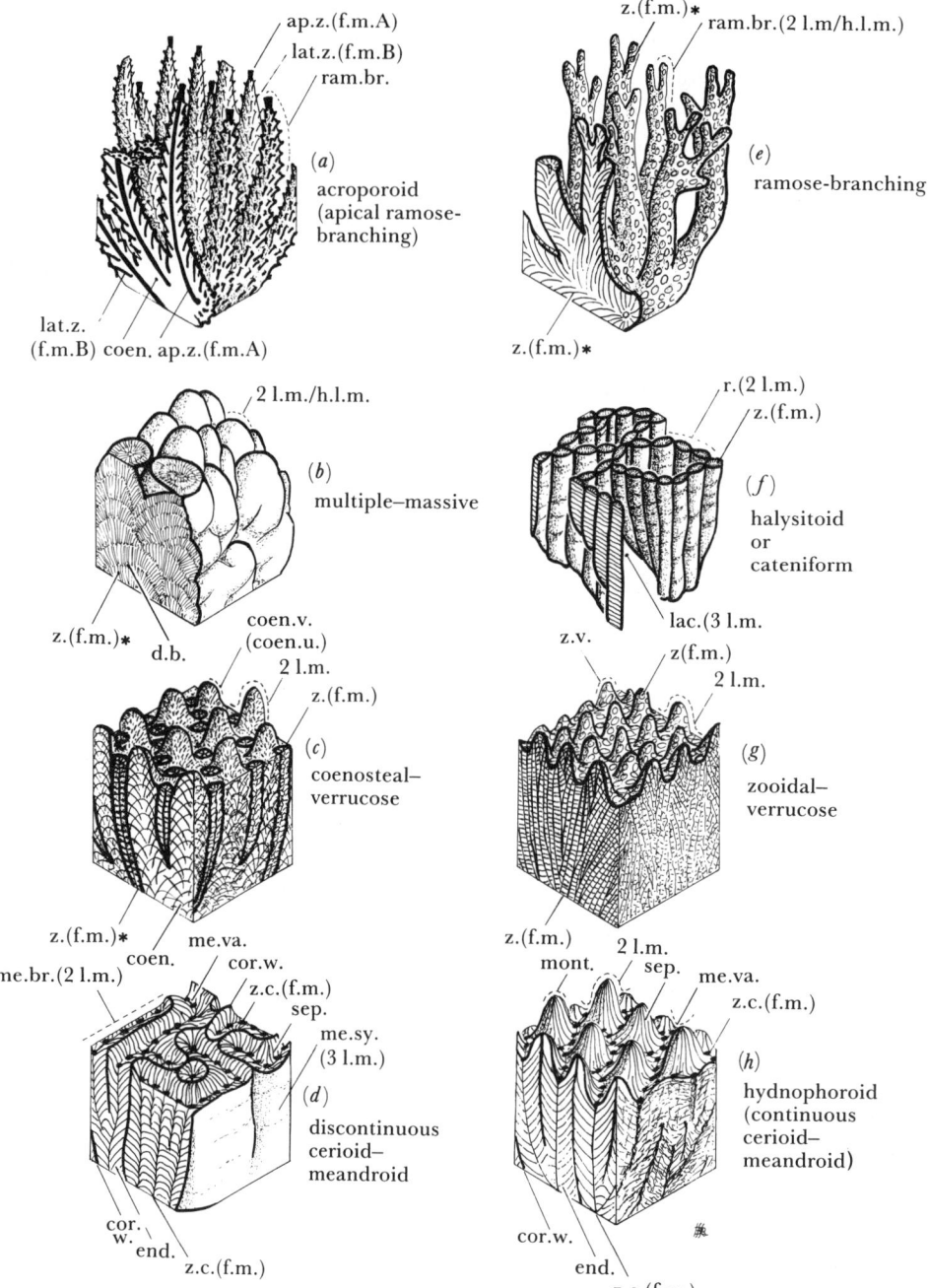

FIGURE 3. Examples of second and higher level modular organization. Diagrams on different scales (for zooidal scale, see figure 1) and not to scale. Each example is based on the organization of a particular coral but is not necessarily accurate in other details. Left-hand faces represent cut longitudinal sections, usually sagittal. Right-hand faces represent more or less broken surfaces (except (a), (b) and (e)). Upper faces represent calicinal surfaces without polyp tissues, sometimes with transverse sections through the higher level modules ((a), (b) and (e)). Note modularity of shape as well as organization in all but (d).

(a) Zooidal organization, a specialized form of figure 2f, but second level modules are heterozooidal–polymorphic reiterations of systems of apical and lateral zooids (based on *Acropora*).

(b) Homomodular reiteration which can occur at different modular levels in different corals, for example, the second level examples consist of homozooidal reiterations of figure 2b–g, and third and higher level examples are reiterations of second or higher level modules like figure 3c, d or h. (Typical of poritids and faviids and common in fossil groups.)

(c) Second level modules consisting of homozooidal reiteration of zooids as in figure 2e, grouped around coenosteal verrucae (typical of *Montipora verrucosa*).

from single-tier modularity as in some species of *Porites* and *Goniastrea*, to multi-tier modularity in meandroid faviids and mussids. Massive colonies, however, often also bear low to conical protuberances as weak shape modules (figure 3b), reflecting second-level homomodular organization. The columns, branches and platey shapes of many corals are, in effect, a further, specialized development of these kinds of protuberances. In this case the branches are described as ramose (figures 3a, e) to distinguish them from the fasciculate branches in other corals (figure 2a). Fasciculate corals are among the most primitive kinds of colony and are especially common in the fossil record. They have only a single modular level, branching zooids, and this corresponds, zooid-for-zooid, with their branching shape-modules (for example, *Siphonodendron*, *Thecosmilia*, *Cladocora*).

The ecological (especially the palaeoecological) literature places an emphasis on shape because corals of particular shapes are broadly associated with particular environmental conditions. Inference of past conditions can, therefore, be made, supposedly, by reference to the shapes of fossil corals, and therefore independently of taxonomic and stratigraphic considerations. Stearn (1982) has discussed some of the problems of this approach, but I would blame some of this on confusion about the modular level at which branching occurs. At the very least, it is necessary to distinguish ramose-branching from fasciculate-branching (Rosen 1971). Hierarchical modular analysis of shape, as outlined here, should also help to provide a finer resolution between shape and environment than exists at present.

Branching is also widely associated with modularity, but in corals this is not universally true. A massive colony of *Montastrea*, for example, has neither branching zooids (figure 2e) nor a branching shape (figure 3b). Branching at the fundamental organizational level is confined to those forms with zooidal walls as asterisked in figure 2. At higher modular levels, branching can occur within meanders (figure 3d), between meanders (figure 3d), and between homomodular (figure 3e) and polymorphic (figure 3a) clusters. Branching organization is not necessarily seen as branching shape: in massive *Favia* it is not (figures 2b and 3b). In ramose-branching *Acropora* (figure 3a) or *Porites* (figure 3e) it is; but their branches are first level shape-modules which correspond to second-level organizational modules. Shape, organization and branching in halysitids is particularly unusual (figure 3f). In *Porites* (*Synaraea*)

---

(d) Heterozooidal–polystatic reiteration of zooids in linear series (second level modules) often in homomodular branching systems (third level modules), typical of many Faviicae and Eusmiliinae. Note new monocentric corallite, front centre. Intermeander organization is cerioid (compare figure 2c), typical of *Platygyra*, but in other meandroid corals may be fasciculate (for example, *Lobophyllia*) or plocoid (for example, *Diploria*).

(e) As for (b) but ramose-branching, and with third or higher level modules based on reiteration of (for example) figure 3d, g or h. Typical of poritids, pocilloporids and many fossil groups (tabulate and scleractinian). Intergrades with platey-ramose forms.

(f) Homozooidal reiteration in ranks (second level modules) grouped in homomodular anastomosing systems enclosing lacunae (third level modules). Typical of halysitid tabulate corals (Ordovician–Silurian).

(g) Homozooidal reiteration of (for example) figure 2e or g grouped into small projections, typical of *Pocillopora*.

(h) Homozooidal reiteration of fundamental modules in continuously interlinked meander-rings (second level modules) which surround isolated portions of wall; typical of *Hydnophora*; like cerioid–meandroid forms (figure 3d), but not polystatic. (Corallite centres not usually as visible in skeleton as shown.)

Abbreviations: ap.z., apical zooids; coen.u., coenosteal upgrowth; coen.v., coenosteal verruca; dens.b., density band; f.m.A, fundamental module of type A; f.m.B, fundamental module of type B; h.l.m., third or higher level module; lac., lacuna; lat.z., lateral zooid; me.br., meander branch; me.sy., meander system; me.va., meander valley; mont., monticule; r., rank or palisade of zooids; ram.br., ramose branch; z., zooid; z.c., zooidal centre; z.v., zooidal verruca; 2 l.m., second level module; 3 l.m., third level module; *, this repeated unit may be a higher level module than fundamental (see captions). Other abbreviations as figures 1 and 2.

*rus*, there is polymorphism of shape modules, with horizontal plates and vertical branch-systems (see Randall & Myers 1983, figures 109 and 110).

## 6. MODULAR ASPECTS OF COLONY GROWTH

### (a) *Statistical aspects*

The question of senescence has attracted attention especially in connection with fecundity and fitness. Thus, Harper & Bell (1979) generalized that 'the earliest modules developed from a zygote are usually vegetative units and later sexual modules are formed in a possibly infinitely extended phase of sexuality and the production of daughter zygotes.' Babcock's (1984) analysis of fecundity confirms this for *Goniastrea aspera* and *G. favulus*, showing that the proportion of sexually mature polyps in a colony approaches 100% only when the colony radius is greater than 3.51–4.00 cm. This is equivalent to about 220 zooids in *G. aspera* (see also Harriott 1983). Babcock also showed that the frequency of eggs per mesentery is greater in polyps of large colonies than of small ones.

Harper & Bell (1979) have also pointed out that Grigg's (1977) work on gorgonian corals shows that fecundity of colonies continues to increase even though their life expectancy decreases. Gorgonian corals are distinct from those treated here, and may differ especially in not showing colony fusion and fission to the same extent as scleractinians. It is difficult to know what life expectancy means in scleractinians other than for juveniles or as a result of catastrophic events. Hughes & Jackson (1985) give data showing that the probability of post-juvenile colony mortality actually decreases with size (and hence, very broadly, with age), and, in general, colonial scleractinian longevity of both colonies and genets is likely to be well in excess of the 70 years maximum shown by Grigg's gorgonian data. (There might be exceptions to Hughes & Jackson's growth generalizations: *Manicina areolata*, perhaps?) Even the relation between colony size and fecundity can be confused by fission and fusion, as Harriott (1983) has inferred for *Porites lutea* and *P. australiensis*.

A related matter has been that of possible decrease of growth rate of colonies with age. Unfortunately there have been almost as many methods of assessing this as papers on this subject. Buddemeier & Kinzie (1976) provided a useful critique of methods, including those used to assess senescence. They concluded that for many corals, growth rates are faster when colonies are small, but thereafter, there is no change with age. Hamada (1973; not discussed by Buddemeier & Kinzie), reviewed earlier growth data on the massive coral *Goniastrea aspera* which purported to demonstrate senescence, and argued that the data could be better explained in terms of the geometrical properties of hexagonal close packing of zooids. These may be a factor, but there may be a simpler explanation. His calculations use the ratio of new zooids to pre-existing zooids per unit time ($I$) as a growth measure, and this decreased at various rates in the observed corals. But $I$ is not independent of the original size of a colony (if the number of zooids is assumed to be directly proportional to the area they occupy). For the same value of growth increment, $I$ is less for large colonies than small ones.

In plants, growth and form have been characterized by the relative numbers, positions and ages of the different kinds of modules present (Harper & Bell 1979; White 1984). Although there may be serious practical difficulties in doing the same for corals, similar studies of corals should be of interest. There is also scope for using the proportionality of the number of modules within one particular hierarchical level to those within the next higher level. Such biometric measures would have taxonomic and ecological value. Wijsman-Best (1974) for instance, counted the number of zooids per meander in some species of *Goniastrea* and *Platygyra* at different depths over about 35 m, and found a decrease of over 80% with increasing depth. (She concluded that 'this may point to the fact that meandering as such is an adaptation to faster growth, since less skeleton has to be formed per unit of living tissue'.) This particular measurement, incidentally, for meandroid corals, indicates the proportion of zooids that develop walls, to those that remain less developed by not doing so (figure 3d), that is, it amounts to a polystasis ratio.

Other measurements of heuristic interest include the number and position of budding zooids at different times in a growing colony, and the same for the sexuality and reproductive states of polyps. The first of these is complementary to the topological growth studies mentioned below. For reproduction, Hughes & Jackson (1985) have suggested that active polyps may generally be concentrated in central areas of colonies.

*(b) Topological aspects: towards architectural models*

Few people have recorded observations on colony growth with particular respect to modules and over a long enough period of time to construct architectural models of corals. Most of the information that is available concerns very juvenile stages or a relatively brief interval in the life of mature corals (notably Stephenson 1931; Manton 1932; Stephenson & Stephenson 1933). There are obvious practical difficulties in carrying out this kind of work because colonies grow slowly and zooids are small and numerous in relation to colony size. Here I review some current knowledge which represents the first steps towards the construction of architectural models of corals.

Stephenson's (1931) observations on *Pocillopora bulbosa* (now regarded as a synonym of *P. damicornis*) showed that the first zooid is surrounded by six subsequent zooids, and that this kind of lateral expansion continues until at some, as yet unobserved stage, branches develop by growth from more than one such centre. The iteration of growth centres must arise from, or correlate with, faster longitudinal growth (figure 1) in these areas. With the onset of branch growth, new zooids appear at or around the tips of branches in a broadly axial manner, but without clearly differentiated axial zooids (Manton 1932). This appears to be a common mode of colony growth in ramose branching corals (figure 3e), with the exception of *Acropora* (figure 3a) with its distinct apical zooids (compare Wood-Jones' (1907, 1910) 'groups' 1 and 2, respectively). In young, but postjuvenile dome-shaped colonies of *Favia* the location of new zooids is much less ordered than in *Pocillopora*, with new zooids appearing di- and tristomodaeally almost anywhere in the colony, but with an extratentacular tendency at the colony margins (Stephenson & Stephenson 1933). This can be seen by shading in new zooids in the authors' figures 5–9. The colony in their figure 5, however, shows restriction of new zooids to an annular zone, but this may just have been fortuitous.

A development from simple direct observation has been manipulation of colonies by transplantation, experimental substrates and artificial damage, probably the best known pioneer work of this kind being that of Wood-Jones (1907, 1910). This kind of work has largely been directed to understanding the influence of environment on colony shape, and also, in *Acropora*, environmental effects on polymorphic zooidal organization. Stephenson & Stephenson (1933), Loya (1976) and Kobayashi (1984) showed that coral regenerative growth from damaged surfaces is faster than from undamaged surfaces (a pattern which is familiar in arboriculture and topiary). Foster (1979) used transplants to analyse variation at the component level of fundamental modules in *Montastrea* and *Siderastrea*. Dauget's (1985) study of *réiteration* has already been mentioned. Although the plasticity of coral morphology is now widely accepted, it was controversial 80 years ago (Wood-Jones 1907, 1910). There is, however, further scope for manipulative studies, especially if, through additional studies of effects on internal as well as external morphology, the plasticity can be specified in modular terms. It is doubtful if there is yet a sufficient body of knowledge to make a synthesis worthwhile.

An alternative approach to making direct observations is to use colony morphology as a growth map. The corallites of many corals are like the petrified wakes of growing polyps (figure 1). Oliver (1968) summarized the generalized growth schemes of a number of extinct corals (see also Krasnov & Preobrazhenskiy 1972) and Randall (1982) has done the same for living *Acropora*. Ideally, there has to be a means of inferring chronology, and, if relevant, the genealogy of zooids. Genealogy is clear in corals whose corallites have a branching organization, but can sometimes be inferred too from mesenterial (hence, sometimes, septal) patterns (figure 2). Matthai's (1914, 1926, 1927, 1948a,b) fundamental contribution to knowledge of colony growth, especially of budding modes in relation to colony form, was based on mesenterial analysis. Generally, however, this is not sufficient on its own to reconstruct the complete colony history of a particular coral. One reason for this is that the mesenterial relationships between mature zooids are often reciprocal or symmetrical (for example, meandroid mussids and faviids) and leave unanswered a question like that of whether new zooids are added at the ends of meanders, or within their length, or both. The figures 11–14 of *Lobophyllia* given by Stephenson & Stephenson (1933) show that both seem to have occurred, though 'the actual birth of these new mouths was not witnessed by anyone'.

Two rather makeshift ways of inferring time from form are (i) the comparison of differently sized colonies of the same species of the same general shape from the same habitat (see, for example, Stephenson 1931); and (ii) assumption that equal distances along or between zooids represent equal lengths of time. On *a priori* grounds (ramose corals, above), as well as evidence from growth banding (see, for example, Buddemeier & Kinzie 1976) the second method, is at best, a hazardous approximation.

Notwithstanding the above reservations, an architectural model should be possible for *Isophyllia dipsacea* (now *I. sinuosa*) by using a combination of the above approaches, and based on Matthai's work (1926, 1927), supplemented, ideally, by evidence from mature colonies. The relatively flat shape of the colony allows it to be read as a two-dimensional map. With suitable evidence for genealogy of zooids, the growth models of other circumoral, two-dimensional corals (figure 3h: especially members of the Agariciidae, Fungiidae and Pectiniidae) should also be readily deduced.

In three-dimensional colonies, sectioning or dissection of the skeleton is usually necessary, except for openly branching fasciculate corals (figure 2a) which are not enclosed by consolidated sediment. An additional assumption that has to be made is that new zooids arise only at or near the living, and therefore outermost, surface of the colony. In fact, this is more likely to have been true of epithecate forms (mostly extinct) than of living scleractinians whose edge-zone usually extends down the exterior of zooidal walls and forms a potential budding site (see Hidaka *et al.* (1982) on the nearly-fasciculate *Galaxea fascicularis*). Fedorowski (1978) illustrated ten examples of growth in the fasciculate, extinct *Heritschioides* sp., but concluded that this coral showed no obvious pattern. Its modular growth model appears to be opportunistic or at least non-determinate. Two other notable fossil examples are the growth studies of *Halysites* and *Cladochonus* by Hamada (1959, 1973).

In theory, serial transverse sectioning of other kinds of three-dimensional corals should provide modular growth information, but there are geometrical difficulties because successive, equally spaced sections only represent equal distances along corallites for corallites which are perpendicular to the planes of the section. Therefore even if growth was at an equal rate throughout the colony, it would be difficult to reconstruct a sequence of events. One solution is to work with very small colonies (see, for example, Beecher's (1893a, b) pioneer work). Another is to use corals which maintained a flat living surface as they grew upward (Scrutton 1983). A growth history inferred from Scrutton's work on *Phillipsastrea nevadensis* suggests that its growth pattern is largely indeterminate, like that of *Heritschioides* sp. and possibly of living *Favia* in the examples already mentioned. Computers might be used for reconstruction from serial sections, but the problem of chronology will remain except where there is additional evidence for it.

Such evidence has been well established for over a decade, but is best seen in sections parallel to the general growth trend, and is therefore less useful in serial transverse sectioning methods. This is density banding (figure 3b), detected by radiography (Knutson *et al.* 1972; see also Buddemeier & Kinzie 1976; Buddemeier 1978 for reviews). These bands represent isochronous surfaces within a colony and reveal very successfully the broad patterns within and between higher level modules. Although the method has been valuable in several respects, its relevance to modular growth has yet to be fully explored. Buddemeier & Kinzie point to this potential in their discussion of Wood-Jones' (1907, 1910) ideas about coral growth. They confirmed that Wood-Jones' three methods of colonial growth were borne out by radiographic work: (i) those in which each polyp is equipotent (for example, *Porites*: figure 3b); (ii) those in which the active site of growth is associated with the youngest polyps (for example, corals like figure 3e); and (iii) those in which it is associated with the oldest [apical] polyp in the colony or branch (for example, *Acropora*: figure 3a). Buddemeier & Kinzie's discussion (pp. 206–207) of this is relevant and stimulating.

Isochronous surfaces within colonies can also be introduced artificially by use of radioisotopes or dyes, and, given enough intervening time between marker bands, can be used to test growth theories (Buddemeier & Kinzie 1976).

A completely different but complementary approach with important potential is computer simulation of growth (Graus 1977). Graus & Macintyre (1976, 1982) showed how colony form of *Montastrea annularis* is related to different patterns of zooidal growth in response to factors of depth and illumination, comparable with their own direct observations of form on reefs, and with Dustan's (1975) experimental results with dye markers. The value of computer simulation in the study of plant growth is well established (Harper & Bell 1979; Bell 1984). There has also been increasing interest in the biomechanical constraints on coral growth with respect to environmental hydrodynamic factors (Graus *et al.* 1977; Schuhmacher & Plewka 1981; Vosburgh 1982; Schuhmacher 1984; and references cited by these authors). Graus *et al.* showed how organization within colonies changes according to hydrodynamic régime. Biomechanical hypotheses, like the illumination hypothesis, above, can also be used in computer simulation.

## 7. MODULES: MEANS OR MEANING?

The idea of modules has been used in two ways: (i) as a means of understanding growth and form; and (ii) as a concept with postulated biological significance for organisms in general. Evolutionary implications can be derived from each, in the first case for relationships between different taxa in a given monophyletic group, and in the second case for evolutionary processes in general. Many of the other papers in this volume have concentrated on this second theme, that of Jackson & Coates in particular being most relevant to corals. Here, I draw on my previous sections to summarize the prospect for continuing with a modular approach to corals, thereby emphasizing the first theme of morphological methodology. The themes are not of course independent and feedback between them will also be mentioned.

### (a) Taxonomy, morphogenesis of colonies and phylogenetics

It is essential that modules be not necessarily convenient, but should also have meaning in terms of homology. To relate one coral to another, through modularity, we must use the same *kind* of unit. The branch of a fasciculate coral (figure 2a) is homologous with a zooid in a non-branching coral like *Montastrea* (figure 2e); the branches of a fasciculate coral are not homologous with ramose branches (figure 3a, e). The concept of a single universal coral zooid modified in various ways in different corals and based on homology of mouth structures, gives a unity to colony analysis and morphogenesis. It will, however, be important to establish a proper basis for this suggested homology, or otherwise refute it. In the meantime, it also allows a second suggestion that the developmental basis of much colony form is paedomorphic arrest of budding.

There follow two implications. First, colony form, hitherto very difficult to characterize either verbally or biometrically, should now be specifiable more precisely and on a proper comparative basis. The modular approach draws attention to colonial organization as a source of new taxonomic characters, and promises to fulfil the belief expressed in the convening of an earlier symposium (Larwood & Rosen 1979) that a way forward through some of the taxonomic difficulties that arise from coral coloniality would be generated by a more broadly based exchange among specialists of different groups of organisms (Rosen 1979). The most useful techniques will probably be radiographic study of marker bands, experimental dye marking and computer modelling.

Second, there are possible patterns of colony evolution alternative to those depicted by Wells (1956) or by Coates & Oliver (1973) but still consistent with Beklemishev's (1969) idea that colony evolution shows a tendency towards increasing integration (that is, mainly loss of discreteness of zooids). Progressively earlier paedomorphic arrest of budding is consistent with Beklemishev's view, but there is also a second trend, that of increasing complexity of modular organization (see below). An important incidental to this, however, will be the need to make further studies of wall structures.

### (b) Ecological influences on growth and form

Improved colony specification should enable us to distinguish more exactly the influence of environmental factors, including hydrodynamic influences, predation, interactions with neighbours and response to damage. Variation might now be analysed not just at the level of zooids and general form, but for each modular level present in a coral, thus expanding on the study framework envisaged by Veron (1982). In this respect Stebbins' (1950) idea, cited by

White (1984), might be tested for corals: 'characters formed by long periods of meristematic activity, being more subject to environmental influences, are likely to be more plastic than those formed rapidly in ontogenesis'. (Concerning meristems, see below.) Foster (1983) has already found evidence that variation patterns at different colony levels are decoupled from each other. Somewhere, between our knowledge of budding modes and the great morphological plasticity of corals, lies the prospect of discovering the inherited architecture of coral colonies.

Conversely, better specification of the effects of environmental variation in corals offers an improved basis for using fossil corals to interpret ancient environments.

### (c) Symbiosis with zooxanthellae, and its age and history

It is widely believed that the symbiosis of corals and zooxanthellae has influenced colony growth, form and evolution (Wells 1954; Coates & Oliver 1973; Rosen 1977, 1981). Actual specification of its morphological consequences has, however, proved elusive. There are both primary effects to consider, as well as those consequent upon the kinds of environment that the symbiosis imposes on corals. The kinds of ecological questions considered above can therefore also be formulated with the additional aim of finding differences between zooxanthellate and non-zooxanthellate corals. Two possibilities are higher levels of integration and complexity in zooxanthellates. Integration (in the morphological sense of loss of zooid discreteness) may have its origin in the faster growth rates shown by zooxanthellate corals as part of a broader tendency to have less dense skeletons (Coates & Oliver 1973; Schuhmacher & Plewka 1981; Schuhmacher 1984). Greater modular complexity may result from the greater number of environmental influences affecting zooxanthellate corals than non-zooxanthellate corals.

It is helpful in this respect to think of colony growth and form in vegetative organisms as the equivalent of behaviour in mobile organisms (Bell 1984). What I have previously referred to as spatial strategies (Rosen 1981) become spatial behaviours. Zooxanthellate corals should have a more complex range of behaviours than non-zooxanthellates on at least three counts: their response (i) to light; (ii) to shallow water hydrodynamic factors; and (iii) to a more crowded and heterogeneous habitat. Extensive fission and fusion of colonies is part of this behavioural response (Jackson & Coates, this symposium). Modular complexity may be another. It might be tested by assessing colonies in terms of the quantity of their hierarchical levels of modules, and of their polymorphic and polystatic forms at each modular level, together with the variety of components present in modules above the zooidal level. A theoretical framework for studying complexity is also provided by Lauder (1981).

Clearly, if correlations can be found between modular form and symbiosis, it might also be applied to the fossil record in order to infer the age and history of the symbiosis. Inference of symbiosis in particular fossil corals should help to provide depth, temperature and latitudinal constraints for the formations in which they occur.

### (d) Modules, meristems and metamers

To broaden the modular approach to other organisms, we can either attempt to extend the homologous basis of module recognition from corals to other organisms, or concentrate on interpreting the significance of repetition on its own terms (that is, regardless of trying to homologize iterative units across phyletic boundaries). The first option is improbable above the level of cells as modules, except in so far as the coral zooid is presumably homologous with other coelenterate zooids. And yet, the alternative, though it attempts to circumvent the homology problem, leads us back to it. This is because White (1984) has explained that Harper modules (of plants) are actually plant metamers, and used zoological criteria to argue this.

To many zoologists, however, metamers convey the serial segmentation typical of annelids and arthropods; and to call coral zooids 'metamers' seemingly implies a close phylogenetic connection between coelenterates and metameric organisms: currently, a very unconventional view. Can we call zooids 'meristems' instead? They cannot be in a botanical sense, but they are meristem-like in that zooids are growth points and developmental foci of reproductive tissue and clonogeny. Occasionally, they are even organized apically: non-reiteratively in circumoral corals like *Echinophyllia* (figure 2*h*), and reiteratively in the case of *Acropora* branches (figure 3*a*). *Acropora*, then, has the equivalent of Prévost modules. The other kinds of zooid are not Prévost modules, but as modules they are rather more tightly envisaged (here) than Harper modules.

Fortunately for this dilemma, the problem of homology of metamers is an old one. Clark (1980) has pointed out that there are several different types of metamerism in the Metazoa, notably annelid and chordate, and that they must have evolved independently. He warns that we should not fall into the trap of supposing that because different groups of animals (to which we might also add plants) are metameric, they are phylogenetically related to each other.

If Clark's argument overcomes the phylogenetic objections, it still leaves some purely descriptive difficulties. Coral zooids do not conform to the general notion of metamers because they are only rarely linear (for example, *Teleiophyllia*, *Thysanus*). Colony and genet growth moreover is largely indeterminate, and zooids are clonogenic. Linear arrangement, however, is related to locomotion (Clark 1964, 1980), and would be less likely in vegetative organisms, whose meres are obviously free to grow in sheet-like, multiple runner or three-dimensional systems. When elongation and determinate growth do occur in colonial corals, however (for example, *Herpolitha*, *Manicina*), it is associated with an unattached, sometimes mobile habit. More significantly, the mobile, pelagic relatives of attached colonial organisms are, in fact, linear and determinate (for example, compare graptoloid with dendroid graptolites, and siphonophores with hydroids, respectively).

Coral modules, as meant here, are therefore meristematic and clonogenic metamers. This is acknowledged by Riedl's solution (1978; especially his figures 10 and 14) in which he regards these colony-meres as features at a higher organizational level than annelid-type segmentation. He refers to them as polymers but 'hypermers' might have avoided the coincidence with chemistry while acknowledging Riedl's point. In the end, the appeal of Harper modules (which would include all the features regarded here as modules of corals), is that they transcend the nuances and constraints of meristems and metamers.

I especially thank Jill Darrell (British Museum (Natural History)) for support and advice throughout this paper. Although we had worked together on unpublished earlier versions and presentations of many of these ideas, she was, unfortunately, not able to contribute directly to preparing this paper. Pat Cook (B.M.(N.H.)), Dr Nancy Foster (University of Iowa), Dr Paul Taylor (B.M.(N.H.)), Professor John Wells (Cornell University) and Rachel Wood (Open University) kindly read and commented on the manuscript. Pat Cook also gave me a great deal of other help. Dr Adrian Bell (U.C.N.W., Bangor) and James White (University College, Dublin) helped me to understand numerous, relevant, broader ideas and, together with Hugh Pearson (B.M.(N.H.)), advised me on botanical matters. I also thank Dr Peter Forey, Dr Dick Jefferies, Dr Colin Patterson (B.M.(N.H.)), and Professor Brian Gardiner (King's College, London) for clarification of certain points, and Professor John Harper, F.R.S. (U.C.N.W., Bangor), for his encouragement.

## References

Babcock, R. C. 1984 Reproduction and distribution of two species of *Goniastrea* (Scleractina) from the Great Barrier Reef province. *Coral Reefs* **2**, 187–195.

Beecher, C. E. 1893a The development of a Paleozoic poriferous coral. *Trans. Conn. Acad. Arts Sci.* **8**, 207–214.

Beecher, C. E. 1893b Symmetrical cell development in the Favositidae. *Trans. Conn. Acad. Arts Sci.* **8**, 215–219.

Beklemishev, W. N. 1969 *Principles of comparative anatomy of invertebrates*, 2 vols, translated by J. M. MacLennan. Edinburgh: Oliver & Boyd. (Translation of 3rd Russian edition, 1964. Moscow: Nauka.)

Bell, A. D. 1984 Dynamic morphology: a contribution to plant population ecology. In *Perspectives on plant population ecology* (ed. R. Dirzo & J. Sarukhán), pp. 48–65. Sunderland, Massachusetts: Sinauer Associates Inc.

Best, M. B., Boekschoten, G. J. & Oosterbaan, A. 1984 Species concept and ecomorph variation in living and fossil Scleractinia. In *Recent advances in the paleobiology and geology of the Cnidaria* (ed. W. A. Oliver, Jr, et al.) *Palaeontogr. am.* **54**, 70–79.

Boschma, H. 1923 Über die Bildung der jungen Kolonien von *Goniopora stokesi* durch ungeschlechtliche Fortpflanzung. *Zool. Anz.* **57**, 284–286.

Boschma, H. 1929 On the postlarval development of the coral *Maeandra areolata* (L.). *Pap. Tortugas Lab. Carnegie Instn* **26**, 129–147.

Buddemeier, R. W. 1978 Coral growth: retrospective analysis. In *Coral reefs: research methods* (ed. D. R. Stoddart & R. E. Johannes), pp. 551–571. Paris: UNESCO.

Buddemeier, R. W. & Kinzie, R., III 1976 Coral growth. *Oceanogr. mar. Biol.* **14**, 183–225.

Buddemeier, R. W., Maragos, J. E. & Knutson, D. W. 1974 Radiographic studies of reef coral exoskeletons: rates and patterns of coral growth. *J. exp. mar. Biol. Ecol.* **14**, 179–200.

Chapman, G. 1981 Individuality and modular animals. *Biol. J. Linn. Soc.* **15**, 177–183.

Chapman, G. & Stebbing, A. R. D. 1980 The modular habit – a recurring strategy. In *Developmental and cellular biology of coelenterates* (ed. P. Tardent & R. Tardent), pp. 157–162. Amsterdam: Elsevier/North Holland Biomedical Press.

Clark, R. B. 1964 *Dynamics in metazoan evolution; the evolution of the coelom and segments.* Oxford: Clarendon Press.

Clark, R. B. 1980 Natur und Entstehungen der metameren Segmentierung. *Zool. Jb. Abt. Zool.* **103**, 169–195

Coates, A. G. & Oliver, W. A. 1973 Coloniality in zoantharian corals. In *Animal colonies: development and function through time* (ed. R. S. Boardman, A. H. Cheetham & W. A. Oliver, Jr), pp. 3–27. Stroudsburg: Dowden, Hutchinson & Ross, Inc.

Connell, J. H. 1978 Diversity in tropical forests and coral reefs. *Science, Wash.* **199**, 1302–1310.

Crossland, C. 1913 *Desert and water gardens of the Red Sea: being an account of the natives and the shore formations of the coast.* Cambridge: University Press.

Dauget, J.-M. 1985 La réaction aux traumatismes: comparaison entre les arbres et les coraux. *Terre Vie* **40**, 113–118.

Duerden, J. E. 1902 West Indian madreporarian polyps. *Mem. natn. Acad. Sci.* **7**, 399–599.

Dustan, P. 1975 Growth and form in the reef-building coral *Montastrea annularis*. *Mar. Biol. Berlin* **33**, 101–107.

Fadlallah, Y. H. 1983 Sexual reproduction, development and larval biology in scleractinian corals. A review. *Coral Reefs* **2**, 129–150.

Fedorowski, J. 1978 Some aspects of coloniality in rugose corals. *Palaeontology* **21**, 177–224.

Fedorowski, J. & Jull, R. K. 1976 Review of blastogeny in Palaeozoic corals and description of lateral increase in some Upper Ordovician rugose corals. *Acta Palaeont. pol.* **21**, 37–78.

Finks, R. M. 1971 Modular structure and organic integration in Sphinctozoa. *Abstr. Progm geol. Soc. Am.* **3**, 764–765. [Abstract.] (Reprinted in *Animal colonies: development and function through time* (ed. R. S. Boardman, A. H. Cheetham & W. A. Oliver, Jr), pp. 585–586. Stroudsburg: Dowden, Hutchinson & Ross. 1973.

Foster, A. B. 1979 Phenotypic plasticity in the reef corals *Montastrea annularis* (Ellis & Solander) and *Siderastrea siderea* (Ellis & Solander). *J. exp. mar. Biol. Ecol.* **39**, 25–54.

Foster, A. B. 1983 The relationship between corallite morphology and colony shape in some massive reef-corals. *Coral Reefs* **2**, 19–25.

Foster, A. B. 1984 The species concept in fossil hermatypic corals: a statistical approach. In *Recent advances in the paleobiology and geology of the Cnidaria* (ed. W. A. Oliver, Jr, et al.), *Palaeontogr. am.* **54**, 58–69.

Graus, R. R. 1977 Investigation of coral growth adaptations using computer modelling. In *Proc. 3rd int. Coral Reef Symp.* (ed. D. L. Taylor), vol. 2, pp. 463–469. Miami: Rosenstiel School of Marine and Atmospheric Science, University of Miami.

Graus, R. R., Chamberlain, J. A. & Boker, A. M. 1977 Structural modification of corals in relation to waves and currents. In *Reefs and related carbonates: ecology and geology* (ed. S. H. Frost, M. P. Weiss & J. B. Saunders), *Stud. Geol. am. Ass. Petrol. Geologists* **4**, 135–153. Tulsa: American Association of Petroleum Geologists.

Graus, R. R. & Macintyre, I. G. 1976 Light control of growth form in colonial reef corals: computer simulation. *Science, Wash.* **193**, 895–897.

Graus, R. R. & Macintyre, I. G. 1982 Variation in growth forms of the reef coral *Montastrea annularis* (Ellis & Solander): a quantitative evaluation of growth response to light distribution using computer simulation. *Smithson. Contr. mar. Sci.* **12**, 441–464.

Grigg, R. W. 1977 Population dynamics of two gorgonian corals. *Ecology* **58**, 278–290.

Haldane, J. B. S. 1927 On being the right size. In *Possible worlds*, pp. 18–26. London: Chatto & Windus.

Hallé, F., Oldeman, R. A. A. & Tomlinson, P. B. 1978 *Tropical trees and forests: an architectural analysis*. Berlin: Springer-Verlag.

Hamada, T. 1959 Corallum growth of the Halysitidae. *J. Fac. Sci. Univ. Tokyo Section 2 Geol.* **11**, 273–289.

Hamada, T. 1973 Caliceal increase in massive corals – a geometrical implication. *Scient. Pap. Coll. gen. Educ. Tokyo* **23**, 53–71.

Hamada, T. 1973 '*Cladochonus*' (tabulate coral from the Red Bed of Malaya). *Geol. Paleont. SE Asia* **13**, 23–37.

Harper, J. L. 1977 *Population biology of plants*. London: Academic Press.

Harper, J. L. & Bell, A. D. 1979 Population dynamics of growth form in organisms with modular construction. In *Population dynamics. 20th Symp. Br. ecol. Soc.* (ed. R. M. Anderson, B. D. Turner & L. R. Taylor), pp. 29–52. Oxford: Blackwell.

Harriott, V. J. 1983 Reproductive ecology of four scleractinian species at Lizard Island, Great Barrier Reef. *Coral Reefs* **2**, 9–18.

Harrison, P. L. 1985 Sexual characteristics of scleractinian corals: systematic and evolutionary implications. In *Proc. 5th int. Coral Reef Congr., Tahiti*, vol. 2, pp. 165–170.

Heyward, A. J. & Collins, J. D. 1985 Fragmentation in *Montipora ramosa*: the genet and ramet concept applied to a reef coral. *Coral Reefs* **4**, 35–40.

Hidaka, M., Uechi, A. & Yamazato, K. 1982 Effects of certain factors on budding of isolated polyps of a scleractinian coral, *Galaxea fascicularis*. In *Reef and man. Proc. 4th int. Coral Reef Symp.* (ed. E. D. Gomez et al.), vol. 2, pp. 229–231. Quezon City: Marine Sciences Center, University of the Philippines.

Highsmith, R. C. 1982 Reproduction by fragmentation in corals. *Mar. Ecol. Prog. Ser.* **7**, 207–226.

Hill, D. 1981 *Treatise on invertebrate paleontology. Part F. Coelenterata. Supplement 1. Rugosa and Tabulata* (ed. C. Teichert), 2 vols. Kansas: The Geological Society of America and the University of Kansas.

Hughes, R. N. 1983 Evolutionary ecology of colonial reef-organisms, with particular reference to corals. *Biol. J. Linn. Soc.* **20**, 39–58.

Hughes, T. P. & Jackson, J. B. C. 1980 Do corals lie about their age? Some demographic consequences of partial mortality, fission and fusion. *Science, Wash.* **209**, 713–715.

Hughes, T. P. & Jackson, J. B. C. 1985 Population dynamics and life histories of foliaceous corals. *Ecol. Monogr.* **55**, 141–166.

Knutson, D. W., Buddemeier, R. W. & Smith, S. V. 1972 Coral chronometers: seasonal growth bands in reef corals. *Science, Wash.* **177**, 270–272.

Kobayashi, A. 1984 Regeneration and regrowth of fragmented colonies of the hermatypic corals *Acropora formosa* and *Acropora nasuta*. *Galaxea* **3**, 13–23.

Krasnov, Ye. V. & Preobrazhenskiy, B. V. 1972 The nature and significance of life forms in tabulates and colonial scleractinians. *Paleont. J.* **6**, 264–268.

Lang, J. 1971 Interspecific aggression by scleractinian corals. 1. The rediscovery of *Scolymia cubensis* (Milne Edwards & Haime). *Bull. mar. Sci.* **21**, 952–959.

Lang, J. 1984 Whatever works: The variable importance of skeletal and of non-skeletal characters in scleractinian taxonomy. In *Recent advances in the paleobiology and geology of the Cnidaria* (ed. W. A. Oliver, Jr, et al.), *Paleontogr. am.* **54**, 18–44.

Larwood, G. P. & Rosen, B. R. (ed.) 1979 *Biology and systematics of colonial organisms*. *Spec. Vol. Syst. Ass.* **11**. London: Academic Press.

Lauder, G. V. 1981 Form and function: structural complexity in evolutionary morphology. *Paleobiology* **7**, 430–442.

Loya, Y. 1976 Skeletal regeneration in a Red sea scleractinian population. *Nature, Lond.* **261**, 490–491.

Manton, S. M. 1932 On the growth of the adult colony of *Pocillopora bulbosa*. *Sci. Rept. Gt Barrier Reef Exped.* **3**, 157–166.

Matthai, G. 1914 A revision of the Recent colonial Astraeidae possessing distinct corallites. *Trans. Linn. Soc. Lond.* (ser 2) **17**, 1–140.

Matthai, G. 1926 Colony-formation in astraeid corals. *Phil. Trans. R. Soc. Lond.* B **214**, 313–367.

Matthai, G. 1927 *Catalogue of the madreporarian corals in the British Museum (Natural History). Vol. VII. A monograph of the Recent meandroid Astraeidae*. London: Trustees of the British Museum.

Matthai, G. 1948 a On the mode of growth of the skeleton in fungid corals and skeletal variation in two large coralla from Tahiti, one of *Pavona varians* (Verrill) and another of *Psammocora haimiana* Milne Edwards and Haime. *Phil. Trans. R. Soc. Lond.* B **233**, 177–199.

Matthai, G. 1948 b Colony formation in fungid corals. I. *Pavona*, *Echinophyllia*, *Leptoseris* and *Psammocora*. *Phil. Trans. R. Soc. Lond.* B **233**, 201–231.

Mori, K. & Minoura, K. 1980 Ontogeny of 'epithecal' and septal structures in scleractinian corals. *Lethaia* **13**, 321–326.

Mori, K., Omura, A. & Minoura, K. 1977 Ontogeny of euthecal and metaseptal structures in colonial scleractinian corals. *Lethaia* **10**, 327–336.

Oliver, W. A., Jr 1968 Some aspects of colony development in corals. *J. Paleont.* (*Mem. paleont. Soc.* 2) **42**, (5 (part II)) 16–34.

Potts, D. C., Done, T. J., Isdale, P. J. & Fisk, D. A. 1985 Dominance of a coral community by the genus *Porites* (Scleractinia). *Mar. Ecol. Progr. Ser.* **23**, 79–84.

Prévost, M. F. 1967 Architecture de quelques Apocynacées ligneuses. *Mém. Soc. bot. Fr.* **114**, 24–36.
Randall, R. H. 1982 Morphologic diversity in the scleractinian genus *Acropora*. In *Reef and man. Proc 4th int. Coral Reef Symp.* (ed. E. D. Gomez et al.) vol. 2, pp. 157–164. Quezon City: Marine Sciences Center, University of the Philippines.
Randall, R. H. & Myers, R. F. 1983 *Guide to the coastal resources of Guam. Volume III. The corals*. Guam: University Press.
Riedl, R. 1978 *Order in living organisms*, translated by R. P. S. Jefferies. Chichester: John Wiley & Sons. (German edition: *Die Ordnung des Lebendigen*. Hamburg: Verlag Paul Parey. 1975.)
Rosen, B. R. 1971 Principal features of reef coral ecology in shallow water environments of Mahé, Seychelles. In *Regional variation in Indian Ocean coral reefs* (ed. C. M. Yonge & D. R. Stoddart), *Symp. zool. Soc. Lond.* **28**, 163–183. London: Academic Press.
Rosen, B. R. 1977 The depth distribution of Recent hermatypic corals and its palaeontological significance. *Mém. Bur. Rech. géol. miniér.* **89**, 507–517.
Rosen, B. R. 1979 Modules members and communes: a postscript introduction to social organisms. In *Biology and systematics of colonial organisms* (ed. G. P. Larwood & B. R. Rosen), *Spec. Vol. Syst. Ass.* **11**, xiii–xxxv. London: Academic Press.
Rosen, B. R. 1981 The tropical high diversity enigma – the corals'-eye view. In *Chance, change and challenge; the evolving biosphere* (ed. P. H. Greenwood & P. L. Forey), pp. 103–129. Cambridge University Press and London: British Museum (Natural History).
Rosen, B. R. & Taylor, J. D. 1969 Reef coral from Aldabra: new mode of reproduction. *Science, Wash.* **166**, 119–121.
Ryland, J. S. 1981 Colonies, growth and reproduction. In *Recent and fossil Bryozoa* (ed. G. P. Larwood & C. Nielsen), pp. 221–226. Fredensborg: Olsen & Olsen.
Sammarco, P. W. 1982 Escape response and dispersal in an Indo-Pacific coral under stress: 'polyp bail-out'. [Abstract.] In *The reef and man. Proc. 4th int. Coral Reef Symp.* (ed. E. D. Gomez et al.), vol. 2, p. 194. Quezon City: Marine Sciences Center, University of the Philippines.
Scheer, G. 1959 Die Formenvielfalt der Riffkorallen. *Ber. naturw. Ver. Darmstadt* (1959), 50–67.
Scheer, G. 1960 Viviparie bei Steinkorallen. *Naturwissenschaften* **10**, 238–239.
Scheer, G. 1974 Investigation of coral reefs as Rasdu Atoll in the Maldives with the quadrat method according to phytosociology. In *Proc. 2nd int. Coral Reef Symp.* (ed. A. Cameron et al.), vol. 2, pp. 655–670. Brisbane: Great Barrier Reef Committee.
Scheer, G. 1978 Application of phytosociologic methods. In *Coral reefs: research methods* (ed. D. R. Stoddart & R. E. Johannes), pp. 175–196. Paris: UNESCO.
Schuhmacher, H. 1984 Reef-building properties of *Tubastraea micranthus* (Scleractinia, Dendrophylliidae), a coral without zooxanthellae. *Mar. Ecol. Prog. Ser.* **20**, 93–99.
Schuhmacher, H. & Plewka, M. 1981 Mechanical resistance of reefbuilders through time. *Oecologia* **49**, 279–282.
Scrutton, C. T. 1983 Astogeny in the Devonian rugose coral *Phillipsastrea nevadensis* from Northern Canada. *Mem. australas. Palaeont.* **1**, 237–259.
Squires, D. F. 1964 Fossil coral thickets in Wairarapa, New Zealand. *J. Paleont.* **38**, 904–915.
Stearn, C. W. 1982 The shapes of Paleozoic and modern reef-builders: a critical review. *Paleobiology* **8**, 228–241.
Stebbins, G. L. 1950 *Variation and evolution in plants*. New York: Columbia University Press.
Stephenson, T. A. 1931 Development and the formation of colonies in *Pocillopora* and *Porites* – Part 1. *Sci. Rep. Gt Barrier Reef Exped.* **3**, 113–134.
Stephenson, T. A. & Stephenson, A. 1933 Growth and asexual reproduction in corals. *Sci. Rept Gt Barrier Reef Exped.* **3**, 167–217.
Stoddart, J. A. 1983 Asexual production of planulae in the coral *Pocillopora damicornis*. *Mar. Biol. Berlin* **76**, 279–284.
Talbot, F. H. (ed.) 1984 *Reader's Digest book of the Great Barrier Reef*. Sydney: Reader's Digest.
van Valen, L. 1978 Arborescent animals and other colonoids. *Nature, Lond.* **276**, 318.
Veron, J. E. N. & Pichon, M. 1976 Scleractinia of eastern Australia. Part I. Families Thamnasteriidae, Astrocoeniidae, Pocilloporidae. *Monogr. Ser. Aust. Inst. mar. Sci.* **1**, 1–86.
Veron, J. E. N. & Pichon, M. 1980 Scleractinia of eastern Australia. Part III. Families Agariciidae, Siderastreidae, Fungiidae, Oculinidae, Merulinidae, Mussidae, Pectiniidae, Caryophylliidae, Dendrophylliidae. *Monogr. Ser. Aust. Inst. mar. Sci.* **4**, 1–422.
Veron, J. E. N. & Pichon, M. 1982 Scleractinia of eastern Australia. Part IV. Family Poritidae. *Monogr. Ser. Aust. Inst. mar. Sci.* **5**, 1–159.
Veron, J. E. N., Pichon, M. & Wijsman-Best, M. 1977 Scleractinia of eastern Australia. Part II. Families Faviidae, Trachyphylliidae. *Monogr. Ser. Aust. Inst. Mar. Sci.* **3**, 1–233.
Veron, J. E. N. & Wallace, C. C. 1984 Scleractinia of eastern Australia. Part V. Family Acroporidae. *Monogr. Ser. Aust. Inst. mar. Sci.* **6**, 1–485.
Vosburgh, F. 1982 *Acropora reticulata*: structure, mechanics and ecology of a reef coral. *Proc. R. Soc. Lond.* B **214**, 481–499.
Wells, J. W. 1954 Status of invertebrate paleontology, 1953. III. Coelenterata. *Bull. Mus. comp. Zool. Harv.* **112**, 109–123.

Wells, J. W. 1956 Scleractinia. In *Treatise on invertebrate paleontology. Part F. Coelenterata* (ed. R. C. Moore), pp. F328–F444. Kansas: Geological Society of America, and University of Kansas Press.

Wells, J. W. 1966 Evolutionary development in the scleractinian family Fungiidae. In *The Cnidaria and their evolution* (ed. W. J. Rees), *Symp. zool. Soc. Lond.* **16**, 223–246. London: Academic Press.

Wells, J. W. 1971 What is a colony in anthozoan corals? *Abstr. Progm geol. Soc. Am.* **3** (7), 748. [Abstract.] (Reprinted in *Animal colonies; development through time* (ed. R. S. Boardman, A. H. Cheetham & W. A. Oliver, Jr), 1973, p. 29. Stroudsburg: Dowden, Hutchinson & Ross, Inc.)

White, J. 1979 The plant as a metapopulation. *Ann. Rev. Ecol. Syst.* **10**, 109–145.

White, J. 1984 Plant metamerism. In *Perspectives on plant population ecology* (ed. R. Dirzo & J. Sarukhán), pp. 15–47. Sunderland, Massachusetts: Sinauer Associates Inc.

Wijsman-Best, M. 1974 Habitat-induced modification of reef corals (Faviidae) and its consequences for taxonomy. In *Proc. 2nd int. Symp. Coral Reefs* (ed. A. M. Cameron *et al.*), vol. 2, pp. 217–228. Brisbane: The Great Barrier Reef Committee.

Wood, E. M. 1983 *Reef corals of the world: biology and field guide.* Neptune City: T.F.H. Publications, Inc. Ltd.

Wood Jones, [*sic*], F. 1907 On the growth-forms and supposed species in corals. *Proc. zool. Soc. Lond.* **2**, 518–556.

Wood-Jones, F. 1910 *Coral and atolls; their history, descriptions, theories of their origin both before and since that of Darwin, the influence of winds, tides and ocean currents on their formation and transformation, their present condition, products, fauna and flora.* London: Lovell Reeve & Co. Ltd. (The text of Part II largely corresponds to that of Wood Jones 1907).

Yonge, C. M. 1930 Studies on the physiology of corals. I. Feeding mechanisms and food. *Sci. Rep. Gt Barrier Reef Exped.* **1**, 13–57.

Yonge, C. M. 1936 Studies on the biology of Tortugas corals. I. Observations on *Maeandra areolata. Pap. Tortugas Lab. Carnegie Instn* **29**, 185–198.

# The simulation of branching patterns in modular organisms

By A. D. Bell

*School of Plant Biology, University College of North Wales, Bangor, Gwynedd LL57 2UW, U.K.*

[Plate 1]

A modular mode of growth in an organism is frequently coupled to the development of a branched structure. Such modular forms may appear to be constructed in a regular manner, exhibiting a high degree of organization, or may be bewilderingly complex and yet not haphazard. These factors, together with the very nature of systems dependent upon the progressive addition of modules, readily promote attempts to model branching processes by means of some sort of constructional simulation. Such models tend to favour visual, graphic presentation. This paper considers aspects of modularity and branching and the different approaches to the modelling of branching organisms, and gives a range of examples. Attention is directed towards the aims of the models and to their different characteristics rather than to a description of the computorial methodology involved. Attempts to simulate branching systems highlight the constructional similarities and differences between various modular organisms, but one key factor is identified for both real and artificial branching patterns: it is the nature of the control of new branch initiation that governs their constructional development.

## 1. Modularity

The recognition of some aspect of modular construction in diverse organisms is not a new concept. It has featured periodically in biological thought but this time its appearance has coincided with the advent of computer technology. The discrete events and entities of modular growth encourage investigation by means of computer simulation, endowed though they are with assumptions regarding 'function' and 'efficiency'.

Modular assemblages develop most frequently into branching forms, and it is the interpretation of the organization behind such complex 'trees' that challenges the morphologist. Frank Lloyd Wright was intrigued by form in the context of his organic architectural designs, and his sentiments apply equally to branched organisms.

> 'Realization of form is always geometrical. That is to say, it is mathematic. We call it pattern. Geometry is the obvious framework upon which nature works to keep her scale in "designing". She relates things to each other and to the whole, while meantime she gives to your eye most subtle, mysterious and apparently spontaneous irregularities in effects.' (Wright 1953)

The implication here is that there must be an intrinsic order to branched growth even in the most elaborate system. An understanding of this sequence demands recognition of the nature of the building blocks for a given type of organism: a 'module' of some kind.

The traditional botanical description of a branching system, the floral diagram, disguises or fails to appreciate the dynamic aspect of branching and its inherent modularity, although such diagrams do convey a great deal of information and can be aesthetically superb (Engler 1876) (figure 1). Dynamic information is obtained if a time sequence of events is presented (for example, Darrow 1929; figure 2) and it is here that a modular approach can be identified. This qualitative manner of monitoring the developmental morphology of plants as they age is directly amenable to simulation model production. An example is seen in Tumidajowicz & Dambski (1984) where sequences of annual rhizome increments are represented as the structural components of a binary 'tree' to represent long term genet growth rather than detail of branching morphology *per se*. The analogy of a computer data structure in the form of a 'rooted tree' and the branching within real trees forms the basis of at least one computer model of tree crown development (Smith & Scoullar 1975).

An organic branching pattern develops by the addition of new components to an existing framework. Depending upon the nature of the organism, these components may be more or

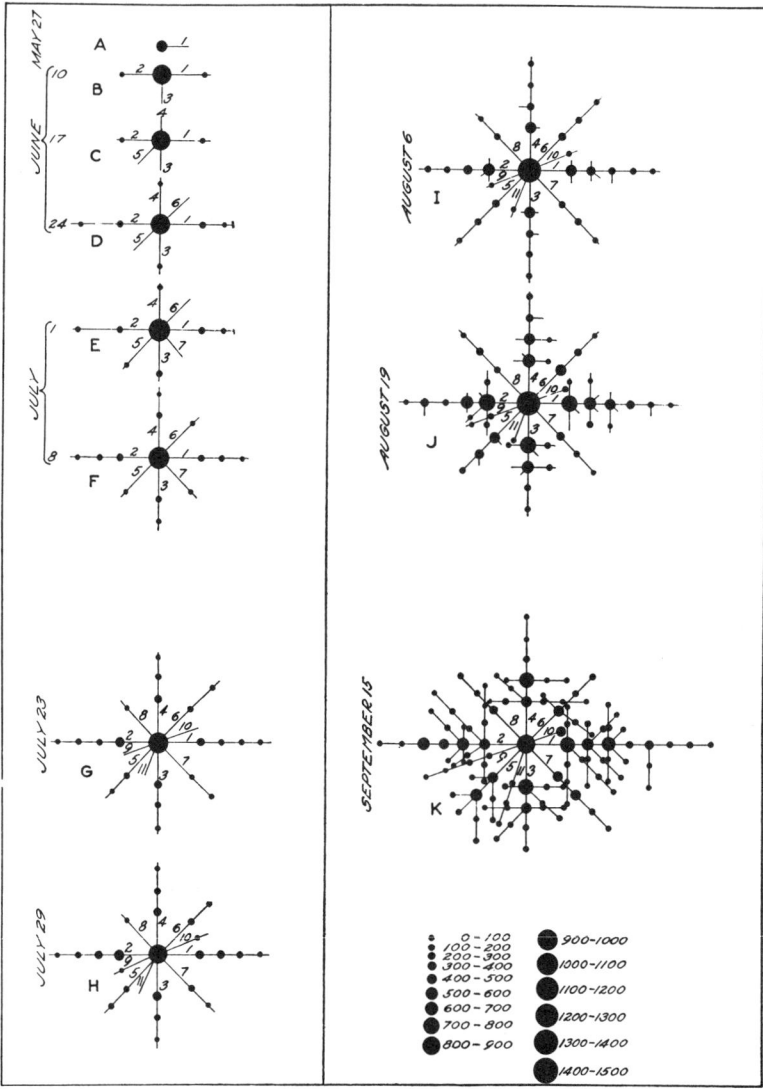

FIGURE 2. A process description of plant form showing a developmental sequence of runner production in strawberry (from Darrow (1929)). In zoological terminology, A represents the ancestrula and the sequence A–K the astogeny of this branching pattern.

FIGURE 1. A 'floral' diagram. A traditional state description of plant form (from Engler (1876)).

less discrete and conform to a greater or lesser extent to one's personal notion of a module. Thus the definition of a modular unit presented by Finks (1973), relating to the uncertain construction of extinct calcareous sponges (Sphinctozoa), has to be deliberately vague: 'a structural unit which makes up, in the aggregate, a given individual or other entity'. Nevertheless an understanding of the development of an organism will help define its modular nature and determine the usefulness of the concept for a particular situation. In bryozoans the individual zooids of the colony readily lend themselves to the epithet 'module' and these modules are aggregated into a branching entity. In many cases, however, application of the modular concept is a matter for debate (outlined in Mackie 1963; White 1984). There is also a temptation, not necessarily fruitful, to look for levels of modular construction within one organism: 'logic can carry one to the conclusion that all colonies are individuals or that all individuals are colonies' (Boardman *et al.* 1973). This paradox may occur whenever modularity is used to analyse a construction. Even the design of complex computer systems involving numerous hardware and software components which can be rationalized by recourse to a modular approach can suffer this complication (Unger & Bidulock 1981).

Branched modular constructions are ubiquitous and, one must assume, successful modes of growth (Mackie 1963). Particularly throughout the plant kingdom, the advent of growth restricted to an apical region represents the morphological key to the branched habit, a potentially efficient mode of development if coupled to the related physiological imperatives of plant architecture (Raven 1984). Apical meristems create building blocks which, in turn, create structure. In the botanical field the soundest level at which to identify modules and their interactions, in my opinion, is that expounded by Prévost (1967) who defined a botanical module as the product of one apical meristem that has a definite, ultimate fate, such as the formation of a terminal inflorescence. The epicotylar meristem of sycamore (*Acer pseudoplatanus*) grows for many years and metres to constitute the trunk of the tree before revealing its determinate nature by terminating in a solitary flower (see also examples in Barthelemy, this symposium). Even the epicotylar meristem of a coconut palm (*Cocos nucifera*) or a monkey puzzle tree (*Auraucaria* species) ceases to grow eventually. It is useful from a constructional point of view to recognize each such component, representing the product of a single meristem, as a module. Modules commonly form the components of a sympodial axis and are often determinate. It seems morphologically logical to extend the concept of the botanical module to include the products of any single shoot apical meristem. The modular components of some sympodial axes are indeterminate and continue to grow as short shoots long after having been evicted to a subsidiary role (Hallé, this symposium). The inception of a new branch in a plant is practically always due to the activation of a single apical meristem. It is the control of such an act that forms the foundation of development in any branching system (see, for example, Thornley 1977).

## 2. BRANCHING

A single line is rarely the shortest distance between three (or more) points; branching and efficiency go hand in hand in the control of space. This universal phenomenon presents a universal problem: how to describe, analyse, and understand the rationale of branching systems, particularly if the system is dynamic and apparently becoming progressively more and more complex. Oxnard (1980) emphasized the need for a diversity of method in the analysis of form and advocates recourse to the new problem-solving techniques in subjects far removed from biology. Branching systems occur in many guises throughout the biological and physical worlds. They are studied by mathematicians (graph theory; see, for example, Biggs *et al.* 1976) and by crystallographers (Mackay 1975). Geographers use the mathematicians' graph theory to devise transport systems (Haggett 1976) as do agronomists interested in root systems (Fitter 1985). Geomorphologists studying river basins have abandoned branch ordering systems as a means of quantification (Jarvis 1977) in favour of the mathematicians' trees (Smart 1978) and neuroanatomists have done likewise (Van Pelt & Verwer 1984). Botanists, lagging behind as usual, apply bifurcation ratios to real trees (see, for example, Oohata & Shidei 1971; Barker *et al.* 1973; but see Tomlinson 1978). The zoologist finds on a crinoid the same branching transport system for harvesting a resource and conveying it to a mouth (Cowen 1981) as a road research laboratory proposes for a banana plantation (Tanner 1967). The very precise branching tunnel system of an extinct ctenostomate bryozoan (Pohowsky 1978) in the shell of a mollusc is uncannily like the branching pattern of many plants or indeed similar to the pattern of veining in their leaves. It is likely that whenever the construction of the form of branching patterns is to be deciphered, the combination of the concept of modularity and computer-aided analysis will be invoked, for example in the analysis of the axon arborizations of a cat's spinal cord (Réthelyi 1981) or the biological control of a clonal weed (Room 1986).

Describing a branching organism in modular terms may not be too difficult, but recognizing interactions of these components, particularly in a developing sequence, presents many problems. It is crucial, however, to find a profitable compromise between simplicity, which has many attributes, and the complexity that will invariably result from attempts to incorporate some hypotheses related to the internal interactions between components.

## 3. Computer simulation

Any object or process can be modelled or imitated and the intentions for so doing vary enormously. Easy access to computer facilities has created a frenzy of simulations in every field of science (Wolfram 1984). Nearly 1500 computer models in environmental biology alone are listed by Kickert (1984), often unwittingly with similar goals and identical names. There is already a taxonomic crisis looming in the modelling world.

'A model represents the isolation of certain features of a complex situation so that their mutual relationships can be seen without the distraction of other features of lesser significance' (Mackay 1975). With this in mind, some simulation models of branching structures have been designed specifically and purely as teaching aids. Typical examples involve the growth and architecture of forests and their management. For example 'CROGRO' (Fellows *et al.* 1983) simulates the growth of crowns of trees as they compete for space and light. Decades of growth are condensed into a few minutes to demonstrate this process. The ability to speed up events is one of the useful byproducts of a growth simulation. 'TIMECOGS' (Welty *et al.* 1985) provides the student with a stand of timber trees portrayed graphically in a manner allowing the observer to walk among them. Individual trees grow better if their neighbours are removed. The student can practice the art of thinning without damaging a forest, and follow the outcome to harvest. An exceedingly convincing and aesthetic three-dimensional simulation of tree shapes is described by Aono & Kunii (1984, figure 21). The applications of this system as envisaged by the authors include graphic design, photogravure production for magazines and landscape gardening. These authors are aware of the modular nature of their trees and use empirical data in some cases. Superb three-dimensional graphic simulations of real tree form based on contemporary interpretations of tree architecture (Hallé & Oldeman 1970; Hallé *et al.* 1978) are currently being produced by Jaeger and co-workers (1985).

The zoological application of computer simulation to form has frequently had a different objective. The assumption is made that change in the simulation 'rules' may mimic an evolutionary change in the genome of the organism. Raup (1966) showed that just four parameters were needed to simulate the gross form of the coiled shell and that actual specimens are not randomly distributed in the total spectrum of theoretically possible forms generated by computer. A more specific simulation, concerned with coiling in ammonoids (Raup 1967) again concluded that during the evolutionary development of ammonoid form only a small part of the geometric range available was represented. This leads to a consideration of the geometric problems of particular theoretical shapes and the conclusion that successful spirals represent compromises that satisfy a number of functional constraints. The computer alone can generate the 'missing' and supposedly unsuccessful shapes: 'In studying the functional significance of the coiled shell, it is important to be able to analyse the types that do not occur in nature as well as those represented by actual species' (Raup & Michelson 1965). Creating a simple simulation programme that can orchestrate a complex structure thus gives some possible indication of the level of complexity of controlling mechanisms. Raup (1968), referring to echinoid growth, stated 'the actual biological system controlling plate patterns need not be more complicated than that used in the computer simulation'. Similar statements appear in Ede & Law (1969) for vertebrate limb morphogenesis, Raup & Seilacher (1969) for foraging patterns of sediment feeders, as a general concept in Cohen (1967) and Gould (1970), and at the cellular level in fossil plants (Niklas 1977).

A simulation of the range of branching morphologies found in fossil colonies of spiral bryozoans (figure 3) again was found to be feasible by using few growth rules (McKinney & Raup 1982). Manipulation of these rules, in a series of replicated simulation runs, highlights the important factors in determining this spiral architecture and shows that quite slight changes of rules produce distinctly different forms (such as the conversion of the form of an extinct spiral bryozoan *Archimedes* into the form of an extant cheilostome bryozoan, *Reliflustra cornea*). One conclusion reached by these authors is that such simulations of branching patterns may indicate forms yet to be realised.

The ability to create ranges of morphological form, surpassing that to be found in extinct or extant organisms, naturally leads to discussion of the course of phylogenetic changes (as for the simulated branching morphologies of early land plants in which the random element of pattern generation is debated (Niklas 1982; Waller 1984; Niklas 1984)). Indeed, an experiment of evolution has been conducted by Papentin (1973) and Papentin & Roder (1975) based on a model of the feeding patterns of sediment worms. A population of 'worms', starting with various 'rules' or 'genes' to control their meandering, compete for food. Some patterns prove to be more successful than others at this activity and these individuals are subsequently granted a greater probability of reproduction involving mutation and recombination of rules. These experiments highlight the pertinent aspects of successful space exploitation by one specific type of pattern (unbranched and meandering). Similar theoretical experiments with more elaborate branched patterns, but lacking the 'gene flow', are described by Bell (1984, 1985).

On occasion, the very process of constructing a computer simulation to reproduce a particular branching structure can be a useful experience in its own right, even without proceeding to

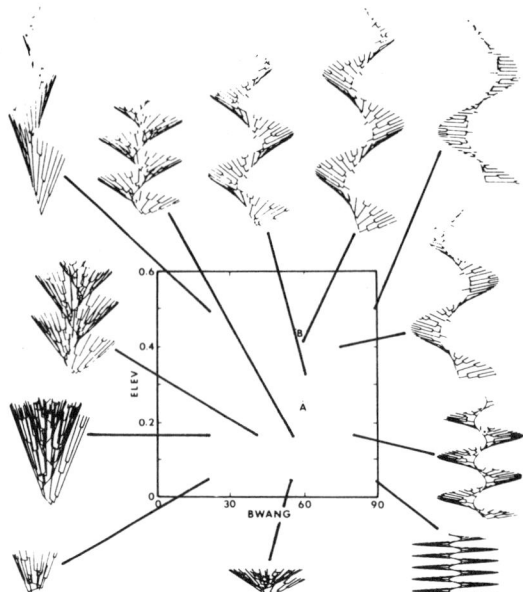

FIGURE 3. Simulated continuum of variation in the architecture of erect spiral bryozoans using few 'growth rules' (from McKinney & Raup (1982)). The extinct *Archimedes* is shown at top right and the extant *Reliflustra* centre bottom (see text).

the use of such a simulation to test an hypothesis. Either the morphology of the organism must be recorded in considerable detail or the underlying features of its developmental architecture fully appreciated, before an attempt can be made to model its growth. Shortcomings in the model will soon become apparent as 'mistakes', which are readily identifiable qualitatively but are not always easy to quantify. A subtle distinction has to be made in interpreting the output of a stochastic simulation between unintentioned teratologies, and the model's version of 'phenotypic' variation between replicates. Van Groenendael (1985) has shown how a model of developmental construction for *Plantago*, based on the modular concept, can be extended and used to construct and predict a wide range of teratological malformations that have been found in plantains. Such comparison of normal and mutant development aided by graphic modelling has likewise proved useful in the study of pea leaf morphogenesis (Young 1983).

In other instances it is the attempt to quantify phenotypic variation in branching pattern produced 'normally' by the organism that motivates the production of a branching model. Nishida (1980) set out deliberately to create tree-like structures based on *Chamaecyparis obtusa* (Japanese cypress) that have a built-in morphological variability similar to that found in nature between trees of the same clone; his intention was to use this controlled situation to quantify the degree of variability. In this manner, the majority of simulation models of branching are developed to test preconceived hypotheses. In this respect a model is not likely to be successful, or valid, unless it is predicted upon sound biological understanding of the growth processes that produce the morphology in question or upon an appreciation of the functional meaning of the morphology (Raup 1970).

## 4. The control of branching in simulation models

A mathematician describes branching patterns in terms of a *junction* (or 'node') where a branch is inserted, and a *link*, the branchless connection between any pair of junctions. A graphical simulation of a branching process must control link length and thus junction location, branching angle, and most importantly branch potential: once the site for a new branch is fixed, will it grow? When will it grow? How will it grow? How will it die? Implicitly, such a simulation will be a process description of branching events and the majority of branching growth models are indeed dynamic. (An exception is the state description model comparing potential productivity in clones of *Populus* with different morphologies (Burk *et al.* 1983).)

Thus a branch simulation model 'grows' according to predetermined 'rules'. A rule controlling branch angle, for example, may be guessed at or based on empirical data and may incorporate a given variability. These aspects of rules of growth have been discussed recently by Lück & Lück (1982) and by Waller & Steingraber (1985). Waller & Steingraber classify growth rules as 'deterministic or stochastic', and 'stationary or non-stationary'. A deterministic branch model operates according to recursive rules that are precise and repeatable during the growth of the pattern and lack built-in random effects. Figure 4 illustrates momentary stages in the astogeny of three type of graptolites (using the simulation procedure of Bell *et al.* (1979)). Rules of growth are deterministic and also stationary, as they do not change (figure 4*a*) or change according to fixed rules (figure 4*b, c*). In these simulations a module is represented by a theca and each new branch is initiated by the production of an extra theca. The model of colony development in encrusting bryozoans produced by Gardiner & Taylor (1982) (figure 5) is likewise deterministic but is, however, non-stationary, branching being modified by interaction between modules, in this case represented by the zooids. The consequences of change in branching angle was investigated in terms of number of zooids per colony and resultant effective feeding area.

The nature of the rules used in the compilation of a branching model, stochastic, non-stationary and so forth, obviously depends on the instigator of the system. The nature of the 'rules' for branching and of their control in the actual modular organism is commonly ignored or unknown. In discussing the validity of such simulation models it is therefore important to identify this key factor: what events occurring within the artificially created organism supplement the control of branching imposed at the overall programme level? Three categories of branching simulation can be identified.

(i) 'Blind' patterns, in which branch initiation is controlled solely by the imposed programme rules, that is from without the 'organism' and its 'environment'.

(ii) 'Sighted' patterns, in which the initiation of a new branch is influenced by factors detected by it in the immediate neighbourhood, such as proximity of other organisms, or parts of the same organism.

(iii) Self-regulatory patterns, in which branch initiation is controlled by the developing simulation itself, using communication via components of the existing framework, whether or not affected by 'environmental' factors.

(i) *'Blind' patterns*

An example of a 'blind' pattern is the graptolite simulation referred to earlier (figure 4) and the complexity of such a pattern depends solely on the complexity of the imposed rules and

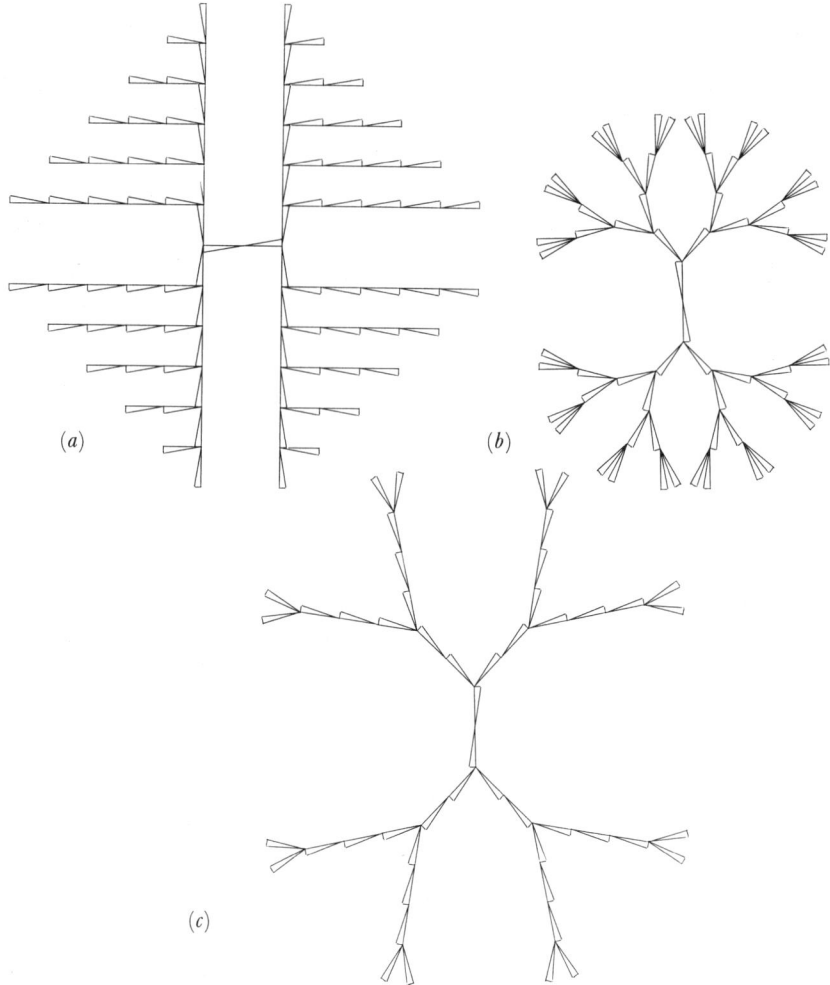

FIGURE 4. Deterministic, stationary simulations illustrating growth form in graptolites. An example of a 'blind' simulation. (a) Graptolite form with stationary rules for branch location and angle. (b) Graptolite form with stationary rules for branch location; branch angle rules also stationary but altering according to fixed rules (decreasing bifurcation in this instance). (c) Graptolite form with stationary rules; both branch location and bifurcation angle changing according to fixed rules.

the duration of the simulation. Hogeweg & Hesper (1973), using the Lindenmayer system for pattern description, were able to generate patterns of considerable complexity exhibiting unexpectedly realistic morphological features such as clustering of modules in short shoot complexes at intervals.

'Blind' patterns with simpler rules produced by Honda (1971), specifically to investigate problems in pattern morphogenesis and recognition, are described as 'tree-like bodies' and demonstrate the considerable variation in form that can result from small changes in branch angles and length ratios. Addition of empirical data to the Honda 'tree' simulations allowed Fisher & Honda (1977) and Honda & Fisher (1979) to study the efficiency of leaf display in the tropical tree *Terminalia* by comparing simulations of actual branch and leaf layouts with theoretical alternatives (see figure 8a–d). A similar simulation study has been conducted by Cheetham *et al.* (1981, 1983) into the efficiency of branching (and thus of feeding by rows of zooids) in adeoniform bryozoans (figure 6).

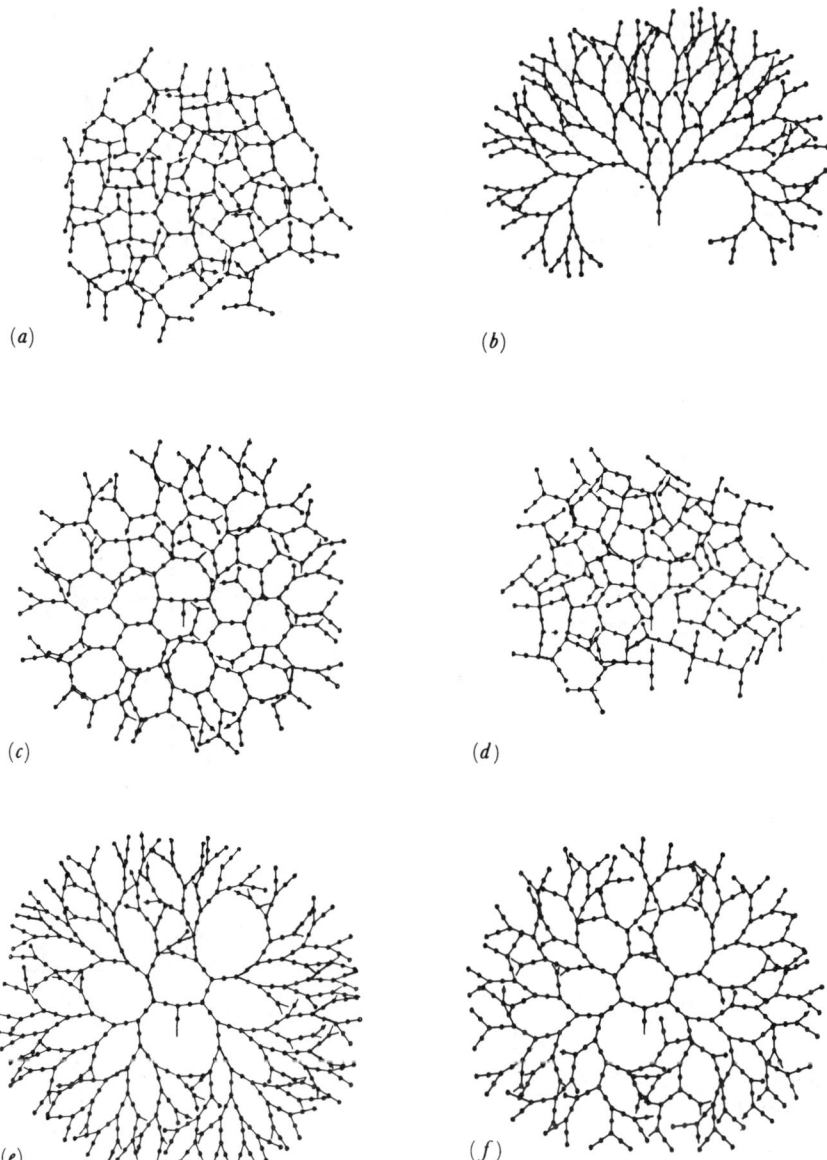

FIGURE 5. Artificial colonies of the fossil encrusting bryozoan (*Stomatopora*, a cyclosome) showing the change in form resulting from change in angle. These are 'sighted' patterns as branch growth is disrupted by proximity of neighbours. (From Gardiner & Taylor (1981)). (a) Constant large bifurcation angle; (b) constant small bifurcation angle; (c) constant decrease in angle; (d) constant increase in angle; (e) arithmetically decreasing angle; (f) exponentially decreasing angle.

(ii) *'Sighted' patterns*

'Sighted' simulations in which branching is influenced by external factors are demonstrated by Gardiner & Taylor (1982) (bryozoans, figure 5) and Cowen (1967). Honda *et al.* (1981) also introduced a branch interaction into their tree branching model by means of 'a horizontal circle of inhibition' located at the current terminal point of any branch, invasion of which influenced subsequent bifurcation (figure 8 *l–o*). A similar detection of available space governs the growth and bifurcation in the lateral branches of the simulation of spiral bryozoan

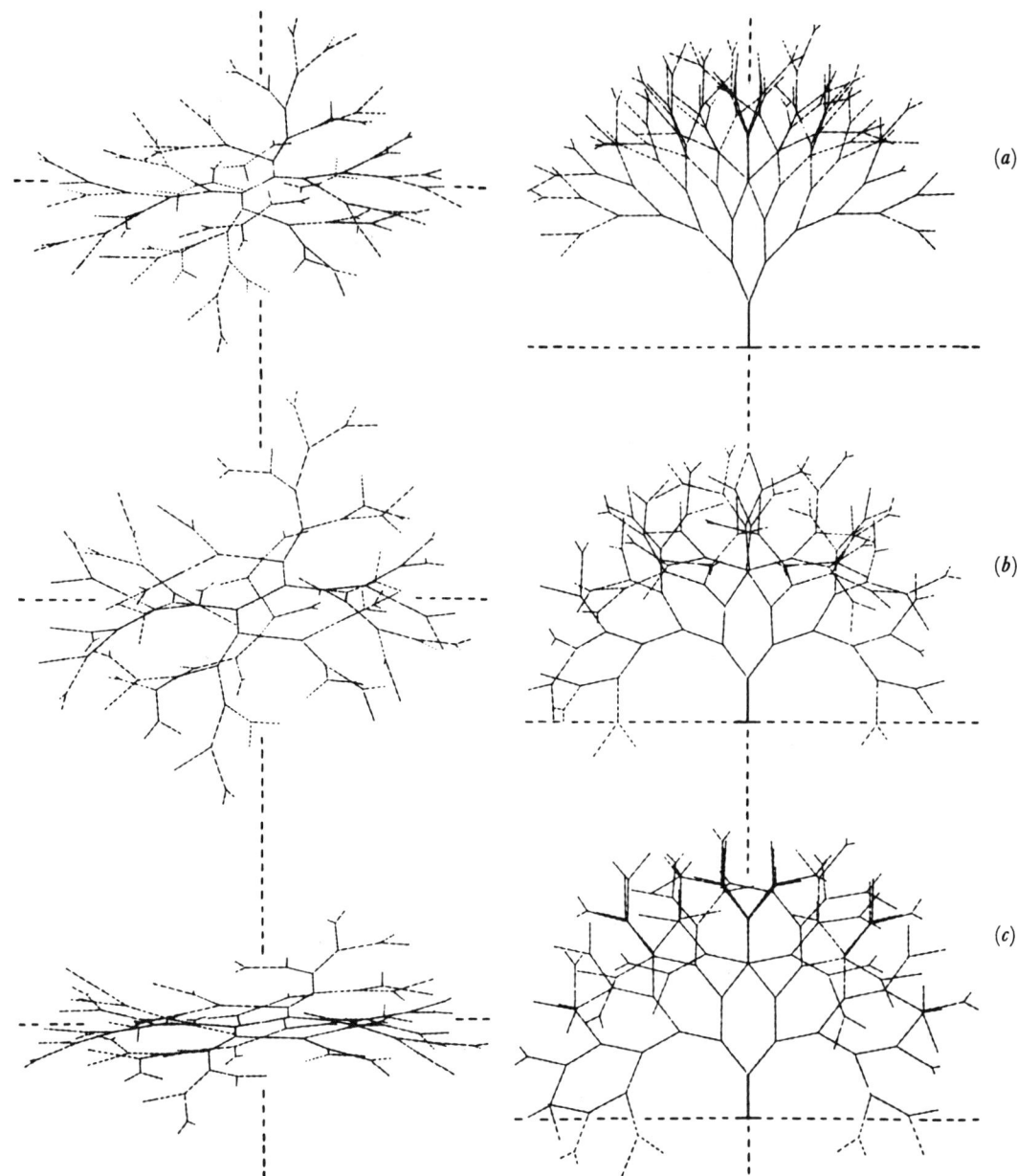

FIGURE 6. Simulation of form in arborescent bryozoans produced to study effectiveness of different geometries in terms of branch interference and zooid feeding efficiency (from Cheetham & Hayek (1983)). Top views on left, corresponding side views on right. (a), (b) and (c) The forms produced by three possible combinations of growth rules governing bifurcation angle and angle of twist between one bifurcation and the next. Combination of angles maximizing colony height do not coincide with those maximizing colony width or radial symmetry. Branching angles also affect the proportion of segments lost by reaching the substrate (shown as dotted lines in (b) and (c)). Angles: (a) bifurcation 50°, twist 50°; (b) bifurcation 80°, twist 50°; (c) bifurcation 80°, twist 20°.

architecture by McKinney & Raup (1982) (see figure 3). Such simulation of interference within developing patterns can demonstrate phenomena that would not otherwise be immediately apparent. Figure 7 (Bell 1985) illustrates that a threshold exists at a branch angle of 23° for this particular sighted pattern, above which a monopodial astogeny is possible, but below which a sympodial system is substituted due to abortion of apices in response to overcrowding.

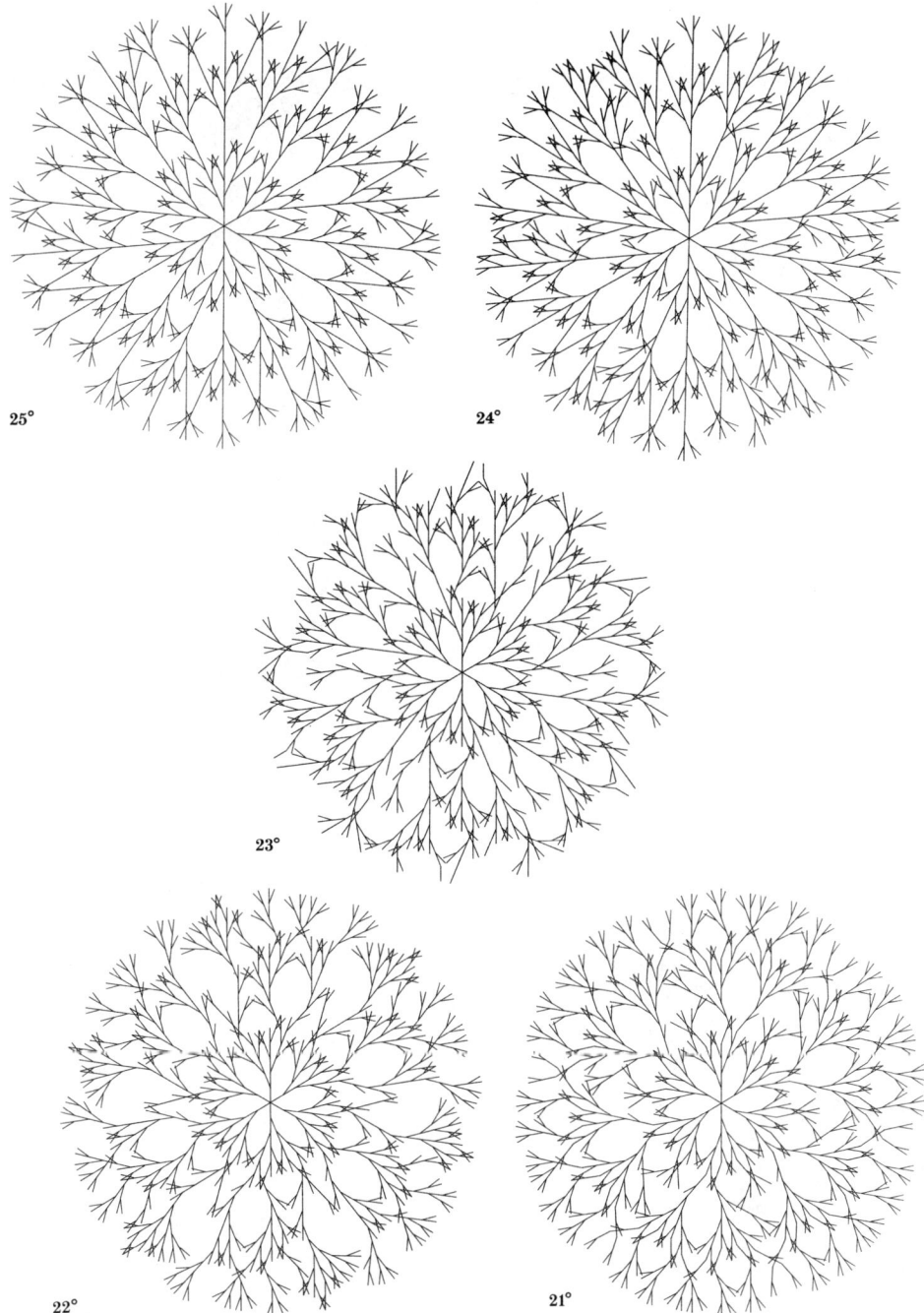

FIGURE 7. Five hypothetical patterns developing radially. Deterministic growth rules are identical in all respects except for lateral branch angles. Rules specify monopodial growth but this is superseded at narrow angles by abortion of crowded apices resulting in the production of sympodial growth. The fundamental switch from monopodial to sympodial branching results at a 1° change of branching angle (24° to 23°). (From Bell (1985).)

The underlying blind 'growth rules' for this simulation call for monopodial growth, but a sympodial development, due to the sighted component, actually materializes.

Sighted simulations incorporating response to interference monitored by a search routine assume that some detection process operates in the real organism, such as a response to shade or touch or release of a chemical. Very little is as yet known about such systems.

FIGURE 8. Simulation of horizontal tiers of branching in a tropical tree, plan view (*Terminalia*).

(*a*)–(*d*) 'Blind' patterns. Change in effective leaf display (represented by a circle centred at each bifurcation point), resulting from change in angle and asymmetry of bifurcation.

(*e*)–(*k*) 'Self-regulatory' patterns. Branching controlled by flow rate of a bifurcation factor within the developing pattern. (*e*) Equal flow to each side of a bifurcation (ratio 1:1). (*k*) No flow to one side (ratio 1:0). Ratios for (*f*)–(*j*) set at 1:0.998, 0.9, 0.66, 0.5 and 0.33, respectively.

(*l*)–(*o*) 'Sighted' patterns with progressively greater radius of search from distal ends for neighbouring branches that will prevent continued growth.

(*a*)–(*d*) From Fisher & Honda (1979); (*e*)–(*o*) from Honda *et al.* (1981).

(iii) *Self-regulatory patterns*

In self-regulatory patterns new branches, although they may be located by deterministic rules, are only initiated in response to events or conditions occurring within the developing simulation. One of the earliest branch simulations, that of Braverman & Schrandt (1965), operated in this manner. Continued stolon growth, hydranth location and type in a branching model of a marine hydroid (*Podocoryne*) was controlled by internal positioning factors based on assumed potential biological mechanisms.

Self-regulatory, internal control is also built into the inflorescence simulations of Frijters (1976, 1978a, b) (using versions of the Lindenmayer branch simulation system; Frijters & Lindenmayer 1974). Frijters recorded that a purely deterministic description of the developmental architecture in *Aster* would become too complex because each branch would require its own separate instructions. Instead, growth of side branches is controlled by information obtained about its neighbours. Similarly in a model for *Hieracium*, branch initiation and flowering sequence are controlled from within the developing simulation by 'flow of auxin'.

The series of studies of tree branch architecture by Honda & Fisher (1979) referred to above have recently been enhanced by recourse to self-regulatory simulation models. Unequal branch lengths at bifurcation points in *Terminalia* can be controlled by unequal flow rates of a 'bifurcation factor' (figure 8) (Honda *et al.* 1981). Some simulated patterns controlled by this means proved to be similar to those observed in actual trees. The simulated geometry and development of branching in another tree, *Tabebuia rosea*, was organized again from within the system by 'flux distribution' (Borchert & Tomlinson 1984; Borchert & Honda 1984). In these simulations self-regulatory mechanisms control branching vigour, potential and asymmetry while feedback systems within the development framework confer the equivalence of apical control.

A final example of a self-regulatory system is illustrated in figure 9 and represents the first stages of a simulation model under development based on the system of Bell *et al.* (1979). Figure 9a shows a developing clover (*Trifolium repens*) stolon at plastochron nine. The apical meristem of the main axis and the apical meristems of buds in the axil of leaves only grow if they receive, via the framework of the simulation, sufficient 'photosynthate' from exporting leaves. Leaves only export photosynthate if they are not being shaded. Figure 9b is an identical simulation in all respects except that the leaves have four exporting leaflets rather than three. The clover in figure 9b will suffer reduction in photosynthate production because of its more congested canopy, and therefore shelf shading, but will enjoy greater photosynthate production because of its larger number of leaflets. The outcome, by plastochron nine, is the earlier activation of specific bud apices as close scrutiny will reveal. This outcome cannot be predicted.

The self-regulatory branch models developed so far are crude, and are based on insecure hypotheses of physiological mechanisms. Nevertheless, the combination of sound morphological branch description, whether computerized or not, and better knowledge of physiological controls must assist the understanding and manipulation of modular growth.

## 5. Conclusion

In different branched modular organisms there is varied correspondence between the identity of the module and the nature of the links in the branch framework. For example, the 'module' in a bryozoan is represented by a zooid but branch initiation in such an organism may begin

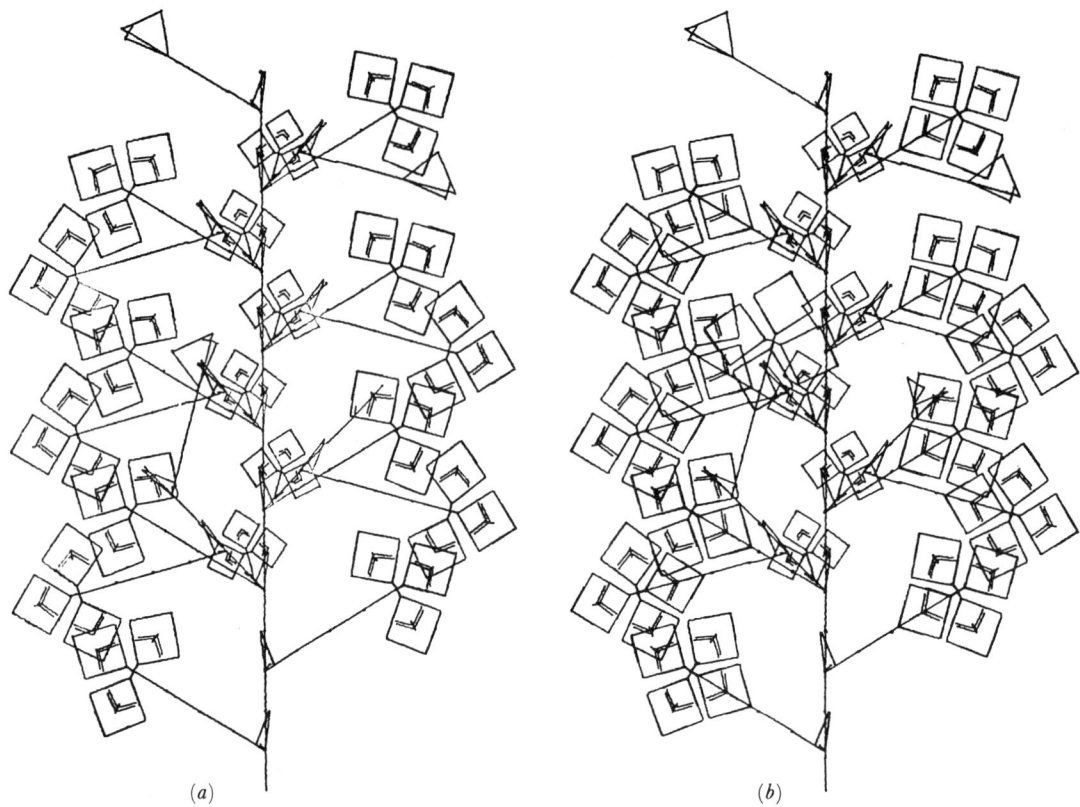

FIGURE 9. A self-regulatory and 'sighted' simulation of white clover stolon growth. Growth of buds is controlled by photosynthate exported from leaflets. Leaflets fail to produce photosynthate if they detect overtopping (shading) by other leaflets. (a) Three leaflets per leaf, (b) four leaflets per leaf, identical simulations otherwise. On balance, greater self-shading in (b) is over-compensated by greater export, and the four-leaflet clover is developing faster.

with the development of a single extra zooid (encrusting cyclostomes), or a group of zooids (arborescent cheilostomes). In higher plants branch initiation is represented by the development of a new apical meristem, and thus new module. Nevertheless, whatever the relationship of module to branch, the lynchpin in branching astogeny is the control of new branch initiation.

Branching architectures are readily simulated by computer graphics. Again, in the manipulation of such developmental forms, it is the control of new branch initiation during the simulation that can distinguish different levels of biological reality. These are described here as 'blind' (simulated organism oblivious of its environment), 'sighted' (each branch initiation influenced by its surroundings but unaware of its connectivity within the organism), and 'self-regulatory' in which astogeny is controlled from within the developing organism using internal and possibly external information. The growth of real modular organisms may be controlled by various combinations of these three factors and simulation may provide an insight into their functioning.

I thank Dafydd Roberts of the Computing Laboratory, U.C.N.W., for continual enhancement and updating of simulation programs.

## References

Aono, M. & Kunii, T. L. 1984 Botanical tree image generation. *Computer graph. applic.* **4**, 10–34.
Barker, S. B., Cumming, G. & Horsfield, K. 1973 Quantitative morphometry of the branching structure of trees. *J. theor. Biol.* **40**, 33–43.
Bell, A. D. 1984 Dynamic morphology: a contribution to plant population ecology. In *Perspectives on plant population ecology* (ed. R. Dirzo & J. Sarukhán), pp. 48–65. Sunderland, Mass.: Sinauer.
Bell, A. D. 1985 On the astogeny of six-cornered clones: an aspect of modular construction. In *Studies on plant demography: a Festschrift for John L. Harper* (ed. J. White), pp. 187–207. London: Academic Press.
Bell, A. D., Roberts, D. & Smith, A. 1979 Branching patterns: the simulation of plant architecture. *J. theor. Biol.* **81**, 351–375.
Biggs, N. L., Lloyd, E. K. & Wilson, R. J. 1976 *Graph theory*, pp. 1736–1936. Oxford: Clarendon Press.
Boardman, R. S., Cheetham, A. H., Oliver, W. A. Jr, Coates, A. G. & Bayer, F. M. 1973 Introducing coloniality. In *Animal colonies: development and function through time.* (ed. R. S. Boardman, A. H. Cheetham & W. A. Oliver), pp. v–ix. Stroudsburg, Pennsylvania: Dowden, Hutchinson & Ross.
Borchert, R. & Honda, H. 1984 Control of development in the bifurcating branch system of *Tabebuia rosea*: a computer simulation. *Bot. Gaz.* **145**, 184–195.
Borchert, R. & Tomlinson, P. B. 1984 Architecture and crown geometry in *Tabebuia rosea* (Bignoniaceae). *Am. J. Bot.* **71**, 958–969.
Braverman, M. H. & Schrandt, R. C. 1965 Colony development of a polymorphic hydroid as a problem in pattern formation. In *The Cnidaria and their evolution. Symp. zool. soc. Lond.* **16**, 169–198.
Burk, T. E., Nelson, N. D. & Isebrands, J. G. 1983 Crown architecture of short-rotation, intensively cultured *Populus* III. A model of first-order branch architecture. *Can. J. For. Res.* **13**, 1107–1116.
Cheetham, A. H., Hayek, L.-A. C. & Thomsen, E. 1981 Growth models in fossil arborescent cheilostome bryozoans. *Paleobiology* **7**, 68–86.
Cheetham, A. H. & Hayek, L.-A. C. 1983 Geometric consequences of branching growth in adeoniform Bryozoa. *Paleobiology* **9**, 240–260.
Cohen, D. 1967 Computer simulation of biological generation processes. *Nature, Lond.* **216**, 246–248.
Cowen, R. 1981 Crinoid arms and banana plantations: an economic harvesting analogy. *Paleobiology* **7**, 332–343.
Darrow, G. M. 1929 Development of runners and runner plants in the strawberry. *U.S.D.A. Tech. Bull.* **122**, 1–27.
Ede, D. A. & Law, J. T. 1969 Computer simulation of vertebrate limb morphogenesis. *Nature, Lond.* **221**, 244–248.
Engler, A. 1876 Vergleichende Untersuchungen über die morphologischen Verhältnisse der Araceae I. Thiel. Naturliches System der Araceae. *Nova Acta der Ksl. Leop-Carol-Deutschen Akademie der Naturforscher* **39**, 135–233.
Fellows, D. M., Sprague, G. L. & Baskerville, G. L. 1983 CROGRO: an interactive forest growth simulator. *Simulation* **41**, 219–228.
Finks, R. M. 1973 Modular structure and organic integration in Sphinctozoa. In *Animal colonies: development and function through time* (ed. R. S. Boardman, A. H. Cheetham & W. A. Oliver). Stroudsburg, Pennsylvania: Dowden, Hutchinson & Ross.
Fisher, J. B. & Honda, H. 1977 Computer simulation of branching pattern and geometry in *Terminalia* (Combretaceae), a tropical tree. *Bot. Gaz.* **138**, 377–384.
Fisher, J. B. & Honda, H. 1979 Branch geometry and effective leaf area: a study of *Terminalia* branching pattern. I. Theoretical trees. *Am. J. Bot.* **66**, 633–644.
Fitter, A. H. 1985 Functional significance of root morphology and root system architecture. In *Ecological interactions in soil: plants, microbes and animals* (ed. A. H. Fitter, D. Atkinson, D. J. Read & M. B. Usher), pp. 87–106. Oxford: Blackwells Scientific Publications.
Frijters, D. 1976 An automata-theoretical model of the vegetative and flowering development of *Hieracium murorum* L. *Biol. Cybernetics* **24**, 1–13.
Frijters, D. 1978a Principles of simulation of inflorescence development. *Ann. Bot.* **42**, 549–560.
Frijters, D. 1978a Mechanisms of developmental integration of *Aster novae-angliae* L. and *Hieracium murorum* L. *Ann. Bot.* **42**, 561–575.
Frijters, D. & Lindenmayer, A. 1974 A model for the growth and flowering of *Aster novae-angliae* on the basis of table $\langle 1, 0 \rangle$ L-systems. In *L-systems* (ed. G. Rosenberg & A. Salomaa) *Lecture notes Comp. Sci.* **15**, 24–52.
Gardiner, A. R. & Taylor, P. D. 1982 Computer modelling of branching growth in the Bryozoan *Stomatopora*. *N. Jb. Geol. Paläont Abh.* **163**, 389–416.
Gould, S. J. 1970 Evolutionary paleontology and the science of form. *Earth Sci. Rev.* **6**, 77–119.
Groenendael, J. M. van 1985 Teratology and metameric plant construction. *New Phytol.* **99**, 171–178.
Haggett, P. 1976 Network models in geography. In *Integrated models in geography* (ed. R. J. Chorley & P. Haggett), pp. 609–668. London: Methuen.
Hallé, F. & Oldeman, R. A. A. 1970 Essai sur l'architecture et la dynamique de croissance des arbres tropicaux. *Collection de monographies de botanique et de biologie végétale*, no. 6. Paris: Masson et Cie.
Hallé, F., Oldeman, R. A. A. & Tomlinson, P. B. 1978 *Tropical trees and forests: an architectural analysis*. Berlin: Springer-Verlag.

Hogeweg, P. & Hesper, B. 1974 A model study on biomorphological description. *Pattern Recog.* **6**, 165–179.

Honda, H. 1971 Description of the form of trees by the parameters of the tree-like body: effects of the branching angle and the branch length on the shape of the tree-like body. *J. theor. Biol.* **31**, 331–338.

Honda, H. & Fisher, J. B. 1979 Ratio of tree branch lengths: The equitable distribution of leaf clusters on branches. *Proc. natn. Acad. Sci. U.S.A.* **76**, 3875–3879.

Honda, H., Tomlinson, P. B. & Fisher, J. B. 1981 Computer simulation of branch interaction and regulation by unequal flow rates in botanical trees. *Am. J. Bot.* **68**, 569–585.

Jaeger, M. 1985 Representation de vegetaux. Rapport de stage. Diplôme d'études approfondies. Traitements graphiques – traitements d'images. Centre de Calcul de L'Esplanade, Université Louis Pasteur, Strasbourg.

Jarvis, R. S. 1977 Drainage network analysis. In *Progress in physical geography*, vol. 1. pp. 271–295.

Kickert, R. N. 1984 Names of published computer models in the environmental biological sciences: a partial list and new potential risks. *Simulation* **43**, 22–39.

Lück, J. & Lück, H. B. 1982 Modellbildungen zur pflanzlichen Verzweigung (eine Übersicht). *Ber. dt. bot. Ges.* **95**, 75–97.

Mackay, A. L. 1975 A metaphor for molecular evolution. *J. theor. Biol.* **54**, 399–401.

Mackie, G. O. 1963 Siphonophores, bud colonies, and superorganisms. In *The lower Metazoa, comparative biology and phylogeny* (ed. E. C. Dougherty). University of California Press.

McKinney, F. K. & Raup, D. M. 1982 A turn in the right direction: simulation of erect spiral growth in the bryozoans *Archimedes* and *Bugula*. *Paleobiology* **8**, 101–112.

Niklas, K. J. 1977 Ontogenetic constructions of some fossil plants. *Rev. Palaeobot. Palynol.* **23**, 337–357.

Niklas, K. J. 1982 Computer simulations of early land plant branching morphologies: canalization of patterns during evolution? *Paleobiology* **8**, 196–210.

Niklas, K. J. 1984 Reply to Waller (1984) *Paleobiology* **10**, 117–120.

Nishida, T. 1980 KOL-system simulating almost but not exactly the same development. Case of Japanese Cypress. *Mem. Fac. Sci. Kyoto Univ. Ser. Biol.* **8**, 97–112.

Oohata, S. & Shidei, T. 1971 Studies on the branching structure of trees 1. Bifurcation ratio of trees in Horton's Law. *Jap. J. Ecol.* **21**, 7–14.

Oxnard, C. E. 1980 Conclusion to the symposium: diversity in the analysis of form. *Am. Zool.* **20**, 721–722.

Papentin, F. 1973 A Darwinian evolutionary scheme. III. Experiments on the evolution of feeding patterns. *J. theor. Biol.* **39**, 431–445.

Papentin, F. & Roder, H. 1975 Feeding patterns: the evolution of a problem and a problem of evolution. *N. Jb. Geol. Paläont Mh.* **H3**, 184–191.

Pelt, J. van & Verwer, R. W. H. 1984 New classification methods of branching patterns. *J. Microsc.* **136**, 23–34.

Pohowsky, R. A. 1978 The boring ctenostomate bryozoa: taxonomy and paleobiology based on cavities in calcareous substrata. *Bull. Am. Paleont.* **73**, 301.

Prévost, M. F. 1967 Architecture de quelques Apocynacées ligneuses. *Mém. Soc. Bot. France* **114**, 23–36.

Raup, D. M. 1966 Geometric analysis of shell coiling: general problems. *J. Paleont.* **40**, 1178–1190.

Raup, D. M. 1967 Geometric analysis of shell coiling: in ammonoids. *J. Paleont.* **41**, 43–65.

Raup, D. M. 1968 Theoretical morphology of echnoid growth. *J. Paleont.* **42**, 50–63.

Raup, D. M. 1970 Modelling and simulation of morphology by computer. *Proc. N. Am. paleont. Conv.*, pp. 71–83.

Raup, D. M. & Michelson, A. 1965 Theoretical morphology of the coiled shell. *Science, Wash.* **147**, 1294–1295.

Raup, D. M. & Seilacher, A. 1969 Fossil foraging behaviour: computer simulation. *Science, Wash.* **166**, 994–995.

Raven, J. A. 1984 Physiological correlates of the morphology of early vascular plants. *Bot. J. Linn. Soc.* **88**, 105–126.

Reffye, Ph. de 1979 Modélisation de l'architecture des arbres tropicaux par des processes stochastiques. Simulation spatiale des modèles tropicaux sous l'effet de la pesantur. Application au *Coffea robusta*. Thèse de doctorat d'Etat Univ. Paris **11**, 1–194.

Réthelyi, M. 1981 The modular construction of the neuropil in the substantia gelatinosa of the cat's spinal cord. A computer aided analysis of golgi specimens. *Acta morph. Acad. Sci. Hung.* **29**, 1–18.

Room, P. M. 1986 Plant architecture and how biological control agents affect the dynamics of weeds. *Proc. VI int. Symp. Biol. Conf. Weeds, Vancouver*.

Smart, J. S. 1978 The analysis of drainage network composition. *Earth Sci. Process.* **3**, 129–170.

Smith, S. & Scoullar, K. 1975 A data structure analogue for modelling tree crowns. *Can J. For. Res.* **5**, 574–579.

Tanner, J. C. 1967 Layout of road systems on plantations. Report LR68 1–14. Road Research Laboratory, Ministry of Transport, Crowthorne, U.K..

Thornley, J. H. M. 1977 A model of apical bifurcation applicable to trees and other organisms. *J. theor. Biol.* **64**, 165–176.

Tomlinson, P. B. 1978 Some qualitative and quantitative aspects of New Zealand divaricating shrubs. *N.Z. J. Bot.* **16**, 299–309.

Tumidajowicz, D. & Dambski, M. 1984 Simulative model of genet growth in an exemplary population of the forest perennial *Dentaria glandulosa* Wetk. Bulletin of the Polish Academy of Sciences. *Biol. Sci.* **32**, 197–297.

Unger, B. W. & Bidulock, D. S. 1981 Modular design of multicomputer systems. *Simulation* **37**, 1–9.

Waller, D. M. 1984 Modelling branching patterns in early land plants. Niklas's simulations of branching in early land plants: canalization of parameters? *Paleobiology* **10**, 115–117.

Waller, D. M. & Steingraeber, D. A. 1985 Branching and modular growth: theoretical models and empirical patterns. In *Population biology and evolution of clonal organisms*. (ed. J. B. C. Jackson, L. W. Buss & R. E. Cook), pp. 225–258. New Haven: Yale University Press.

Welty, J. J., Moser, J. W., Jr & Bailey, M. J. 1985 TIMECOGS: an educational computer graphics timber-marking simulator. *Computer graph. applic.* **5**, 61–67.

White, J. 1984 Plant metamerism. In *Perspectives on plant population ecology* (ed. R. Dirzo & J. Sarukhán), pp. 15–47. Sunderland, Massachusetts: Sinauer.

Wolfram, S. 1984 Computer software in science and mathematics. *Scient. Am.* **251**, 188–203.

Wright, F. L. 1953 *The future of architecture*. New York: Horizon Press.

Young, J. P. W. 1983 Pea leaf morphogenesis: a simple model. *Ann. Bot.* **52**, 311–316.

# Physiological consequences of modular growth in plants

By R. C. Hardwick

*National Vegetable Research Station, Wellesbourne, Warwick CV35 9EF, U.K.*

Modular plants are structurally stable systems. Perturbations in the genome that arise during growth (somatic mutations) are, it is conjectured, detected and eliminated by diplontic selection at the apical meristem. This requires accurate control, which, it is suggested, results in the accuracy of module arrangement, or phyllotaxy. During differentiation vegetative modules tend to retain autonomy and totipotency; the differentiated state is maintained by a constant flux of signals between parts. This enables damage to be repaired and differentiation to be adjusted to the availability of resources. The differentiation of modules to form flowers, on the other hand, is achieved by loss of totipotency, by hierarchical organization of the genotype, and by tissue-specific signals between parts. Elaborate, but fixed-function, structures can be produced in this way.

The physiology of modular plants is best described in terms of cooperation, not competition, between modules. The theory of multicomponent systems predicts that as plants increase in size, structural stability of growth will be lost unless the connectance between modules is kept below a critical value. Experiments confirm that the exchanges of assimilate between modules are limited, but not fixed (the system can adapt to damage). The distribution system is vulnerable to exchanges that might benefit individual modules but that would reduce the inclusive fitness of the genome. Such exchanges are controlled by the organized senescence of branches, leaves, fruit and ovules.

## Introduction

Within the walls of each plant cell three genomes (nuclear, mitochondrial and chloroplast) cooperate for their mutual survival. For life in the open sea, the optimal phenotype can be constructed from a single planktonic cell, whose activities can be controlled by the diffusion of mRNAs from a single nucleus. In most other environments, small phenotypes are not optimal, and here multicellularity evolved. Small propagules were still required for dispersal, and since their parents were sessile, they needed mechanisms for autonomy. Modular growth enables the young plant to attain autonomy while it is still small, and yet, by repetition of parts, to reach a large final size. This paper considers the mechanisms involved in (i) initiation, (ii) differentiation, (iii) regulation, and (iv) death, of modules. The analysis is developed through a series of propositions.

## Initiation of modules

I use the term module to refer to a repetitive unit of construction, for example, a single cell in a filamentous alga, leaf plus stem plus axillary bud in a higher plant.

Proposition. *An organized, modular, phenotype is the product of a set of organizing principles or rules.*

Lindenmayer (1982) has shown that the growth of various species of filamentous alga, such as *Chaetomorpha* and *Ulothrix*, can be simulated by a computer provided with a set of rules for 'cell' division, polarity sensing and differentiation. The rule for cell division requires that the two daughter cells develop differently, one remaining mitotic, the other differentiating and becoming non-mitotic. This implies the existence in filamentous algae of a mechanism to determine the polarity of division and differentiation. Laboratory studies have identified such a mechanism; it involves electric fields (Jaffé & Nuticelli 1977; see also Trinci & Cutter, this symposium).

In the higher plants, cell division is largely confined to specialized regions, which constitute only a small proportion of the plant body. The meristematic growth habit enables delicate thin-walled dividing cells to be protected against drought, predators and mechanical collapse, at the cost of a substantial reduction in growth rate. While it is still in the ovary, the number of cells in an embryonic plant of *Daucus carota* L. (carrot) doubles every 24 h (Gray *et al.* 1984). If cell division continued to occur throughout the plant's body, an embryo would grow to maturity (about $2^{28}$ cells; C. C. Hole, personal communication) in 28 days. In another fortnight it would contain $2^{42}$ cells and weigh about 2.5 t. Localizing cell division to meristematic areas limits the plant's potential growth rate, a handicap in variable environments, but it opens the possibility of modular growth. Modular growth required, however, the evolution of further mechanisms.

PROPOSITION. *In multicellular plants, modular growth requires the repetitive establishment of autonomy and a secondary axis of polarity among the cells leaving the apical meristem.*

Division of labour can be achieved without such mechanisms. In the roots of modern Gymnosperms and Angiosperms, different functions are assigned to different ages of tissue; in the dimorphic root systems of some amphibious plants such as mangroves, *Typha* and *Ludwigia*, each meristem is assigned permanently to the production of a particular set of tissues. But such patterns are relatively inflexible. They have proved to have only a limited potential compared with the repetitive production of parts from a single apical meristem, that is, modular growth.

In modular plants the repetitive processes at the apical meristem take place with remarkable precision and regularity, resulting in accurate phyllotaxies. Two approaches have been used to study the rule systems of phyllotaxy: geometry for the state-rules and computer simulation for the process-rules. Geometric analysis of the properties of regular lattices shows that the set of regular (stable) arrays on the surface of a cone or cylinder can be described by a function of two parameters, $\theta$ and $h$ (Erickson 1983). $\theta$ is the angle subtended between two successive elements of the genetic spiral, $h$ is the vertical distance between them. Phyllotaxies resemble a Feigenbaum system (Feigenbaum 1980) where $h$ is the bifurcation parameter, and $\theta$ is the dependent variable. Stability is only attainable within rather narrow limits; for example, a regular 3:5 phyllotaxy can be maintained only between the points ($h = 0.11$, $\theta = 135.9°$) and ($h = 0.29$, $\theta = 142.1°$) (Erickson 1983). Thus close control is required of the parameters $\theta$ and $h$.

The computer simulations use reaction–diffusion equations to model the processes in shoot apical meristems. The apical meristem is represented as a two- or three-dimensional net of communicating cells, continually increasing in size. Some sort of morphogenetic signal, either mechanical (strain) or chemical (activator or inhibitor), is assumed to be generated in each cell and to be communicated across the net with a characteristic decay function. Local minima arise in the resulting morphogenetic field, and as growth continues it is possible, by appropriate choices of parameter values and boundary conditions, to make the successive minima adopt a regular configuration (Ridley 1982). Again, strong control of the parameters is a pre-requisite.

Phyllotaxy is a good taxonomic character; it is under strong genetic control and is little affected by perturbations in the environment. The value of $\theta$ is maintained with remarkable precision; for example, in *Epilobium* the standard error in $\theta$ was 0.26° (Meicenheimer 1981). In a meristem of 100 μm diameter an angular deviation of 0.26° corresponds to a distance of only 0.2 μm, about the diameter of a single cell. This is probably an underestimate of the precision of the initiation process, for the fluctuations in $\theta$ are in part systematic, not random (Kumazawa & Kumazawa 1971).

In discussing the 'mysteries' of phyllotaxy, Thompson (1917) remarked that 'I, for my part, see no subtle mystery in the matter other than what lies in the steady production of similar growing parts similarly situated at similar successive intervals of time'. What can be the adaptive value of such steadiness? Since leaves can twist on their petioles, I suggest that if an accurate phyllotaxy has an adaptive value it is more likely to relate to conditions during the initiation of primordia, than to their subsequent life; it seems unlikely that accurate phyllotaxy is simply a means for optimizing light-interception. Now apical meristems, in addition to their main function of initiating modules, have various second-order functions:

PROPOSITION. *Extensive modular growth requires long-lived apical meristems. These require a mechanism for preserving the integrity of the genome, by removal of mutant cells from the germ line as they arise (that is, diplontic selection).*

The shoot apical meristem is the principal source of new nuclei within the phenotype. It is therefore largely responsible for maintaining the integrity of the genome. In tissue culture the plant genome is remarkably labile (Scowcroft 1984), but in the intact plant it is, with few exceptions (Durrant 1962; Whitham & Slobodchicoff 1981; Cherfas 1985) very stable; many cultivars that are vegetatively propagated appear to maintain their genetic integrity indefinitely. How is this stability achieved? Klekowski & Kazarinova-Fukshansky (1984) suggest that mutant cells are removed from the germline by a mechanism of diplontic selection, which operates at the apical meristem. Diplontic selection requires a means of determining the type (same or mutant) of young cells, for example by comparing the local value of a morphogenetic field with an internal reference value (Wolpert 1981). This leads to the following conjecture.

PROPOSITION. *The patterning of primordia at the apical meristem is the by-product of systems that regulate the value of $h$ and $\theta$ with time. The primary function of these mechanisms is in diplontic selection; they maintain a stable morphogenetic field at the shoot apical meristem.*

According to this conjecture diplontic selection requires stable morphogenetic fields; and the occurrence of tightly regulated phyllotaxies in the shoots of modular plants reflects this stability. Since leaves, roots, etc., rarely contribute to the next generation they would not be expected to exhibit diplontic selection or accurate control of cell lineages.

Plants and crops can be considered as populations of modules (Bazzaz & Harper 1977; Hardwick & Milbourn 1967). In the analysis of the growth processes of such populations 'birth' comprises module initiation and emergence from the apical bud. The rate of initiation of primordia sometimes varies with age (Romberger & Gregory 1977) but it generally responds rather little to variations in the external environment (Berg & Cutter 1969). The rate of emergence of modules from the apical bud does vary with the environment (Milford *et al.* 1985). Hence the following:

PROPOSITION. *Accommodation of plant growth rate to the supply of environmental resources is achieved by variation in the rate of module growth and in the number of functional meristems, rather than in the rate at which modules are initiated at each apical meristem.*

Thus variation in, for example, nutrient availability results in variation in the size of the plant (that is, size and number of modules) at flowering and harvest; the dates of flowering and harvest (a function of the rate of module initiation) remain relatively invariant. This is in accordance with predictions from models of life-cycle characteristics (Cohen 1971).

### DIFFERENTIATION OF MODULES

In stable environments large phenotypes are at an advantage. A small phenotype can be enlarged by the dichotomous branching of one axis into two, each daughter being an identical copy of the parent axis. As the copying process continues, the dimensions of the overall structure will increase linearly; but the number of growing points will increase exponentially. So the growing points on the periphery of the plant will become increasingly, and impossibly, tightly packed. This leads to the following:

PROPOSITION. *A mechanism for differentiation, that is, a mechanism that ensures that the daughters of a branching do not all behave similarly, is an essential requirement in the production of large, branched, modular structures.*

A mechanism for meristem 'dormancy' meets this requirement. It also confers the ability to reconstruct damaged structures (see below). Growth of a modular structure without differentiation imposes other problems. A variety of analyses suggest the following.

PROPOSITION. *Module differentiation is organized in such a way as to optimize a cost–benefit function.*

This proposition is supported, for example, by cost–benefit analyses of branch disposition and light interception (Honda & Fisher 1978), of canopy height and light interception (Iwasa *et al.* 1984), of rhizome branching angle and soil exploration (Bell 1979), of the costs and benefits of sun or shade leaves (Bazzaz 1979), of vegetative or floral branches (Smith 1984), of long and short shoots (Wilson 1966), and of elastic versus geometric similarity in trees (MacMahon & Kronauer 1976).

Cost–benefit analyses have severe limitations: for example, it is mathematically impossible to maximize for more than one variable at the same time; in natural selection, history and chance and other unknowns also enter the equations. Nevertheless, it has proved easier to analyse the costs, benefits and trade-offs of various differentiation patterns than it has to identify the mechanisms by which they are achieved. In trees, two broad groups can be distinguished. The first group, typified by, for example, some species of *Araucaria*, are stoutly constructed trees; the tissues are long-lasting or 'evergreen' and the developmental sequences appropriate to each position in the structure are determinate and almost invariant. Invariant differentiation may be achieved by mechanisms such as topophysis (patterns of differentiation determined by position) and by hydraulic dominance (patterns of xylem flows determined by xylem resistance) (Zimmermann 1978).

The developmental processes that characterize the second group appear to have arisen later in evolution (Corner 1967). These plants rely less on mechanical strength and invariant patterns of development, and more on stand-by or 'dormant' buds and on conditional patterns of development. Dormant buds enable damage to be repaired (reiterative growth), and variation in the abundance of resources such as mineral nutrients or light to be accommodated ('opportunistic' growth (Tomlinson 1982)). Evidently a signalling system is involved, but the detailed mechanisms of 'apical dominance' and correlative inhibition are still obscure (Hillman 1984).

Studies of the abstract rules for form generation in higher plants are not as well advanced as they are in filamentous algae, although Lindenmayer (1984) has derived rules for simulating the growth of certain complex inflorescences. The rules are position-dependent and require signalling between branches.

The experimental work on plant signalling systems can be summarized as follows.

PROPOSITION. *Discrete or 'pulse' signals for differentiation may pass between modules. Once differentiation has been achieved, it is maintained by a constant flux of signals from neighbouring cells or modules.*

The signalling systems must match, in terms of information content or variety, the processes they control. It seems unlikely that the recognized plant growth substances can provide this variety; polysaccharide moieties may also be involved (Canny 1985). But morphogenetic signals are not necessarily chemical. Changes in electrical membrane potential accompany many developmental processes. Strain, which is a function of the mass of distal modules, is a signal for the differentiation of reaction wood, and hence for the adjustment of branch angle (Fisher & Stevenson 1981). Other mechanical stimuli such as wind loading and plant–plant contact, which elicit adaptive morphological responses, also appear to be sensed as mechanical strain (Biddington 1986). Mechanical strain generated by cell expansion determines the direction of the division plane in cytokinesis (Lintilhac & Vesecky 1984). The red:far-red ratio in the light at the base of the canopy is a function of presence of upper leaves as light filters (Holmes 1983), and is a signal for the release or initiation of axillary buds (Deregibus *et al.* 1985; see also Franco, this symposium). Short range 'pulse' signals between modules occur in the differentiation of successive perianth components (Heslop-Harrison 1964). A long-range, continuous flux of signals (auxin) is maintained from shoot apex to root tip, overriding local fluctuations in availability of nutrients and by its logarithmic dose–response characteristics 'smoothing out' the processes of differentiation (Trewavas 1982). Local variations in this flux of auxin appear to be responsible for the characteristic pattern of xylem differentiation along the stem (Ewers & Zimmerman 1984), and hence for the differential growth of buds (Salleo *et al.* 1985) and the development of characteristic branching patterns (Honda *et al.* 1981).

A continuous flux of signals also passes upwards from root to shoot, indicating the current availability of mineral nutrients (de Wit & Penning de Vries 1983) and of water (Blackman & Davies 1985). The root-signals are a function of the root's genotype; the growth of the shoot systems of fruit trees can be regulated by appropriate choice of rootstock (Rogers & Beakbane 1957).

In a number of cases the signal for differentiation remains obscure. The branches of some gymnosperms exhibit topophysis, each retaining its characteristic angle of growth even when the others have all been removed, or even when the branch is taken from the plant and grown on as a cutting (Worrall 1984). Some plants, for example peas (*Pisum sativum*), switch from initiating vegetative modules to initiating flowers after a precisely counted number of nodes (Hardwick 1985). Diffusates, probably gibberellins, from the cotyledons are implicated (Murfet & Reid 1985). In other species the following is possible.

PROPOSITION. *Meristem size is an autonomous signal for differentiation.*

The size of the shoot apical meristem gradually increases during ontogeny. It has been suggested that this provides a signal for the mechanisms that control branching (Thornley 1977) and the switch to floral growth (Battey & Lyndon 1984). We do not know how the parameter $h$, and meristem size, are controlled (Charles-Edwards 1984), but the apical meristem is a potential site of accumulation for various materials that are carried in the vascular system (Sheldrake 1973) and it has been suggested that the tendency of materials such as boron to accumulate at the shoot apex of primitive plants may have provided the necessary opportunity for localizing cytokinesis and building an apical meristem there, by the evolution of a mechanism that made cytokinesis critically dependent on the ambient level of boron (Lovatt 1985). Elaborations of such mechanisms may be involved in the control of meristem size.

During vegetative growth, plant cells tend to retain totipotency; irreversible determination is uncommon (Sussex 1983). This contrasts with the situation in many animals, where totipotency is rare, and differentiation seems to reflect a hierarchical organization of the genome (Britten & Davidson 1969). It seems that:

PROPOSITION. *Modular organization of the phenotype imposes a non-hierarchical organization on the genotype; totipotency is a consequence of modular growth.*

In unitary organisms (such as vertebrates and insects), the phenotype is reliably entire and even-aged, and differentiation can be achieved by a hierarchically organized genome, dependent on tightly controlled flows of signals between parts. The benefit is that very complex patterns of differentiation can be achieved, but the cost is a loss of totipotency. In the development and differentiation of modular, uneven-aged, organisms, on the other hand, a degree of local autonomy and totipotency is valuable as insurance against accident and because it enables the size, type and number of modules to be adjusted to the resources available. Such adjustments are appropriate during vegetative growth, when the function of each module is the exploitation of resources. For modules concerned with reproduction (that is, sepals, petals, etc.) function and therefore structure are fixed. Here we tend to find a hierarchical organization of the genome, tissue-specific signals, and loss of totipotency.

Studies of abnormal patterns of differentiation of the inflorescence of *Plantago lanceolata* by van Groenendael (1984) suggest that in this plant differentiation of reproductive structures occurs by commitment to a defined sequence of events. Various abnormal inflorescences of *Plantago* can be explained in terms of the substitution at a late stage in the sequence of 'instructions' appropriate to an earlier stage. The system is reminiscent of the homeotic box system in animal segmentation. In other species gene transfer experiments demonstrate that the expression of at least some plant genes is tissue-specific. For example, the gene for production of the seed protein phaseolin has been transferred from beans to tobacco. In its new host the gene is preferentially expressed in the appropriate tissue, that is, the seed (Netzer 1984). The nature of the tissue-specific signal or signals is not known.

In animal differentiation, morphogenetic signals may be amplified by the cAMP–adenyl cyclase mechanism. In the higher plants the mechanism appears to be less well developed (Brown & Newton 1981). Evidence is accumulating that:

PROPOSITION. *Stable (canalized) patterns of differentiation can be established in a tissue of totipotent cells by driving reactions to completion by autocatalysis.*

Instances of autocatalytic (or positive feedback or flow facilitation) mechanisms in plant morphogenesis include, for cytokinin, the habituation of tobacco pith cells (Meins & Foster 1985), and, for auxin, the two-cell developmental model of Stange (1984), auxin enhancement of the synthesis or activity of its own binding sites (Starling 1984), and the auxin-enhanced differentiation of auxin-transporting (vascular) tissues in stems (Sachs 1981). A computer model of autocatalytic processes in a laminar tissue provides a convincing simulation of the development of leaf venation (Mitchison 1980).

The vegetative parts of modular plants are, then, at least potentially, totipotent and autonomous. The ways in which relationships between such units are regulated is discussed in the next section.

## Relationships between modules

The cell plate is not entire but perforate, so that syncytial connections extend throughout the plant body, linking cells to one another in a three-dimensional network. Modules are linked to one another by a 'superapoplast' (xylem) and 'supersyncytium' (phloem) (Raven 1977). As the number of modules, $n$, in the system increases, the potential number of flows of substrate between modules increases rather rapidly (as $n^2$). Yet the growth of higher plants is characterized by 'harmonious relationships' (Münch 1938), or 'functional equilibria' (Brouwer 1983). How is this achieved?

Consider a simple two-component system, for example, shoot and root. We are interested in the rules for allocating increments in mass between components, that is in the form of the relations $dS/dt = f(S, R, t)$; $dR/dt = f'(R, S, t)$, where $R$ is mass per root, $S$ is mass per shoot and $t$ is time. A particularly simple case occurs when $f$ and $f'$ are defined by (1) and (2):

$$dS/dt = g_{ss}S + g_{sr}R, \tag{1}$$

$$dR/dt = g_{rs}S + g_{rr}R, \tag{2}$$

where the values of the coefficients $g_{ss}$ and $g_{rr}$ are fixed and where $g_{sr} = g_{rs} = 0$. Then

$$S = bR^a,$$

and growth is allometric. Allometry has often been claimed, but usually on somewhat inadequate data. Critical analysis of a particularly large and well structured data set (Currah & Barnes 1979) confirmed that the estimated value of the coefficient $a$ ($= g_{ss}/g_{rr}$) was more or less constant over time and over treatments. A physiological explanation of this constancy, which has often been observed, is not as yet available. The estimated value of the coefficient $b$ was constant with plant spacing, but its value decreased with time: that is, growth in time was not allometric. This too appears to be a general phenomenon (for an explanation see Barnes (1979)).

Equations (1) and (2) could, in principle, be extended to describe the growth in mass of a plant of $n$ modules. This would require $n$ differential equations and $n^2$ coefficients. The analysis of such equation systems shows that in large, randomly connected sets of $n$ interacting components, stability is critically dependent on the degree of connectance (the percentage of non-zero values of the coefficients $g_{ij}$). If connectance is greater than about 15% the system will almost certainly be unstable to perturbations (Gardner & Ashby 1970). Large values of connectance strength will also cause instability (May 1972). Two propositions follow from these theoretical studies. The first is:

PROPOSITION. *To avoid instability of growth, the phloem connectance between modules must be less than about* 15%.

The proposition is in accordance with 'Canny's rules' (Canny 1984). Labelled assimilates can move along single strands of phloem for hundreds of centimetres and all connections between modules appear to be potentially available, but (unless the system is perturbed) it seems that each module serves as 'source' to, or draws as 'sink' from, only a few of its nearest neighbours; interaction strength ($g_{ij}$) decreases sharply as the distance between source and sink increases (Cook & Evans 1983). Experiments with labelled assimilates confirm that in modular plants, phloem connectance is maintained at a low value.

The mechanical equivalent of connectance is cross-bracing. Trees, unlike most man-made structures of similar size, are not cross-braced. It might seem that cross-bracing would offer a substantial increase in mechanical strength for little extra cost. But the connectance result predicts that in an artificially cross-braced tree (which could be constructed by making a large number of approach grafts between branches), patterns of growth would not be stable. The same is predicted to occur if connections develop between stilt roots and natural roots in plants such as *Ficus benghalensis*. Where connections develop between individual plants, such as in the hemiparasite *Bartsia* (which parasitizes its own species) the result of more than 15% connectance would be expected to be a loss of stability of growth of the ensemble and the emergence of a few very large individuals.

The open vascular system and primitive eustele of the stems of late Devonian plants soon evolved into a closed system, allowing lateral transfer between leaf traces (Beck *et al.* 1982). Lateral connections allow the system to compensate for damage and to smooth out variations in fluxes between modules. A significant percentage of fixed nitrogen imported by a module (a leaf) from the roots tends to be re-exported in the phloem to the roots, and the same appears to be true of carbon assimilates passing from shoot to root (Lambers 1983). These mechanisms may help confer functional stability on the system.

The second stability condition was that the average strength of interactions between modules should not exceed a critical value. This requirement is met relatively easily if all the modules of a plant have the same genotype but:

PROPOSITION. *If connections develop between modules of different genotypes, strong interactions may ensue, threatening the stability of growth of the system as a whole.*

In angiosperms, functional mixtures of modules of two (or more) different genotypes occur (i) in somatic mosaics; (ii) in periclinal chimeras; (iii) in plants attacked by a vascular parasite; (iv) in plants joined by root or shoot grafts; and (v) in the angiosperm ovary after outcrossing.

Somatic mosaics may arise by amplification of gene copy during growth or by the failure of diplontic selection during growth; the phenomena do not seem to be widespread and there have been few physiological investigations on the growth of mosaic plants, or of plant chimeras. The physiology of vascular parasites and their hosts, on the other hand, has received detailed attention. Modules of two markedly different genotypes develop intimate (but apoplastic) connections, and the resulting imbalance is such that the parasite modules increase at the expense of those of the host. Mistletoes (*Viscaceae*) maintain lower transpiration resistances, and higher transpiration fluxes, than in the neighbouring tissues of the host and thereby obtain a disproportionate share of the solutes dissolved in the host's xylem sap (Ehleringer *et al.* 1985). Dodders (*Cuscutaceae*) maintain a low concentration of organic solutes in the apoplast around their haustoria; the result is an enhanced rate of phloem unloading and of phloem translocation by the host to that area (Wolswinkel *et al.* 1984). A dramatic variant of vascular parasitism occurs in some naturally root-grafted plants: Epstein (1978) discusses a case in which 22 years after a forest of Douglas firs was selectively felled, 23% of stumps were still alive, preserved by translocates reaching them via root grafts to the remaining trees.

These examples suggest that the system of connections between modules, both xylem and phloem, is not strongly protected against the development of strong interactions and exploitation by 'selfish' individuals. Why have such vulnerabilities not been eliminated during evolution?

PROPOSITION. *A parsimonious strategy by every module towards its neighbours (that is, minimizing the values of $g_{ij}$) would change the general nature of the interactions in the internal environment from 'mutualistic' to 'antagonistic'.*

Law (1985) suggests that an environment of 'antagonistic' relationships is less stable than one of 'mutualistic' relationships. The proposition suggests that the same is true of interactions within the plant. A non-antagonistic rule system allows modules to cooperate in exploiting a patchy distribution of resources (for example, of sun flecks, or mineral nutrients) as in *Linnea borealis* (Antos & Zobel 1985) and to support injured modules (tillers of the tundra grass *Dupontia fischeri* that are injured by grazing lemmings appear to draw assimilates from less severely damaged neighbours (Mattheis *et al.* 1976)). However,

PROPOSITION. *Strong pressures for selection against 'selfish' interactions would be expected where these pose a consistent threat to the inclusive fitness of the genome, that is, where they are strongly expressed (large values of the parameters $g_{ij}$) and occur in every generation.*

Attack by a phloem parasite is a relatively rare event: as a result phloem parasites do not exert any great selective pressures for change in the phloem unloading system. The mechanism of unloading across the apoplast to the parasite *Cuscuta* is, however, very similar to the physiology of phloem unloading to ovules (Wolswinkel 1984); ovules have a different genotype to the mother plant, and are produced in every generation. It has been suggested that the result of these pressures has been the evolution by the seed plants of the integuments and the endosperm, whose function, it is suggested, is to control phloem unloading from parent to offspring, offset the trend towards 'selfishness' in the offspring, and organize the death of 'surplus' ovules (Queller 1984). Thus to avoid large values of $g_{ij}$, a module may be 'aborted'. This leads to the following proposition.

### DEATH OF MODULES

PROPOSITION. *The threat of 'selfish' interactions is controlled by organized senescence of meristems or modules.*

In tree stumps supported by root grafts, a disadvantaged individual can apparently be maintained by phloem translocates indefinitely (Epstein 1978). Interactions between advantaged and disadvantaged modules of the same plant are organized more strongly. The first analyses of the development of the leaf canopy assumed that shaded leaves at the bottom of the canopy would be retained indefinitely, and that the increasing drain of resources would set an upper limit to the effective leaf area index. But observation showed that leaves that have been shaded do not persist: their maintenance respiration reduces, and eventually they senesce and die. Similarly, the lower branches of forest trees do not persist indefinitely; they 'age', in response to the shifting pattern of distribution of mineral nutrients within the tree (Moorby & Wareing 1963), itself a function of the tree's hydraulic architecture. Eventually, branches die, in an organized fashion, falling from a pre-existing collar and leaving an articulation zone or socket. Branches of a large tree may die in light intensities in which small trees of the same species stay alive (Went 1973) (this establishes a clean trunk below the morphological inversion point (Hallé *et al.* 1978)). In *Lolium multiflorum*, dependent tillers are maintained only for a short time, then they die (Ong & Marshall 1979). Climbing plants continually move their substance upwards so as to reach the light, the lower leaves dying in an organized fashion. In these and other cases of organized senescence there may appear to be a link with other processes, such as the development of self-shading, or of internal shortages of resources; but the correlation is not necessarily causal. It seems rather that module death is a canalysed and adaptive feature, conferring fitness on the genome, and that any correlation with other events is coincidental (Hardwick 1983).

From the point of view of plants as populations of modules (Bazzaz & Harper 1977), the death of meristems and modules is an important determinant of growth. Plant structure and agricultural yield in species as diverse as *Coffea* and *Pisum* can be described as a function of meristem and module death rates (Reffye 1981; Hardwick 1985). Leaf senescence is under nuclear control (Woolhouse 1982), but is subject to internal regulation; for example, the rate of leaf senescence can be accelerated or retarded by exogenous application of growth regulators, calcium and polyamines. In many species leaf longevity is either invariant or varies smoothly along the axis, so that zones of senescing and abscissing leaves tend to progress smoothly along each axis (sequential senescence). In monocarpic senescence, on the other hand, all leaves tend to senesce more or less simultaneously. Here a close correlation has often been demonstrated between leaf longevity (leaf area duration) and yield. As in other aspects of modular growth, we have only a poor understanding of the mechanisms involved.

## Discussion

In this paper I have attempted to explore the rule systems, and derive some principles, of modular growth. Many details are still lacking, so that at present the propositions have the status of conjecture or hypothesis, rather than axiom or law. A recurring feature has been that modular plants exhibit 'structural stability' (Thom 1972). The result is that plants grown in the same environment are 'similar but not the same' (Nishida 1980). If the modular plant is a physiologically stable system, how should interactions between modules be described? Many plant physiologists have applied the notion of competition and competitiveness to describe relationships between modules. But successive modules have the same genome and are means to the same end, that is, the propagation of that genome. It may therefore not be helpful to imply that modules 'strive' against one another. Even where one module appears to obtain material benefit at the expense of a second (for example an upper leaf benefits, a lower one abscisses; a proximal fruit grows, a distal fruit aborts) it is arguable that the interaction should be described in terms of 'cooperation' rather than 'competition' (cooperative interactions maximize the inclusive fitness of the two modules involved).

'Modularizing the small detached house', concluded the visionary architect Albert Farwell Bemiss (1936) 'will give not only the benefits of mass production but those of superior architectural talent to the hundreds of thousands who will live in these homes... Mass productive methods have come to stay, because they are simply the further development of the division of labour'. In higher plants, as in houses, it seems there were benefits to be gained from 'superior architectural talent', and from 'division of labour'; mass production through modularity of construction arose early in evolution and has evidently come to stay. To the question *cui bono?* (who profits by it? who is most likely to have brought it about?), the answer is that modular growth is brought about by the genome, and the profits of modular architecture accrue to the genome. Physiological mechanisms provide the means by which the genome achieves and maintains a modular phenotype; the end is evolutionary fitness.

## References

Antos, J. A. & Zobel, D. B. 1985 Ecological implications of below ground morphology of nine coniferous forest herbs. *Bot. Gaz.* **145**, 508–517.

Barnes, A. S. 1979 Vegetable plant part relationships. II. A quantitative hypothesis for shoot/storage root development. *Ann. Bot.* **43**, 487–499.

Battey, N. H. & Lyndon, R. F. 1984 Changes in apical growth and phyllotaxis on flowering and reversion in *Impatiens balsamina* L. *Ann. Bot.* **54**, 553–567.
Bazzaz, F. A. 1979 The physiological ecology of plant succession. *A. Rev. Ecol. Syst.* **10**, 351–371.
Bazzaz, F. A. & Harper, J. L. 1977 Demographic analysis of the growth of *Linum usitatissimum*. *New Phytol.* **78**, 193–208.
Beck, C. B., Schmid, R. & Rothwell, G. W. 1982 Stelar morphology and the primary vascular system of seed plants. *Bot. Rev.* **48**, 691–815.
Bell, A. D. 1979 The hexagonal branching pattern of rhizomes of *Alpinia speciosa* L. (Zingiberaceae). *Ann. Bot.* **43**, 209–223.
Bemiss, A. F. 1936 *The evolving house*, vol. 3, *Rational design*. Cambridge, Mass.: The Technology Press.
Berg, A. R. & Cutter, E. G. 1969 Leaf initiation rates and volume growth rates in the shoot apex of *Chrysanthemum*. *Am. J. Bot.* **56**, 153–159.
Biddington, N. L. 1986 Mechanically induced stress in plants: a review. *Pl. Growth Regul.* (In the press.)
Blackman, P. G. & Davies, W. J. 1985 Root to shoot communication in maize plants of the effects of soil drying. *J. exp. Bot.* **36**, 39–48.
Britten, R. J. & Davidson, E. H. 1969 Gene regulation for higher cells: a theory. *Science, Wash.* **165**, 349–357.
Brouwer, R. 1983 Functional equilibrium: sense or nonsense? *Neth. J. agric. Sci.* **31**, 335–348.
Brown, E. G. & Newton, R. P. 1981 Cyclic AMP and higher plants. *Phytochemistry* **20**, 2453–2463.
Canny, M. J. 1984 Translocation of nutrients and hormones. In *Advanced plant physiology* (ed. M. B. Wilkins), pp. 277–296. London: Pitman.
Canny, M. J. 1985 Ashby's law and the pursuit of plant hormones: a critique of accepted dogmas, using the concept of variety. *Aust. J. Pl. Physiol.* **12**, 1–7.
Charles-Edwards, D. A. 1984 On the ordered development of plants. I. An hypothesis. *Ann. Bot.* **53**, 633–707.
Cherfas, J. 1985 When is a tree more than a tree? *New Scient.* **106**, 42–45.
Church, A. H. 1920 *On the interpretation of phenomena of phyllotaxis*. Botanical memoirs number 6, pp. 1–58. Oxford University Press.
Cohen, D. 1971 Maximising final yield when growth is limited by time or by limiting resources. *J. theor. Biol.* **33**, 299–307.
Cook, M. G. & Evans, L. T. 1983 The roles of sink size and location in the partitioning of assimilates in wheat ears. *Aust. J. Pl. Physiol.* **10**, 313–327.
Corner, E. J. H. 1967 On thinking big. *Phytomorphology* **17**, 24–28.
Currah, I. E. & Barnes, A. 1979 Vegetable plant part relationships. I. Effects of time and population density on the shoot and storage root weights of carrot (*Daucus carota* L.). *Ann. Bot.* **43**, 475–486.
Deregibus, V. A., Sanchez, R. A., Casal, J. J. & Trlica, M. J. 1985 Tillering responses to enrichment of red light beneath the canopy in a humid natural grassland. *J. appl. Ecol.* **22**, 199–206.
Durrant, A. 1962 The environmental induction of heritable change in *Linum*. *Heredity, Lond.* **17**, 27–61.
Ehleringer, J. R., Schulze, E. D., Ziegler, H., Lange, O. L., Farquhar, G. D. & Cowan, I. R. 1985 Xylem-tapping mistletoes: water or nutrient parasites? *Science, Wash.* **227**, 1479–1481.
Epstein, A. H. 1978 Root graft transmission of tree pathogens. *A. Rev. Phytopathol.* **16**, 181–192.
Erickson, R. O. 1983 The geometry of phyllotaxis. In *The growth and functioning of leaves* (ed. J. E. Dale & F. L. Milthorpe), pp. 53–88. Cambridge University Press.
Ewers, F. W. & Zimmerman, M. H. 1984 The hydraulic architecture of eastern hemlock (*Tsuga canadensis*). *Can. J. Bot.* **62**, 940–946.
Feigenbaum, M. J. 1980 Universal behaviour in nonlinear systems. *Los Alamos Science*, no. 1 (summer 1980), pp. 4–27.
Fisher, J. B. & Stevenson, J. W. 1981 Occurrence of reaction wood in branches of dicotyledons and its role in tree architecture. *Bot. Gaz.* **142**, 82–95.
Gardner, M. R. & Ashby, W. R. 1970 Connectance of large dynamic (cybernetic) systems: critical values for stability. *Nature, Lond.* **228**, 784.
Gray, D., Ward, J. A. & Steckel, J. R. A. 1984 Endosperm and embryo development in *Daucus carota* L. *J. exp. Bot.* **35**, 459–465.
Groenendael, J. M. van 1985 Teratology and metameric plant construction. *New Phytol.* **99**, 171–178.
Hallé, F., Oldeman, R. A. A. & Tomlinson, P. B. 1978 *Tropical trees and forests: an architectural analysis*. Berlin: Springer-Verlag.
Hardwick, R. C. 1983 Why do seed legumes lose their leaves? In *Interactions between nitrogen and growth regulators in the control of plant development* (ed. M. B. Jackson), pp. 61–74. British Plant Growth Regulator Group, monograph 9, 1983.
Hardwick, R. C. 1985 Yield components and processes of yield production in vining peas. In *The pea crop: a basis for improvement* (ed. P. D. Hebblethwaite, M. C. Heath & T. C. K. Dawkins), pp. 317–327. London: Butterworth.
Hardwick, R. C. & Milbourn, G. M. 1967 Yield analysis in the vining pea. *Agric. Prog.* **42**, 24–31.
Heslop-Harrison, J. 1964 Sex expression in flowering plants. *Brookhaven Symp. Biol.* **16**, 109–122.

Hillman, J. R. 1984 Apical dominance. In *Advanced plant physiology* (ed. M. B. Wilkins), pp. 127–148. London: Pitman.

Holmes, M. G. 1983 Perception of shade. *Phil. Trans. R. Soc. Lond.* B **303**, 503–521.

Honda, H. & Fisher, J. B. 1978 Tree branch angle: maximizing effective leaf area. *Science, Wash.* **199**, 888–890.

Honda, H., Tomlinson, P. B. & Fisher, J. B. 1981 Computer simulation of branch interaction and regulation by unequal flow rates in botanical trees. *Am. J. Bot.* **68**, 569–585.

Iwasa, Y., Cohen, D. & Leon, J. A. 1984 Tree height and crown shape, as results of competitive games. *J. theor. Biol.* **112**, 279–297.

Jaffé, L. F. & Nuticelli, R. 1977 Electrical controls of development. *A. Rev. Biophys. Bioeng.* **6**, 445–476.

Klekowski, E. J. & Kazarinova-Fukshansky, N. 1984 Shoot apical meristems and mutation: fixation of selectively neutral cell genotypes. *Am. J. Bot.* **71**, 22–27.

Kumazawa, M. & Kumazawa, M. 1971 Periodic variations of the divergence angle, internode length and leaf shape, revealed by correlogram analysis. *Phytomorphology* **21**, 376–389.

Lambers, H. 1983 'The functional equilibrium', nibbling on the edges of a paradigm. *Neth. J. agric. Sci.* **31**, 305–311.

Law, R. 1985 Evolution in a mutualistic environment. In *The biology of mutualism: ecology and evolution* (ed. D. H. Boucher). London: Croom Helm.

Lindenmayer, A. 1982 Developmental algorithms: lineage versus interactive control mechanisms. In *Developmental order: its origin and regulation* (ed. S. Subtelny & P. B. Green), pp. 219–249. New York: Liss.

Lindenmayer, A. 1984 Positional and temporal control mechanisms in inflorescence development. In *Positional controls in plant development* (ed. P. W. Barlow & D. J. Carr), pp. 461–486. Cambridge University Press.

Lintilhac, P. M. & Vesecky, T. B. 1984 Stress-induced alignment of division plane in plant tissues grown *in vitro*. *Nature, Lond.* **307**, 363–364.

Lovatt, C. J. 1985 Evolution of xylem resulted in a requirement for boron in the apical meristems of vascular plants. *New Phytol.* **99**, 509–522.

MacMahon, T. A. & Kronauer, R. E. 1976 Tree structures: deducing the principle of mechanical design. *J. theor. Biol.* **59**, 443–466.

Mattheis, P. J., Tieszen, L. L. & Lewis, M. C. 1976 Responses of *Dupontia fischeri* to simulated lemming grazing in an Alaskan Arctic tundra. *Ann. Bot.* **40**, 179–197.

May, R. M. 1972 Will a large complex system be stable? *Nature, Lond.* **238**, 413–414.

Meicenheimer, R. D. 1981 Changes in *Epilobium* phyllotaxy induced by $N$-1-naphthylphthalamic acid and $\alpha$-4-chlorophenoxyisobutyric acid. *Am. J. Bot.* **68**, 1139–1154.

Meins, F. & Foster, R. 1985 Reversible, cell heritable change during the development of tobacco pith tissues. *Devl Biol.* **108**, 1–5.

Milford, G. F. J., Pocock, T. O. & Riley, J. 1985 An analysis of leaf growth in sugar beet. 1. Leaf appearance and expansion in relation to temperature under controlled conditions. *Ann. appl. Biol.* **106**, 163–172.

Mitchison, G. J. 1980 A model for vein formation in higher plants. *Proc. R. Soc. Lond.* B **207**, 79–109.

Moorby, J. & Wareing, P. F. 1963 Ageing in woody plants. *Ann. Bot.* **27**, 291–308.

Münch, E. 1938 Investigations on the harmony of tree shape. *Jb. wiss. Bot.* **86**, 581–673.

Murfet, I. C. & Reid, J. B. 1985 The control of flowering and internode length of *Pisum*. In *The pea crop: a basis for improvement* (ed. P. D. Hebblethwaite, M. C. Heath & T. C. K. Dawkins), pp. 67–80. London: Butterworth.

Netzer, W. J. 1984 Appropriate bean gene expression reported. *Biotechnology* **2**, 596.

Nishida, T. 1980 KOL system simulating almost but not exactly the same development. Case of Japanese cypress. *Mem. Fac. Sci. Kyoto Univ. Ser. Biol.* **8**, 97–122.

Ong, C. K. & Marshall, C. 1979 The growth and survival of severely-shaded tillers in *Lolium perenne* L. *Ann. Bot.* **43**, 147–155.

Queller, D. C. 1984 Models of kin selection on seed provisioning. *Heredity, Lond.* **53**, 151–165.

Raven, J. A. 1977 The evolution of vascular land parts in relation to supracellular transport processes. *Adv. Bot. Res.* **5**, 153–219.

Reffye, Ph. de 1981 Modèle mathématique aléatoire et simulation de la croissance et de l'architecture du caféier robusta. 2. Étude de la mortalité des méristèmes plagiotropes. *Café–Cacao–Thé* **25**, 219–230.

Ridley, J. N. 1982 Computer simulation of contact pressure in capitula. *J. theor. Biol.* **95**, 1–11.

Rogers, W. S. & Beakbane, A. B. 1957 Stock and scion relations. *A. Rev. Pl. Physiol.* **8**, 217–236.

Romberger, J. A. & Gregory, R. A. 1977 The shoot apical ontogeny of the *Picea abies* seedling. III. Some age-related aspects of morphogenesis. *Am. J. Bot.* **64**, 622–630.

Sachs, T. 1981 The control of the patterned differentiation of vascular tissues. In *Advances in botanical research* (ed. H. W. Woolhouse), pp. 152–255. London: Academic Press.

Salleo, S., Lo Gullo, M. A. & Oliveri, F. 1985 Hydraulic parameters measured in 1-year-old twigs of some Mediterranean species with diffuse-porous wood: changes in hydraulic conductivity and their possible functional significance. *J. exp. Bot.* **36**, 1–11.

Scowcroft, W. R. 1984 *Genetic variability in tissue culture: impact on germplasm conservation and utilisation*. Rome: International Board for Plant Genetic Resources.

Sheldrake, A. R. 1973 Do coleoptile tips produce auxin? *New Phytol.* **72**, 433–447.

Smith, B. H. 1984 The optimal design of a herbaceous body. *Am. Nat.* **123**, 197–211.
Stange, L. 1984 Cellular interactions during early differentiation. In *Cellular interactions* (ed. H. F. Linskens & J. Heslop-Harrison), pp. 424–452. Berlin: Springer Verlag.
Starling, R. J. 1984 The question of plant hormone binding sites. *Trends biochem. Sci.* **9**, 48–49.
Sussex, I. M. 1983 Determination of plant organs and cells. In *Genetic engineering of plants: an agricultural perspective* (ed. T. Kosuge, C. P. Meredith & A. Hollaender), pp. 443–451. New York: Plenum.
Thom, R. 1972 *Stabilité structurelle et morphogénése*. New York: Benjamin.
Thompson, W. D'A. 1917 *On growth and form*. Cambridge University Press.
Thornley, J. H. M. 1977 A model of apical bifurcation applicable to trees and other organisms. *J. theor. Biol.* **64**, 165–176.
Tomlinson, P. B. 1982 Chance and design in the construction of plants. In *Axioms and principles of plant construction* (ed. R. Sattler), pp. 162–183. The Hague: Martinus Nijhoff/Dr W. Junk Publishers.
Trewavas, A. J. 1982 The regulation of development and its relation to growth substances. *What's new Pl. Physiol.* **13**, 41–43.
Went, F. W. 1973 Competition among plants. *Proc. natn. Acad. Sci. U.S.A.* **70**, 585–590.
Whitham, T. G. & Slobodchikoff, C. N. 1981 Evolution by individuals, plant–herbivore interactions, and mosaics of genetic variability: the adaptive significance of somatic mutations in plants. *Oecologia* **49**, 287–292.
Wilson, B. F. 1966 Development of the shoot system of *Acer rubrum* L. *Harvard Forest Paper* **14**, 1–21.
Wit, C. T. de & Penning de Vries, F. W. T. 1983 Crop growth models without hormones. *Neth. J. agric. Sci.* **31**, 313–323.
Wolpert, L. 1981 Positional information and pattern formation. *Phil. Trans. R. Soc. Lond.* B **295**, 441–450.
Wolswinkel, P. 1984 Phloem unloading and 'sink strength': the parallel between the site of attachment of *Cuscuta* and developing legume seeds. *Pl. Growth Regul.* **2**, 309–317.
Wolswinkel, P., Ammerlaan, A. & Peters, H. F. C. 1984 Phloem unloading of amino acids at the site of attachment of *Cuscuta europaea*. *Pl. Physiol.* **75**, 13–20.
Woolhouse, H. W. 1982 Leaf senescence. In *The molecular biology of plant development* (ed. H. Smith & D. Grierson), pp. 256–284. Oxford: Blackwell.
Worrall, J. 1984 Predetermination of lateral shoot characteristics in grand fir. *Can. J. Bot.* **62**, 1309–1315.
Zimmermann, M. H. 1978 Hydraulic architecture of some diffuse-porous trees. *Can. J. Bot.* **56**, 2286–2295.

# From aggregates to integrates: physiological aspects of modularity in colonial animals

By G. O. Mackie

*Biology Department, University of Victoria, Victoria, British Columbia, Canada V8W 2Y2*

[Plate 1]

Physiological adaptations of colonies in the areas of resource-sharing, defensive behaviour and locomotion are reviewed and discussed.

Exchange of nutrients between zooids is a fundamental attribute of true colonies, although the transport systems involved vary and, in some cases, complex junctional structures appear to regulate passage of materials between zooids, as in the parallel case of interconnected cell systems. It is not known if regulatory molecules, as well as nutrients, pass via these interzooidal pathways.

Almost all colonies show some form of behavioural coordination but, again, the pathways employed vary considerably. Examples of coordination by nerves, by conducting epithelia and by photic signals, are given. Zooids typically retain a high degree of local autonomy and the interzooidal pathways serve more as labile links between the action systems of the zooids than as centres where behaviour is initiated. Siphonophores are exceptional in having nerves, in their stems, which initiate as well as relay behaviour.

Defensive responses are the commonest form of coordinated activity in colonies, but some colonies show coordinated locomotion. The most striking examples are those salps and siphonophores that regulate the activity of their locomotory modules either in terms of frequency of pulsation or in terms of direction of locomotion, or both. Modularity has theoretical mechanical advantages in animals swimming by jet propulsion.

The requirements of pelagic locomotion have led planktonic colonies to diverge markedly from the pattern seen in sessile colonies. Pelagic colonies, unlike sessile ones, are typically linear and unbranched, show an anterior–posterior axis, are determinate in form, at least at the anterior pole, and lack much of the plasticity and regenerative ability that characterize sessile colonies.

The significance of coloniality has been much discussed in the past and is once again the focus for research and debate, chiefly among population biologists. A new way of looking at clones and colonies has emerged, and along with it a new language, and those of us who participated in the earlier debate must come to terms with these developments. I will approach the topic, therefore, by reviewing some of the prevailing ideas from my standpoint as a classically trained zoologist and as a behavioural physiologist with a special interest in gelatinous marine macroplankton.

## 1. The nature of colonies: old and new ideas

### (a) Individuals and modules

Where now we speak of modules, the operative word used to be 'individuals', and discussions of coloniality often centred around levels and degrees of individuality (see, for example, Beklemishev 1969). Going up the scale of complexity from cell to tissue to multicellular organism to colony or society we find individuality expressed at each successive level, along with a tendency for the individualities of the lower order units to be suppressed or submerged. The cell (itself probably derived from a symbiotic association of microorganisms) tends to lose its individuality in multicellular associations, just as the individuality of the zooid is subordinated to that of the colony of which it forms a part, or that of the unitary animal to that of the society.

These ideas go back at least to the time of Haeckel, who recognized six levels of morphological integration (see White 1984). Most writers have couched the argument in terms of an evolutionary progression from lower to higher organizational levels. As Dendy (1924), for instance, expressed it 'evolution consists to a very large extent, if not mainly, in the progressive merging of individualities of a lower order in others of a higher order'. One might quibble about 'if not mainly', but most people would probably accept Dendy's comment as a valid expression of an important biological truth. The idea is useful also because it encourages comparison of processes occurring at different organizational levels, a point touched upon below (§3) in reference to intercellular and interzooidal communication.

If, then, as aggregates progressed along the road toward integration they began to acquire a new individuality of their own, we must still agree with Herbert Spencer (1898) who remarked that 'it would be impossible to say where the lower individualities ceased and the higher individualities commenced', and that 'there is no definition of individuality that is unobjectionable' (he went on to define it anyway!). Attempts to classify animals in order of structural complexity according to their supposed level of individuality can produce bizarre results. In Cattaneo's (1879) scheme, for example, tapeworms are put above molluscs. Many other difficulties surround use of the word 'individual' (Mackie 1963), not least its failure to distinguish between the products of replicative growth and the products of sexual reproduction (Harper & Bell 1979). Thus, while the general concept (individuality) is useful, we run into problems when it comes to calling specific objects individuals.

The same is true with modularity and modules. A module is a product of iterative growth. The term is happily free of evolutionary and metaphysical overtones. Originally used in a restricted sense to describe the products of a plant apical meristem, it has now acquired a much broader range of possible meanings, especially in reference to animals. 'Module' could be used to describe cells, developmental compartments, segments, metamers, tagmata, cormidia, cyclosystems, chimney systems and members of clones, colonies and societies. Modules need not be attached to one another physically and they need not even be genetically identical. Rosen (1979), for instance, regards insect colonies as 'discontinuous modular societies'. Thus, while modularity is valuable as an expression of a mode of growth, use of 'module' itself is hazardous unless the identity of the particular module under discussion is made unmistakeably clear. White (1984) would like to see a return to a more restrictive definition of the word. This seems a forlorn hope. 'Module' is firmly established in a general sense and cannot now be used to replace any of the old descriptive terms applying to specific morphological entities.

The modules I will be referring to in this article are the zooids in animal colonies. Zooids have these characteristics:

(i) They arise by budding from single oozooids and are therefore parts of single genets. Fusion of colonies representing different genets may subsequently occur producing a chimeric colony, no different in principle from a chimera made by combining blastomeres from sheep and goat embryos (Fehilly et al. 1984).

(ii) Zooids are organically linked in such a way that direct exchange of metabolites can occur. A problem here is lack of information about exactly what if any exchange takes place in specific cases. Some compound acidians (for example, *Botryllus*, *Perophora*) are truly colonial, being linked by a common blood vascular system. Others show no such obvious pathways of metabolic exchange and might better be regarded as mere aggregates of unitary organisms.

(iii) With few exceptions, zooids in colonies are coordinated behaviourally by nerves, conducting epithelia, or both.

(iv) Zooids evolved from single unitary organisms and are equivalent to them in terms of primary individuality, however specialized they may have become, and however subordinate to the emergent colonial individuality. Thus 'zooid' has a strong evolutionary connotation. It is not always possible to decide if a particular structure is a true zooid. Bracts and pneumatophores in siphonophores and spines in bryozoans may be colonial neomorphs, but are often regarded as zooids.

Metamerically segmented animals are probably not colonies or the derivatives of colonies, and it is doubtful if metamers are or ever were zooids in the sense defined above, unless the corm theory of annelid evolution is correct. Accordingly, this discussion will not cover metamerically segmented animals. Though still problematical in botany (White 1984) the term metamer has found a stable niche in zoological usage meaning essentially the same as 'segment'. Little is to be gained by thinking of zooids as metamers, and we must hope that Harper's dalliance with this idea (Harper 1984) fails to develop into a full-blown affair.

### (b) *Benthic versus pelagic colonies: do the same rules apply?*

An important literature has grown up on the adaptive significance of animal modularity, which is concerned almost exclusively with plant-like sessile colonies: sponges, cnidarians, bryozoans and ascidians in particular (see, for instance, Buss 1979; Chapman & Stebbing 1980; Hughes & Jackson 1985; Jackson 1977, 1979). These colonies share a number of common characteristics, which distinguish them from solitary forms.

(i) They often show indeterminate growth, with the potential for exponential increase in numbers of modules, and hence for rapid expansion over available substrates. Rapid exploitation of substrate is also presumably the rationale for coloniality in various parasitic worms (*Echinococcus granulosus*, *Taenia crassiceps*, etc.) as discussed by Beklemishev (1969). The phenomenon of polyembryony in parasites comes in the same category, along with such oddities as *Thompsonia*, the colony-forming parasitic barnacle.

(ii) Following from (i), many colonial genets achieve large size, wide dispersal and extreme longevity. Corals, for example, can be far larger than solitary anthozoans. They have a high surface area:volume ratio, as they consist of many small zooids covering a skeleton which they secrete. Size increase is not accompanied by changes in this ratio as would be the case, along with various attendant problems (see Strathmann & Chaffee 1984) in solitary forms. The ability of corals and other sessile colonies to fragment and for the fragments to fuse and/or regenerate imparts the potential for extensive dispersal and for indefinite survival. Some corals living today may be over 1000 years old, and corals living in the past when sea levels were more stable may have lived many times longer (Potts 1984).

(iii) Modular growth typically produces a branching colony structure. Branching ensures that the modules are spaced out or packed together in some way that automatically 'regulates competition between zooids' (Knight-Jones & Moyse 1961), and is presumably 'the most effective arrangement for feeding' (Bayer 1973). This may of course be assisted by active orientation movements of the zooids, rather than being merely a product of patterned or tropistic growth processes.

(iv) Plasticity is an obvious characteristic of many sessile colonies. They can assume a variety of sizes and shapes depending on spatial and temporal variations in the substrate. The ectoproct *Stomatopora*, when displaced from its original substrate by a bivalve, grows up and over the shell of the bivalve (Jackson 1977). *Electra pilosa*, which grows on *Fucus* fronds, finds refuge at the tip of the frond, keeping up with the latter's growth by extending runners in that direction (Stebbing 1973).

Not only size and shape vary, but the disposition of different zooids can sometimes be adjusted to local conditions, as in *Hydractinia*, where spiral zooids are produced only around the opening of the shell it lives on. Some ascidian colonies and calyptoblastic hydroids resorb and replace their zooids according to a regular cycle, a process of continual rejuvenation (Gordon 1977) which also permits rapid changes in zooid numbers and distribution (Sabbadin 1979).

If a sessile colony hits hard times or suffers partial predation it can contract, expand again in new directions, replace missing modules or pursue a number of other strategies made possible by its plastic properties. Degenerate zooids are not wasted, but provide nutrients for new growth (Werner 1979).

(v) The existence of tissue connections between the zooids in colonies nearly always allows transmission of behavioural signals warning of danger or disturbance. Thus, zooids far from the site of impact can protect themselves in advance.

When we come to look at pelagic colonies we find a somewhat different picture. As with sessile colonies, defensive reactions are highly developed in pyrosomes, salps and siphonophores but these often take the form of escape locomotion, because all pelagic colonies (cystonectid siphonophores excepted) are highly specialized for swimming by jet propulsion. In keeping with their locomotory style of existence, pelagic colonies are linear and polarized. Branching is minimal (doliolids, siphonophores) or absent (salps, pyrosomes). Doliolid chains are budded from the rear end of the oozooid ('nurse') stage, salp blastozooids from a stolon produced by the oozooid; siphonophores bud their nectophores from one budding zone and their other appendages from another, but both budding zones are incorporated into a single linear stem, which is derived from the oozooid (figure 1). Pyrosomes are cylinders closed at one end, open at the other. Streamlining sometimes reaches a high pitch of perfection, as in diphyid siphonophores. Some salp chains and siphonophores show bilateral symmetry. There is always a front and a back end.

Thus we find that one end of the colony has a well-defined shape and may consist of one or a fixed number of specialized zooids. Indeterminate growth, to the extent that it occurs, is seen at the other (posterior) end where modules are added progressively. Even here, however, detachment of modules may be balanced by production of new ones so that overall body form is maintained. The body proportions of most siphonophores change little as the animal grows even where stem groups detach to become free eudoxids. Calycophores and physonects seem to reach and maintain a certain size, characteristic for each species. Physonects cannot really be said to show indeterminate growth at either end, except possibly in some poorly understood cases (for example, *Apolemia*, which may be over 10 m long).

Pelagic colonies show little plasticity. Siphonophores, doliolids and salps can bud new modules only from their one or two localized budding zones. The only fragment of a physonectid

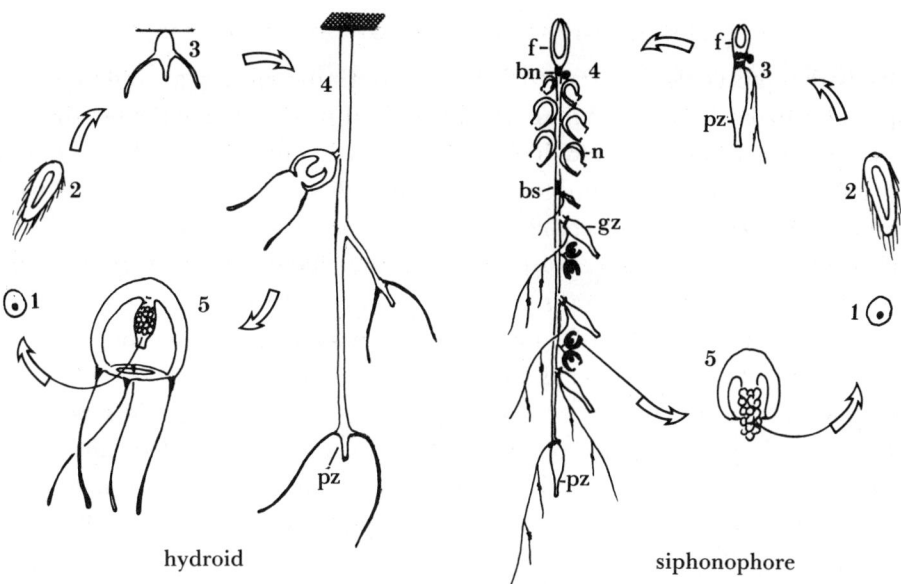

FIGURE 1. Comparison of the life cycle of an agalmid siphonophore with that of an athecate hydroid. In the hydroid cycle, following the egg (1) and planula (2) stages the larva becomes a tentaculated actinula (3) which settles by its aboral pole. The actinula forms the primary polyp or protozooid (pz) of the hydroid colony (4), from whose elongated base secondary polyps and medusae are budded. The medusae (5) are retained in some species, set free in others, as here. They produce and release the gametes. In the siphonophore cycle, the siphonula larva (3) can be regarded as a type of actinula but it forms a protozooid with only one tentacle, and the aboral end, instead of attaching to a substrate, becomes invaginated as a float (f). Elongation of the middle section of the larva produces a long stolon, the stem of the siphonophore colony (4), with the protozooid at its distal tip. The stem buds off sexual medusae (5) (which are usually retained, but are sometimes set free) and sterile locomotory medusoids, or nectophores (n). Several sorts of polyp are also budded off, of which only gastrozooids (gz) are shown. The stem elongates and buds from two blastogenic zones, one for the nectophores (bn) and one for the gastrozooids and other siphosomal appendages (bs). Most specialists now regard the float as a neomorphic structure, not as zooid of medusoid origin. Tentacles in siphonophores are parts of gastrozooids or palpons, not separate 'dactylozooids'.

siphonophore that could regenerate to produce a perfect colony would be a piece of stem containing both budding zones as well as the float. Isolated zooids cannot produce new colonies. Pelagic colonies do not fuse, vary their shapes, shrink and re-expand or switch their zooids around.

Salps and certain pelagic polychaetes (for example, *Trypanosyllis*) come closest to fitting the paradigm established for benthic colonies. In both cases, though linear and unbranched, they show unipolar indeterminate growth, producing chains of zooids which separate either as shorter chains or as solitary zooids. Phenomenal rates of increase are reported for salps under phytoplankton bloom conditions (Heron 1972). The Portuguese man-of-war 'is distinguished from every other siphonophore by the complexity of branching by budding of the cormidia' (Totton 1960). It thus somewhat resembles sessile, benthic colonies. However, it is not truly pelagic, and cannot swim. It lives at the air–water interface and can be regarded as sessile at the surface. Thus, its resemblance to sessile benthic colonies is not so surprising. The peculiar branching syllid, *Syllis ramosa*, likewise is not truly pelagic, but parasitizes sponges.

## 2. Physiological aspects of coloniality

It seems likely that most colonies evolved from clone-forming solitary ancestors (see, however, Silén (1944) on bryozoan origins). Where the rate of bud production exceeded the rate of bud separation, temporary colonies would have been formed, much as in *Hydra*, *Loxosoma* and *Cephalodiscus* today. Having thus stumbled upon coloniality, the ancestral forms would have found advantages in remaining interconnected, perhaps to monopolize the substrate. The retention of primary tissue connections between the zooids must have had important consequences. By making possible transfer of metabolites between the zooids it would have allowed some colonies to evolve non-feeding zooids which could become specialized for other tasks. Secondly, it would have provided pathways for the direct transmission of behavioural responses, so that certain functions, particularly defensive responses, could spread across the colony. Even in highly polymorphic colonies such as siphonophores, where the physiological functions are distributed among several different zooid types, we still see a high degree of behavioural coordination in the colony as a whole, thanks to the presence of a colony-wide network of nervous connections.

Beyond this point, it is hard to generalize about colonial evolution. The picture is not a tidy one. Each group has gone its own way. Not all colonies became polymorphic, and of those that did, not all are strikingly well integrated. The most dramatic examples of behavioural integration are to be found in certain free-swimming pelagic colonies, but the actual mechanisms of coordination involved differ from group to group. We can only review the facts as they apply to different present-day colonies and consider how their zooids share resources, communicate with one another and collaborate in common tasks.

### (a) Sharing of resources

Feeding may be carried on independently by individual zooids or be organized on the colonial level, for instance in the formation of chimneys in ectoproct bryozoans (Lidgard 1981) and in the collaborative activity of several different sorts of zooid in stylasterine cyclosystems (de Kruijf 1977). The mere ability to deploy large numbers of relatively small feeding zooids within a confined space may enable predatory colonies to capture very large prey. This is seen in the Portuguese man-of-war, which can capture and digest whole mackerel. Although it cannot ingest the fish in the usual sense, it can envelop it completely in the expanded mouths of numerous gastrozooids, which press up tightly to one another, thus creating an improvised external stomach wherein extracellular digestion takes place (Hardy 1956) (figure 2). Biggs (1977) reports similar observations for *Rhizophysa*.

In cnidarian colonies the gut cavity is continuous throughout the colony and transport of nutrients is simple and rapid. Rees (1971) found that $^{14}$C-labelled food spread through small *Pennaria* colonies within 30 min. Rinkevitch & Loya (1983) showed that labelled metabolites underwent translocation from a donor branch of the coral *Stylophora* to a recipient branch after fusion of the two. The label appeared in highest concentrations at the tips of the recipient branches, presumably the most metabolically active sites.

In siphonophores, only the gastrozooids feed, and all nutrients must therefore be transported to the other regions, sometimes over distances of many centimetres or even metres (*Physalia* has tentacles over 6 m long). Mackie & Boag (1963) used carmine-infiltrated food to trace movement of digested material in a siphonophore. Five minutes after feeding, carmine appeared

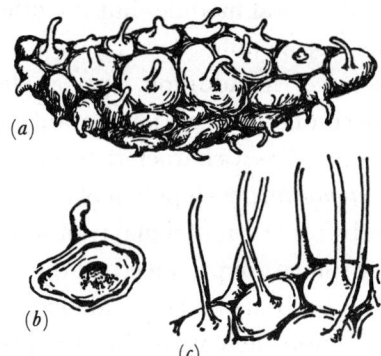

FIGURE 2. Evidence of collaborative feeding in the Portuguese man-of-war (from Hardy 1956). A partly digested fish, forcibly removed from the siphonophore, is shown in (a), from a preserved specimen. The gastrozooids have broken off near their distal ends, but their expanded mouths still envelop the prey. A single such expanded 'mouth' is shown in (b). The condition in life approximates to (c).

in the stem canal and was flushed up and down the stem by rhythmic pumping movements of the gastrozooids and palpons. Cilia lining the canal assist the process. Valves at the bases of the zooids regulate exchange of fluids with the stem. Peristaltic pumping movements assist food dispersal in many hydrozoan colonies (reviewed by Mackie 1973).

Colonial styelid ascidians and *Perophora* have common colonial blood-vascular systems, which presumably serve for metabolite exchange. Pumping structures (ampullae) are present in the styelids. Their contractions are coordinated over large areas by propagated electrical signals (Mackie & Singla 1983a). When two colonies fuse, one of the first indications of fusion is that the ampullary rhythms come into synchrony.

Interzooidal communication in ectoprocts is less straightforward than in the above cases. The polypides of phylactolaemates arise as buds within a common coelom initially, but partitions develop later in some species; the common coelom presumably guarantees dispersal of food metabolites. In some fossil stenolaemates the zooids were entirely walled off from one another but in existing cyclostomes the zooids communicate via holes in the septal walls. The funicular system does not pass through these pores, but it is thought that coelomic fluid can be exchanged. In gymnolaemates pores are again present, but they are plugged by funicular and other tissues. Bobin's (1977) histological studies on *Bowerbankia* show evidence of lipid transport from degenerating autozooids to the stolon, from which new zooids are budded. There are few experimental studies on interzooidal transport in bryozoans, but Best & Thorpe (1985) recently fed $^{14}$C-labelled flagellates to *Membranipora* and found spread of the label to adjacent zooids within 15 min. Spread to a distance of 9 cm (equivalent to about 90 zooids) took place over 24 h. The label accumulated at the growing edge of the colony. It seems likely that the funicular system is the transport pathway.

### (b) *Distant early warning systems*

A solitary organism can receive warning of the approach of a predator or the presence of some other threat only if it has sense organs capable of detecting the danger at a distance; otherwise, it may be mortally injured before it has had time to protect itself. Most marine invertebrates are poorly equipped with respect to such sense organs. Colonies, on the other hand, frequently have built-in signalling systems that allow the zooids to react in anticipation.

Damage at one point may wipe out several modules but the others protect themselves and the genet survives, often to regain the lost modules.

Highly polymorphic colonies have, in addition, special defensive zooids (for example, avicularia). Solitary animals have similar devices (for example, pedicellaria) and possession of defensive structures cannot be regarded as a characteristic of coloniality *per se*. In at least one case, however, the ectoproct *Membranipora*, development of defensive structures (spines) appears to be organized on a colonial basis by some sort of signal transmitted within the colony following attack. Harvell (1984) shows that spines are produced all the way around the periphery even following strictly local attack by a nudibranch. The evidence is most readily explained on the assumption of internal signalling within the colony, but the precise mechanism has not been determined.

In the remaining cases, to be discussed below, we are dealing with behavioural responses involving conventional neuromuscular or other neuroeffector actions transmitted through colonies by electrical impulses in nerves or excitable epithelia. A range of examples illustrating the variety of such responses is provided in table 1.

TABLE 1. EXAMPLES OF PROPAGATED DEFENSIVE RESPONSES IN MODULAR ORGANISMS

| organism | response | conduction system | conduction velocity $mm\ s^{-1}$ | source |
|---|---|---|---|---|
| *Rhabdocalyptus* (hexactinellid sponge) | ciliary arrest | trabecular tissue | 3 | Mackie & Singla 1983*b* |
| *Mimosa* (vascular plant) | leaves fold up | vascular tissue | 10 | Sibaoka 1966 |
| *Membranipora* (ectoproct) | lophophore retraction | nerve net | 10 | Thorpe *et al.* 1975 |
| *Botrylloides* (ascidian) | siphon closure, ciliary arrest | vascular epithelium | 20 | Mackie & Singla 1983*a* |
| *Cordylophora* (hydroid) | polyp retraction | stolon epithelium | 30 | Josephson 1961 |
| *Renilla* (pennatulid) | bioluminescence, polyp withdrawal | nerve net | 60 | Anderson & Case 1975 |
| *Pegea* (young salp chain) | change in direction of locomotion | nerves and epithelia in alternation | 125 | Mackie & Bone 1977 Anderson & Bone 1980 |
| *Chelophyes* (siphonophore) | stem contraction | exumbrellar epithelium | 500 | Mackie & Carré 1983 |
| *Forskalia* (siphonophore) | escape locomotion | giant nerve axon | 4000 | Mackie 1978 |

*(i) Muscular evasive actions*

With few exceptions (for example, *Syncoryne*) both thecate and athecate hydroids show hydranth retraction responses after strong stimulation of the stolon or of other hydranths. Neuronal mediation has been demonstrated in several cases (reviewed by Spencer & Schwab 1982).

In siphonophores, the most obvious propagated defensive response is escape swimming (see below) which is mediated by the twin nerve nets of the central stem of the colony. The reaction can, however, be set off by tactile stimulation of certain areas lacking nerves, such as the epithelia covering the bracts and nectophores. It was this observation that led to recognition of the important role of conducting epithelia in the mediation of protective reactions in colonial hydrozoans and pelagic tunicates (reviewed by Anderson 1980).

Siphonophores also show protective contractions of their zooids, shortening of the stem, elimination of gas bubbles from the float (causing sinking) and involution of the velum and bell margin (in hippopodiids). These responses spread for varying distances depending on the strength and duration of stimulation. Labile neural or epithelioneural couplings link the stem and its attached zooids (figure 11a). Habituation is rapid. Thus the colony as a whole is not obliged to maintain a defensive posture if the warning signals are not of exceptional duration and magnitude. Giant axons in the stems of many siphonophores enable them to respond with lightning speed to damaging stimuli.

Coordinated retractions of coral polyps have been thoroughly studied and are reviewed elsewhere by Horridge (1957) and Shelton (1979, 1982). Local stimulation may cause responses only in the immediate vicinity (*Porites*) or the response may spread through the whole colony (*Tubipora*). In *Porites*, repeated stimuli may be needed to procure maximum spread of the response. In most cases, spread is thought to be mediated by the colonial nerve net.

It is worth stressing that polyp retraction responses in cnidarian colonies operate on a similar basis to retractions in solitary species. Similar muscles and nerve pathways are involved. The colony provides interzooidal conducting pathways that determine *how* the response spreads: locally, generally, incrementally, decrementally, at what threshold, with what fatigue properties, etc. Only in the siphonophores do we find completely new coordinating elements introduced at the colonial level, such as giant nerve axons. The capacity for autotomy has also been evolved, apparently *de nouveau*, as a defensive measure in physonectid siphonophores. Bracts and nectophores have autotomy joints at their junctions with the stem and shed themselves when strongly stimulated.

A colony-wide nerve plexus (Hiller's plexus) is present in cheilostome ectoprocts, confirmed in recent work on *Electra*, summarized by Lutaud (1977), but is absent in the ctenostome *Alcyonidium* (Lutaud 1981). Information is lacking on cyclostomes. Phylactolaemates have long been known to have a colonial nervous system.

The system in cheilostomes consists of fine neurites which run from zooid to zooid via the mural pores, connecting the cerebral ganglia of each zooid. In *Membranipora*, stimulation elicits bursts of electrical events, presumably action potentials in the colonial nerve net (see Thorpe 1982). Giant axons are thought to mediate lophophore retraction and associated muscular events. Lophophore retraction is a very rapid process, and Thorpe's data suggest that the retractor may be one of the fastest contracting muscles known. Surprisingly, it is said to be a smooth muscle, but I have observed the retractors of *Plumatella* and *Fredericella* to be striated. Operculate forms close the operculum as the lophophore is pulled in.

Contrary to earlier reports, Thorpe (1982) finds no good evidence for coordinated lophophore retractions in phylactolaemates. The colonial nerve net might be concerned with other functions, for example, locomotion in the case of *Cristatella*.

Interzooidal communication has been proposed to explain the apparent coordination of vibracular sweeping movements in the cheilostomes *Caberea* and *Cupuladria*. The phenomenal *Selenaria* can uncover itself when covered with sand, and right itself when overturned, by seemingly coordinated movements of its vibracular setae (Cook & Chimonides 1978). The animals are sensitive to light in the blue–green range of the spectrum and move towards it (Cook & Chimonides 1981). Light increases the frequency of small electrical signals resembling nerve impulses, which spread over the surface of the colony, and may be indirectly responsible for the setal movements of locomotion (Berry & Hayward 1984).

(ii) *Ciliary arrest*

Organisms that filter their food from water are generally sensitive to water quality and may arrest or reverse the flow when stimulated by 'bad' water. For a colonial animal, there are clear advantages, not open to solitary forms, in being able to signal the presence of noxious stimuli from zooid to zooid within the colony, so that zooids remote from the stimulus can protect themselves in advance.

*Rhabdocalyptus* and other hexactinellid sponges are well-individualized solitary animals which bud asexually, forming temporary colonies consisting of a parent and several buds in various growth stages. The buds eventually separate. While still attached to the parent the buds are physiologically integrated with it. Hexactinellids, unlike other sponges, show a rapidly propagated ciliary arrest response when stimulated. Excitation spreads through the whole sponge within a few seconds, and in the case of bud colonies, through the whole colony. Sponges are very susceptible to clogging of their canals by suspended sediments and arrest of the feeding current probably helps prevent this. Any local disturbance can set the response off (Mackie *et al.* 1983).

In *Botryllus* and related colonial styelid ascidians, the blood vessels that interconnect the zooids are lined with excitable epithelium, which provides a non-nervous conduction pathway for electrical impulses triggering ciliary arrests in the gills of the zooids (figure 3). On entering the zooid, these signals excite the normal nerve reflexes of squirting, siphon closure and ciliary arrest common to both colonial and solitary ascidians. All that has been added is an interzooidal conduction pathway (Mackie & Singla 1983a).

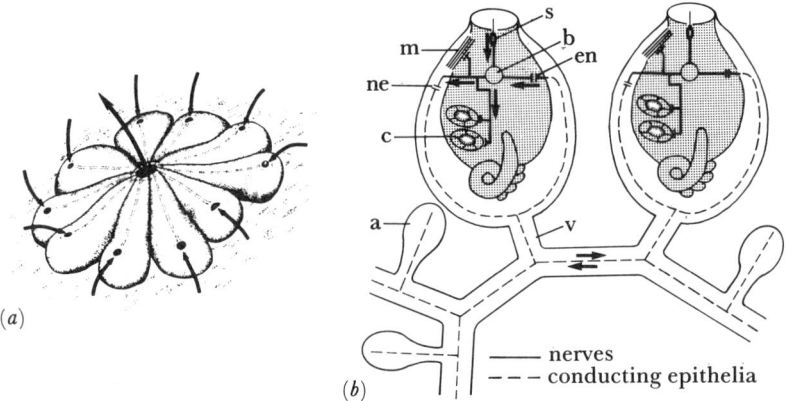

FIGURE 3. *Botryllus*. (*a*) Cyclosystem, with current flow indicated. The zooids share a common cloacal opening. From Vogel 1981. (*b*) Action systems and conduction pathways mediating defensive responses (squirting, and ciliary arrest). Nervous links within zooids are shown as solid lines, conducting epithelia between zooids as broken lines. From Mackie & Singla 1983a. Abbreviations: a, ampula; b, brain; c, ciliated cells of branchial basket; en. epithelioneural coupling; m, muscle of mantle wall; ne, neuroepithelia coupling; s, sensory cell; v, vascular network.

*Pyrosoma* (figures 4 and 7) coordinates siphon closures and ciliary arrests in a completely different and unique manner. The zooids are organized around a central cavity into which they empty their effluent water streams (figure 4*a*). The streams combine to provide the locomotory jet which propels the colony. The zooids are not linked by nerves or conducting epithelia but they are situated side by side in close proximity, as in the sessile, compound ascidians from which they doubtless evolved. Each zooid has a light-producing organ and an eye capable of detecting flashes of light produced by the light organs of neighbours. Under adverse conditions, a zooid closes its siphons, arrests its cilia and gives off a flash of light. Neighbouring zooids detect the flash and respond in turn by ciliary arrests and flashes. In this way it is thought that a photic signal can be relayed from zooid to zooid along the colony causing them all to arrest their feeding currents (figure 4b). The flash is, in fact, so bright that nearby colonies could also be affected. The system only works in dim light for obvious reasons, but *Pyrosoma* lives in mesopelagic waters where light levels are low. Each zooid's flash would represent a signal saying 'beware/bad water ahead/turn off feeding current/pass message on'. An incidental effect of cessation of feeding is that locomotion also stops and the colony, being somewhat heavier than water, starts to sink and could thus enter a different, more suitable water layer (Mackie & Bone 1978; Bone & Mackie 1982).

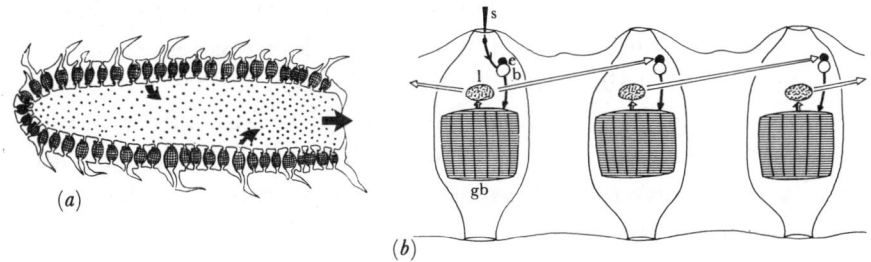

FIGURE 4. *Pyrosoma*. (a) Sagittal section of the whole colony from Barnes (1974) after Grobben. Water currents (arrows) from individual zooids enter the common cloacal cavity and are expelled as the locomotory current to the rear. (b) Scheme showing serial photic excitation of zooids following stimulation of the left hand zooid, from Bone & Mackie (1982). Abbreviations: b, brain; e, eye; gb, gill basket; l, luminous organ; s, stimulus.

(iii) *Light emission*

Waves of luminescence accompany spreading polyp retractions in various colonial hydroids and pennatulids (summarized in Morin 1974). The conduction system is probably an excitable epithelium in the hydroids, and a nerve net in the pennatulids. The functions of these responses have not been convincingly demonstrated but speculation favours a protective role of some sort.

In the siphonophore *Hippopodius*, following sharp or damaging stimulation, waves of luminescence spread between the nectophores, and are accompanied by protective involution of the nectophore margins. At the same time, the siphonophore, which is normally transparent, becomes opaque owing to the formation of light-scattering granules in the mesogloea (figures 5 and 6, plate 1), and endodermal secretory cells of unknown function discharge their product. Opacification by day is thought to function like luminescence at night, making the organism loom up terrifyingly out of nowhere, so discouraging molesters. The response is spread by excitable epithelia in the nectophores but passage from one nectophore to another involves a labile nervous intermediate step (Mackie & Mackie 1967; Bassot *et al.* 1978).

(iv) *Locomotion*

Locomotory ability has evolved in several benthic colonies, including some pennatulids, the acrobatic coral *Diaseris* (Hubbard 1972), ectoprocts such as *Cristatella* and *Selenaria* (discussed by Thorpe 1982) and various colonial ascidians (Carlisle 1961; Ryland *et al.* 1984). However, these are exceptional: the vast majority of benthic colonies are firmly attached to their substrates. It is the pelagic colonies that most clearly illustrate the possibilities of modularity in the area of locomotion.

In brief, a serial arrangement of interconnected propulsive units can provide smoother and more economical locomotion than separate units and, provided that the modules can communicate behaviourally, there need be no lessening of the ability of the colony to execute coordinated escape movements: the colony can behave like a well-integrated unitary organism. Rapid changes in direction are possible in some cases.

### (a) *The locomotion of salps*

In salps such as *Salpa fusiformis*, the blastozooids fit snugly together in streamlined chains. The zooids can also separate and live freely. Bone & Trueman (1983) show that feeding and swimming of chains is a more economical process than that of zooids operating individually. This is because the chain experiences relatively less drag than the isolated zooid. Drag depends on (i) total exposed surface area; (ii) frontal area; and (iii) greatest projected area (Alexander 1968). The last two factors will be about the same for colonies as for separate zooids in the case of linear, 'tubular' salp chains like *S. fusiformis*. Exposed surface area, however, will be proportionately less in the chain than in the single zooid because of the way the zooids fit together in the chain, with large parts of their outer surfaces pressed together. This then results in reduced drag. A salp chain can swim as fast as a solitary zooid with less effort per zooid. Only a fraction of the zooids need swim at all for the chain to achieve acceptable cruising velocities. Water passes through swimmers and non-swimmers alike, so the non-swimmers can feed without expenditure of energy in locomotion.

Unlike single zooids, which swim in jerks, filling and emptying in a regular cycle, the chain glides along in a smooth manner owing to the fact that the zooids contract asynchronously. Bone & Trueman (1983) suggest that this steady-state swimming further reduces drag owing to the establishment of stable boundary layers. In addition, smooth locomotion probably favours food capture. Salps use a mucous net to filter particles from the water. This is a delicate structure which could be disorganized by strong contractions or sudden movements.

Salps are extremely efficient swimmers in mechanical terms (Bone & Trueman 1983) and perform extensive diurnal vertical migrations. Wiebe *et al.* (1979) report that *Salpa aspersa* ascends and descends daily over a range of 800 m. To do so, it must swim at velocities of 5–10 m min$^{-1}$ (8.3–10.7 cm s$^{-1}$). Fedele (1923) further demonstrated that salps can swim both forwards and backwards.

A salp blastozooid can be thought of as a contractile tube, closable by valves at either end (figure 9a). If, when the tube contracts, the front valve closes first, water will be jetted backwards, and the salp will move forwards.

Salps swim backwards by the simple expedient of changing the order in which the valves close. The brain produces the locomotory rhythm, and determines the timing of the opening and closing of the valves. In a solitary salp swimming forwards, collision or a touch at the anterior end generates impulses, in sensory nerves, which travel to the brain and cause either a brief locomotory arrest or an arrest followed by reverse locomotion. If the rear end of the swimming salp is touched, the response is an acceleration of foward swimming. Fedele (1923) showed that chains perform exactly like individual salps. The mechanism has now been worked out (Bone *et al.* 1980; Anderson & Bone 1980).

Each zooid in the salp chain is linked to each of its neighbours by a pair of attachment plaques

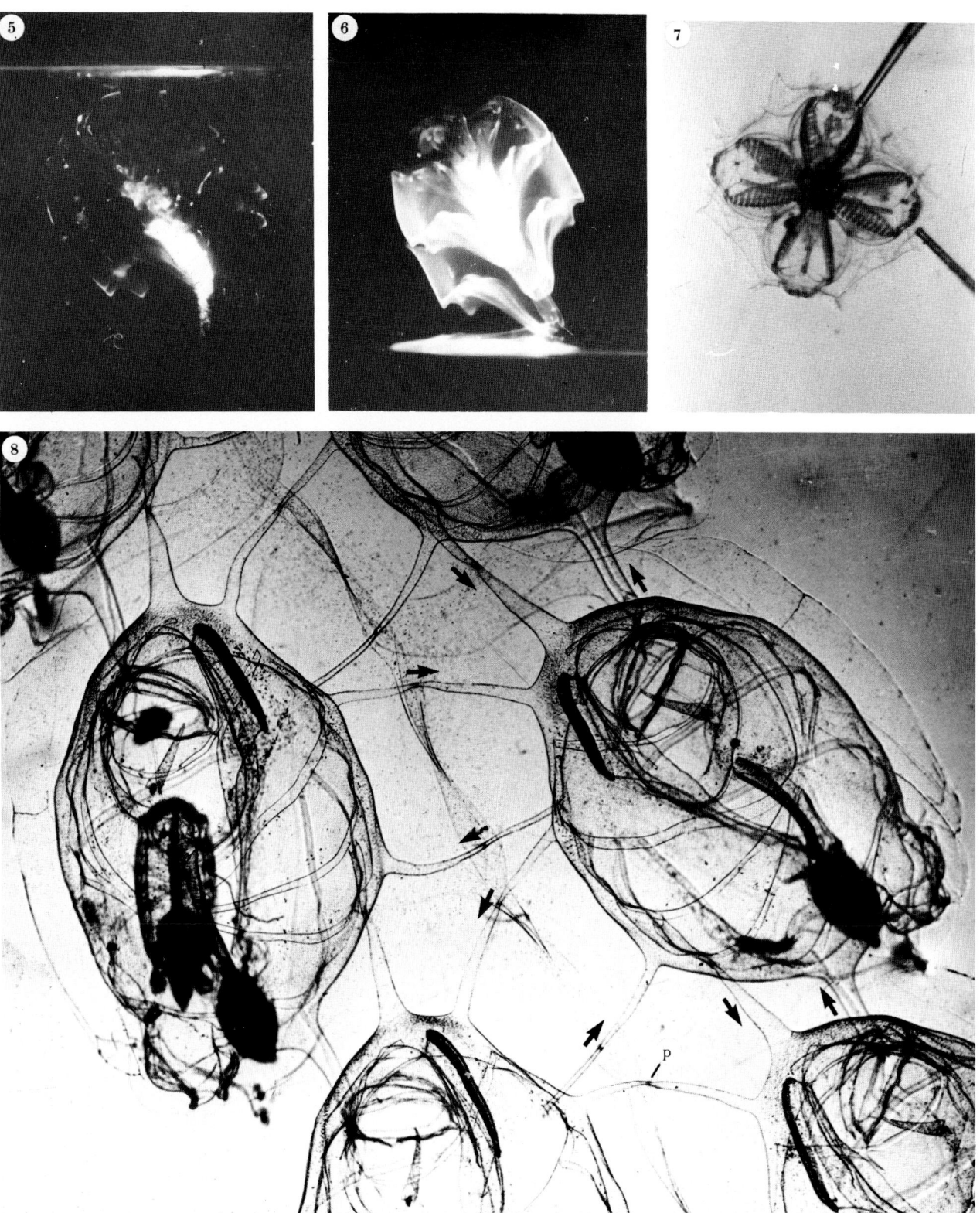

FIGURE 5. The siphonophore *Hippopodius hippopus* in its normal, transparent state (Mackie & Mackie 1967).

FIGURE 6. *Hippopodius* in opaque state following a sharp tap to one of the nectophores (Mackie & Mackie 1967).

FIGURE 7. *Pyrosoma atlanticum*, tetrazooid larva held by suction tubes which also serve as external electrodes for recording ciliary arrest potentials (Mackie & Bone 1978).

FIGURE 8. Part of a salp chain, *Thalia democratica*, (photo by Claude Carré), showing connections between blastozooids. An adhesion plaque is visible at p. Arrows show the presumed polarities of the eight plaques by which a single blastozooid is attached to its four neighbours, as in *Salpa fusiformis* (see figure 9b).

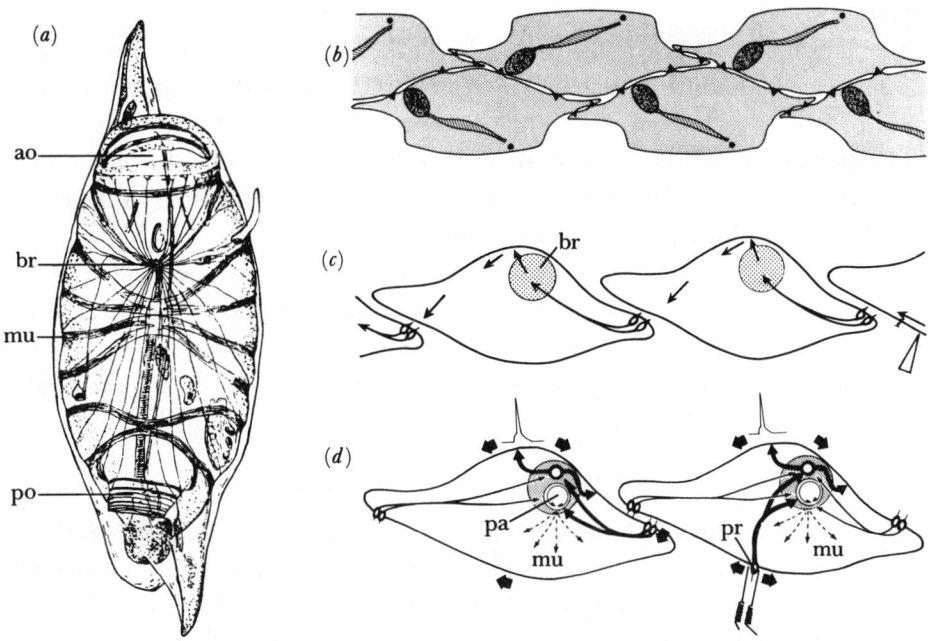

FIGURE 9. Coordination in salp chains. (a) A dorsal view of the sinistral form of a blastozooid of *Salpa maxima*. The dextral form is a mirror image of the sinistral (Fedele 1923). Abbreviations: ao, anterior opening; br, brain; mu, muscle band; po, posterior opening. (b) Positions of attachment plaques in *S. fusiformis*. Triangles indicate the innervated side of each plaque. Conduction is polarized toward that side (Bone *et al.* 1980). (c and d) Routes whereby epithelial signals (outer skin pulses) spread between zooids in the chain (Anderson & Bone 1980). (c) A simplified diagram showing generation of outer skin pulses (short arrows) at a site of mechanical or electrical stimulation (arrowhead right). Outer skin pulses enter sensory neurons at the plaque and pass as nerve impulses to brain (br) of the next salp, where they trigger outer skin pulses which pass on as before. (d) A diagram as in (c) above, but showing additional features: pacemakers (pa) in the brain which generate rhythmic output to the swimming muscles (mu); peripheral receptors (pr) in the skin which feed into the brain evoking the same responses as sensory signals arriving via the plaque receptors; and electrical traces of outer skin pulses recorded from the skin (shown above the drawing).

(figures 8 and 9b). Sensory neurons are located on one side of each plaque, their processes extending across to the other side. Of the two plaques joining a pair of zooids, one plaque is polarized so as to transmit forwards, the other backwards. Excitation entering the nerve at the front end of a zooid passes to the brain and causes the switch to reverse swimming. At the same time, impulses invade the skin, which is an excitable epithelium propagating impulses in all directions (Mackie & Bone 1977). These epithelial impulses travel to the plaque at the rear of the animal and are picked up by sensory processes projecting across the plaque from the salp behind. Thus, the signal to change direction is relayed through a series of neural and epithelial pathways from salp to salp down the chain (figure 9c, d).

Likewise, acceleration of forward swimming requires signals to be transmitted through plaques polarized in the anterior direction. The sensory nerves receiving these signals enter the brain and cause an increase in the output frequency of the locomotory pacemaker. At the same time, skin impulses are generated and pass to the front, where they are picked up by the sensory cells of the salp next in line ahead. The interposition of polarized nerve pathways in the system ensures that, even though the skin pulses are propagated to all parts of the skin, directional information is conserved, and the whole chain swims as a unit either forwards or backwards.

When no information is transmitted, the salps swim forward at their own rhythms, or coast along passively.

Strong stimulation causes the zooids to detach from one another. Fedele (1923) showed that a strong stimulus to the middle of a chain causes separation at that point. The front half then swims forwards at an accelerated rate while the back half swims backwards in the reverse response. The mechanism worked out by Bone and his colleagues is both elegant and simple. The separated zooid continues to respond exactly as it did while a member of the colony, using the same conduction pathways and brain circuitry.

Another group of pelagic tunicates, the doliolids, also alternate between free and attached stages. Free stages show a capacity for bidirectional locomotion and also for arresting their branchial cilia (Bone & Mackie 1977) but it is not known if these responses are coordinated between zooids during the colonial parts of the life cycle, when the zooids are held together by peculiar placenta-like structures.

### (b) *The locomotion of siphonophores*

As shown in figure 1, siphonophores can be viewed as pelagic, athecate hydroids in which the primary medusa stage is somewhat reduced, and which have produced several polymorphically specialized secondary zooids, some derived from polyps, some from medusae. Polyps are basically designed for food capture and for budding other zooids, medusae for swimming and for gamete production. By combining within a single organism altered versions of two zooid types which originally followed in sequence in the life cycle the siphonophore has gained the advantages of both. It gets feeding modules (gastrozooids) and digestive modules (palpons) from polyps, and locomotory modules (nectophores) from medusae.

Superficial appearances notwithstanding, nectophores retain all the essential medusan locomotory features. They have lost their tentacles, mouth and gonads. They cannot feed and depend on the colony for their nutrition. They are stripped down to the bare locomotory essentials: striated swimming muscles arranged circularly in the wall of the subumbrellar cavity, a velum to direct the water jet, and marginal nerve rings containing the pacemakers driving locomotion. Their strange shapes are related to streamlining, or to articulation with other nectophores.

In a physonectid siphonophore such as *Nanomia* (figure 10) the nectophores are deployed in two rows, back to back. They fit together into a 'solid' column by means of grooves and flanges. As in a salp chain, the exposed surface area, and hence drag, is presumably greatly reduced by the packing arrangement. The frontal area is small, offering little water resistance. When swimming normally (figure 10$g$), *Nanomia* allows its nectophores to pulsate at their own natural frequencies, which results in smooth, gliding motion, despite some tendency to veer from side to side. Steady-state conditions, which further reduce drag, doubtless apply here, as in salps. *Nanomia* is a vertical migrator, capable of ascending and then descending through hundreds of metres every 24 h.

When abruptly stimulated at the rear end, *Nanomia* darts forward at high velocity (20–30 cm s$^{-1}$ compared with 8–10 cm s$^{-1}$ in normal swimming). The nectophores contract in unison during this response (figure 10$a$). The escape reaction is mediated by giant axons in the stem, which contract at velocities in the range 1.0–3.0 m s$^{-1}$, about an order of magnitude faster than non-giant axons in typical hydrozoans. Nerves run from the stem to the marginal nerve rings of the nectophores, coordinating the swimming contractions. These nerves, and the

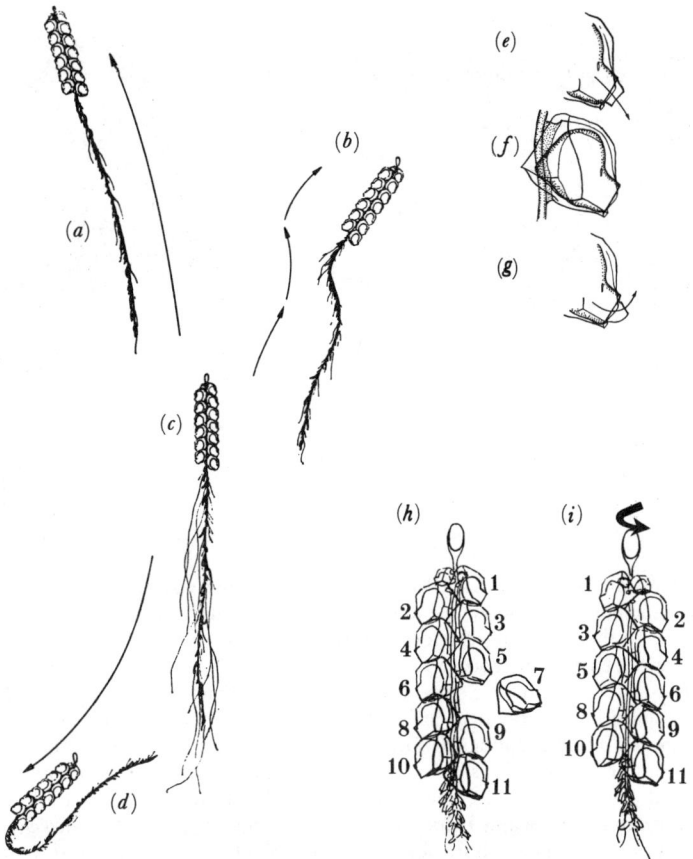

FIGURE 10. *Nanomia cara*, a physonectid siphonophore, from Mackie (1964). (*a*) Colony performing escape forwards swimming; (*b*) slower, serpentine swimming (normal locomotion); (*c*) colony at rest; (*d*) reverse escape swimming; (*e*) side view of nectophore showing jet direction during forward swimming; (*f*) the same, at rest; (*g*) the same, in reverse swimming; (*h*) following autotomy of the seventh nectophore, the upper part of the colony rotates through 180° (*i*) restoring symmetry.

giant axons, are clearly special components evolved for purposes of the colonial escape response (Mackie 1984).

When the anterior end of the colony is stimulated, giant axons are excited as before, but this time the colony swims backwards (figure 10*d*). The mechanism involves contraction of bundles of radial muscle fibres located at either side of the velar opening. These contract simultaneously with the swimming contraction and, because of their asymmetrical placement, change the direction of the water jet (figure 10*e*, *f*, *g*). The colony shoots violently backwards, as all the nectophores perform reverse swimming in unison. Conducting epithelia, not nerves, prove to be the peripheral pathway for colonial control over the radial muscles. A similar, indeed homologous, system mediates protective involution of the margin (crumpling) in hippopodiid siphonophores and in many free medusae. Apart from concentrating the radial muscles into bundles and preserving the primary conducting epithelial links with the stem, *Nanomia* has simply had to convert one protective response (crumpling) into another (escape locomotion).

Clearly, for effective swimming, the nectophore column must be bilaterally symmetrical. In normal growth, nectophores are inserted to left and right in an alternating pattern. Loss of nectophores from within the column would weaken the whole structure and introduce

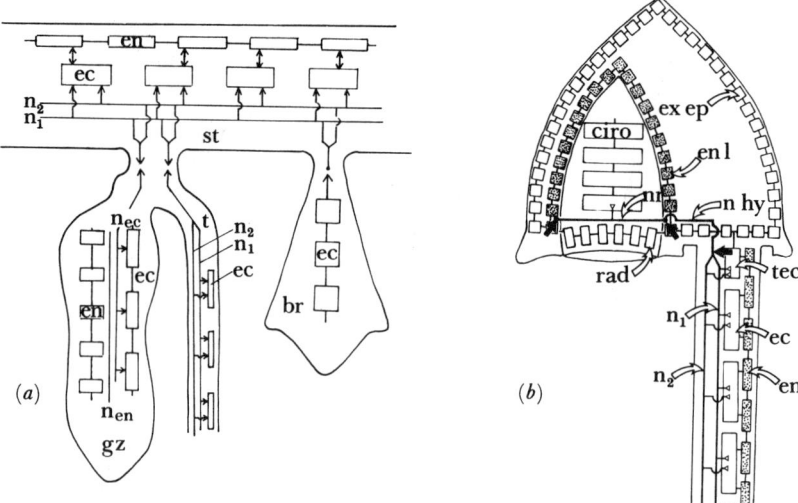

FIGURE 11. Action systems and conduction pathways in siphonophore colonies. (a) *Nanomia*, showing conducting pathways (arrows) and effector muscles (rectangles and squares) in the stem (st), in a gastrozooid (gz), a tentacle (t) and a bract (br). In the stem, the ectodermal muscle (ec) receives dual excitatory input from the two stem nerve nets ($n^1$, $n^2$) and is electrically coupled to the endoderm (en), which is a conducting epithelium. Activated via these two routes, the stem muscles show fast and slow contractions respectively. In the gastrozooid both ectoderm and endoderm have nerve nets ($n^{ec}$, $n^{en}$). The ectodermal net is physiologically connected with the stem nervous systems by labile couplings. Contractions can spread in either direction, but habituation is rapid. The two nerve nets in the tentacle also show labile functional connections with the stem. The bract lacks nerves, but its covering epithelium is excitable, and conducts impulses to the stem, where they evoke nervous activity and resulting contractions or escape behaviour (from Mackie 1978). (b) *Chelophyes*, showing nervous and non-nervous links between a nectophore and the stem. Black arrows symbolize epithelioneural excitation, triangles show neuromuscular excitation. Lines between excitable units represent electrical coupling. Stimulation of the exumbrellar epithelium (ex ep) evokes impulses that propagate to the margin where they excite nerves in the nerve ring (nr) that excites the circular swimming muscle (circ). Epithelial impulses also propagate directly to the radial muscles (rad) of the velum, causing them to contract during swimming. On reaching the stem, epithelial impulses pass directly to ectoderm cells in a transitional zone (t ec) that has the property, when excited, of exciting the stem nervous systems ($n^1$, $n^2$), both of which contain a giant axon. Depolarizations of the transitional ectoderm also pass directly to the endoderm (en) where they propagate as slow (s) events. Throughout the length of the stem, activity in $n^1$ and $n^2$ causes twitch depolarizations in the ectodermal muscles (ec) but impulses do not propagate within this tissue. Two-way interactions occur between the ectoderm and the endoderm. Activity in the stem nerves leads to twitch depolarizations in the transitional ectoderm that, if sufficiently large, propagate to the exumbrellar epithelium as exumbrellar impulses. On reaching the nectophore margin, they cause swimming as before. The hydroecial nerve (n hy) is implicated as an alternate, but faster, pathway mediating escape swimming responses following stem stimulation (from Mackie & Carré 1983).

asymmetries. This danger is minimized by a torsional response of the stem which takes place whenever a nectophore is lost (figure 10 h, i).

Calycophoran siphonophores cannot swim in reverse but many of them, especially the streamlined diphyids, are extremely agile and powerful swimmers, capable of a wide range of swimming velocities. *Chelophyes*, for example, can swim at 1 cm s$^{-1}$ using only its posterior nectophore, or at any intermediate velocity up to 30 cm s$^{-1}$ during escape behaviour, with both nectophores active. Here too, giant axons mediate escape reactions (Bone & Trueman 1982; Mackie & Carré 1983). Figure 11 b shows the main action systems and conduction pathways linking the nectophore of a diphyid with the stem, as established by neurophysiological investigation.

It seems clear that much of the success of siphonophores is due to their possession of highly efficient locomotory mechanisms. Swimming is a function of the colony. The individual swimming modules are subordinated to what Vogt (1853) called *la volunté commune*. Interesting parallels exist between the swimming of salps and physonectid siphonophores in the serial arrangement of their propulsive units, which interlock in configurations that minimize drag, in their ability to swim slowly and smoothly in normal forward swimming, in their ability to swim forwards rapidly in escape, and in their ability to perform escape locomotion in the reverse direction.

### 3. The wisdom of the colony: gaps in our understanding of chemical communication

Cannon (1932) talked of the 'wisdom of the body' as the sum of its homeostatic mechanisms, which we would now consider to include mechanisms involving hormones, neuromodulators, growth factors, morphogens and other mysterious regulatory substances. Plants, which are modular organisms, also communicate internally by means of hormones, growth regulators, flowering regulators, etc. Altogether, these substances fall into some nine categories (Devlin 1975). It would therefore be surprising if animal colonies were completely devoid of any capacity for internal chemical communication but to date no evidence for this has come forth.

In considering this problem in relation to colonial forms, it is apparent that major differences exist between the two principal polymorphic, colonial groups, the cnidarians and the bryozoans. All the zooids in a cnidarian colony are in communication by open gastrovascular canals. These serve to distribute digestive products and appear quite unsuited to the more subtle forms of chemical communication now under consideration. It is hard to see how chemical gradients could be maintained in such a system. If colony morphogenesis or homeostasis does involve chemical or electrical gradients, these must be set up in the epithelia themselves, or perhaps in the nerve cells running in the intraepithelial spaces. The complete lack of gastrovascular compartmentalization in cnidarian colonies has not prevented them from becoming polymorphically specialized. The Portuguese man-of-war, for example, has at least seven, possibly eight, different zooid types.

Ryland (1979), by contrast, holds that compartmentalization was a necessary prerequisite for polymorphic specialization in ectoproct evolution. Phylactolaemates, which are the least compartmentalized, show least polymorphism, whereas cheilostomes, which are highly compartmentalized, show 'spectacular zooidal evolution and diversification'. Compartmentalization has, however, prevented bryozoans from evolving a true colonial individuality. It is in the least compartmented genus, *Cristatella*, that colonial individuality is most clearly expressed.

It would seem, then, that in the evolution of ectoprocts, if not of cnidarians, specialization of zooids for different functions could not occur in a continuum and that partial metabolic isolation by means of compartmental divisions was necessary for this to take place. A need for isolation on the one hand had to be reconciled with a need for intercommunication on the other, and the result was a set of compromises. This is all very reminiscent of a similar problem in cell biology, discussed by Loewenstein (1984). Like interzooidal communication, intercellular communication is 'a compromise between cell connectivity and cell individuality' and Loewenstein is 'inclined to view the evolution of cellular communication systems as an endless maneuvering for achieving cell orchestration without loss of cell individuality'.

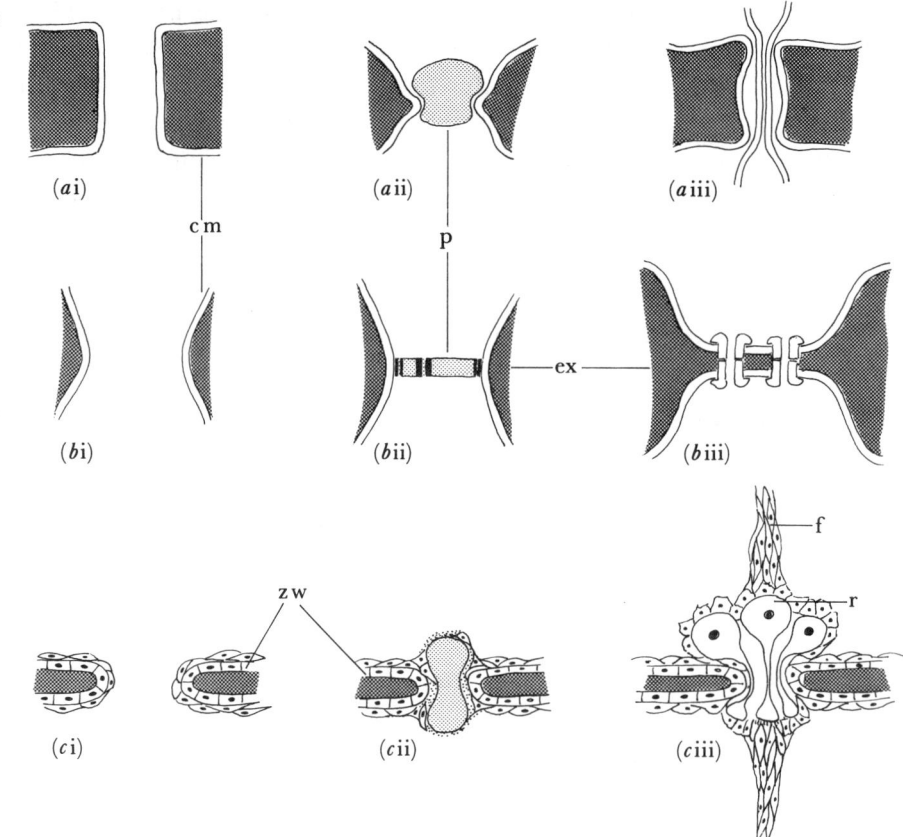

FIGURE 12. Comparison of intercellular junctions of plants (ai–aiii) and animals (bi–biii) with interzooidal junctions in ectoproct bryozoans (ci–ciii). (ai) Simple intercellular bridge, such as a septal pore of an ascomycete. (aii) Plugged 'transfer' connection as in *Polysiphonia*. (aiii) Plasmodesma of higher plant. (bi) Simple syncytial bridge as found between follicular nurse cells and oocytes in insects. (bii) Plugged bridge with pores, in a hexactinellid sponge. (biii) Gap junction of higher animal. (ci) Open mural pore between the coelomic compartments of adjacent zooids as in cyclostome stenolaemates. (cii) Immature pore with mucus plug, as in *Bowerbankia*. (ciii) Mature pore of *Bowerbankia* with rosette complex and funicular connections. The pore is polarized toward the stolon side (downwards in the drawing). Abbreviations: cm, cell membrane; ex, extracellular material (dark stipple); f, funiculus; p, plug; r, rosette cell; sw, zooid wall. Freely adapted from Bobin (1977) (cii, ciii); Mackie & Singla (1983b) (bii); Robards (1976) (ai, aiii); Telfer (1975) and Wetherbee (1980) (aii).

In figure 12 I show three grades of connectivity as expressed in plant cells, animal cells and bryozoan colonies. Unfortunately, bryozoologists have encountered the same sorts of difficulties as students of cellular communication when it comes to deciding what, if any, morphogenetic and regulatory substances are actually translocated via the junctions in question. Most plant physiologists assume that plasmodesmata mediate translocation of natural regulatory substances but, although experiments with fluorescent and radioactive markers show what sizes of molecules could pass, evidence for translocation of specific metabolites is still scant (Gunning & Overall 1983; Erwee & Goodwin 1985). Students of gap junctions in animals have been equally poorly placed. Now at last, with the development of antisera to the gap junction protein, the way seems open for research into what actually goes through these junctions (Warner *et al.* 1984).

In bryozoans, Ryland (1979) lists a number of forms of coordination that could be due to movement of regulatory factors between zooids, but in no case is there clear evidence for such processes and the 'coordination' could simply be a matter of the working out of ageing rates, that is part of the built-in developmental programme. However, the progression towards complex junctional structures seen alike in cellular and colonial systems suggests that junctions such as $c$iii in figure 12 are highly selective filtering devices which allow certain molecules to pass, but block others, and are therefore potentially well suited to playing a role in the establishment of chemical gradients and for translocation of key morphogenetic and regulatory factors.

## 4. General conclusions

We have looked at physiological adaptations of colonies in the areas of resource sharing, defensive behaviour, locomotion and chemical coordination.

Resource sharing can be regarded as a fundamental attribute of all true colonies, which sets them apart from mere aggregates. The pathways by which digestive products move between zooids may be blood vessels, extensions of the gut or shared portions of the body cavity. Restrictions on the free exchange of metabolites may be inferred from the complex morphology of the interzooidal connections in cheilostome bryozoans but the precise significance of these specializations is unknown. The same pathways that mediate transport of nutrients might also allow chemical regulators, morphogens, hormones, etc., to spread between zooids, but this is another unexplored area.

The possession of a colonial nervous system or of excitable epithelia functioning like nerves in conduction of electrical signals is an attribute of almost all true colonies. The parsimony principle applies here. Zooids typically show a high degree of local autonomy and use the same, or modified versions of, basic neuroeffector action systems found in unitary organisms or in the forms from which they are derived. Interzooidal pathways of behavioural communication function, so far as we know, in a rather restricted way, serving more as labile links between the action systems of the zooids than as centres where colonial behaviour is initiated. The siphonophores are exceptional in having nervous systems in their stems which not only relay responses between zooids but initiate activities such as swimming.

While defensive responses are the commonest, and often the only, known type of coordinated behaviour, some colonies show coordinated locomotion. Coordination may simply mean that the swimming modules are all active or all quiescent at the same time, but in some cases there is overall control of the frequency of the swimming contractions and even of the direction in which swimming occurs. Physonectid siphonophores and salp chains can swim either forwards or backwards in a coordinated manner.

The requirements of pelagic locomotion have led planktonic colonies to diverge from the pattern seen in the sessile colonies from which they presumably evolved. Most pelagic colonies are linear and unbranched, show an anterior–posterior polarity, are determinate in form, at least at the anterior end, and lack much of the plasticity and regenerative ability characteristic of sessile colonies.

It seems clear that the evolution of colonies occurred many times, in many different ways and according to diverse selective pressures. Beklemishev (1969) sees a progression in colonial evolution marked by increasing polymorphism, increasing zooidal physiological interdependence, reduced local autonomy, etc., culminating in the siphonophores where the emergence of a truly colonial individuality can be recognized. Beklemishev's 'progression' is, however, essentially a pedagogical formulation and it does not necessarily reflect actual directions of evolution, certainly not in all colonial groups. Many very simple colonies have been highly successful. Aggregates have flourished as well as integrates. There is no one consistent trend in colonial evolution that would allow us to define 'higher' and 'lower'; accordingly, attempts to categorize and compare various groups in such terms are rather academic and should not be mistaken for descriptions of evolutionary processes.

Of those who have helped with the preparation of this paper I wish to express my particular thanks to Dr Drew Harvell for critically reading the manuscript and for her many valuable suggestions.

## REFERENCES

Alexander, R. M. 1968 *Animal mechanics*. London: Sedgwick and Jackson.
Anderson, P. A. V. 1980 Epithelial conduction: its properties and functions. *Prog. Neurobiol.* **15**, 161–203.
Anderson, P. A. V. & Bone, Q. 1980 Communication between individuals in salp chains. II. Physiology. *Proc. R. Soc. Lond.* B **210**, 559–574.
Anderson, P. A. V. & Case, J. F. 1975 Electrical activity associated with luminescence and other colonial behaviour in the pennatulid *Renilla* Köllikeri. *Biol. Bull. Mar. biol Lab., Woods Hole* **149**, 80–95.
Barnes, R. D. 1974 *Invertebrate zoology*, 3rd ed. Philadelphia: W. B. Saunders.
Bassot, J. M., Bilbaut, A., Mackie, G. O., Passano, L. M. & Pavans de Ceccatty, M. 1978 Bioluminescence and other responses spread by epithelial conduction in the siphonophore *Hippopodius*. *Biol. Bull. mar. biol. Lab., Woods Hole* **155**, 473–479.
Bayer, F. M. 1973 Colonial organization in octocorals. In *Animal colonies: development and function through time* (ed. R. S. Boardman, A. H. Cheetham & W. A. Oliver, Jr), pp. 69–93. Stroudsberg, Pennsylvania: Dowden, Hutchinson & Ross.
Beklemishev, W. N. 1969 *Principles of comparative anatomy of invertebrates*, vol. 1, *Promorphology*. (Transl. from the Russian J. M. MacLennan; ed. Z. Kabata) Edinburgh: Oliver and Boyd.
Berry, M. S. & Hayward, P. J. 1984 Nervous and behavioural responses to light in colonies of the free-living bryozoan *Selenaria maculata* (Busk). *Experientia* **40**, 108–110.
Best, M. A. & Thorpe, J. P. 1985 Autoradiographic study of feeding and the colonial transport of metabolites in the marine bryozoan *Membranipora membranacea*. *Mar. Biol.* **84**, 295–300.
Biggs, D. C. 1977 Field studies of fishing, feeding and digestion in siphonophores. *Mar. Behav. Physiol.* **4**, 261–274.
Bobin, G. 1977 Interzooecial communications and the funicular system. In *Biology of bryozoans* (ed. R. M. Woollacott & R. L. Zimmer), pp. 307–333. New York: Academic Press.
Bone, Q., Anderson, P. A. V. & Pulsford, A. 1980 The communication between individuals in salp chains. I. Morphology of the system. *Proc. R. Soc. Lond* B **210**, 549–558.
Bone, Q. & Mackie, G. O. 1977 Ciliary arrest potentials, locomotion and skin impulses in *Doliolum* (Tunicata: Thaliacea). *Riv. biol. norm. patol.* **3**, 181–191.
Bone, Q. & Mackie, G. O. 1982 Urochordata. In *Electrical conduction and behaviour in simple invertebrates*. (ed. G. A. B. Shelton), pp. 473–534. Oxford: Clarendon Press.
Bone, Q. & Trueman, E. R. 1983 Jet propulsion in salps (Tunicata: Thaliacea). *J. Zool., Lond.* **201**, 481–506.
Bone, Q. & Trueman, E. R. 1982 Jet propulsion of the Calycophoran siphonophores *Chelophyes* and *Abylopsis*. *J. mar. biol. Ass. U.K.* **62**, 263–276.
Buss, L. W. 1979 Habitat selection, directional growth and spatial refuges: why colonial animals have more hiding places. In *Biology and systematics of colonial organisms*. (ed. G. Larwood and B. R. Rosen), pp. 459–497. The Systematics Association Special vol. no. 11. London: Academic Press.
Cannon, W. B. 1932 *The wisdom of the body*. New York: W. W. Norton.
Cattaneo, G. 1879 Le individualità animali. *Atti. Soc. Ital. Sci. Nat.* **22**, 1–71.
Carlisle, D. B. 1961 Locomotory powers of adult ascidians. *Proc. Zool. Soc. Lond.* **136**, 141–146.
Chapman, G. & Stebbing, A. R. D. 1980 The modular habit – a recurring strategy. In *Developmental and cellular biology of coelenterates* (ed. P. Tardent & R. Tardent), pp. 157–162. Amsterdam: Elsevier/North Holland Biomedical Press.
Cook, P. L. & Chimonides, P. J. 1978 Observations on living colonies of *Selenaria* (Bryozoa, Cheilostomata). I. *Cah. Biol. mar.* **19**, 147–158.
Cook, P. L. & Chimonides, P. J. 1981 Observations on living colonies of *Selenaria* (Bryozoa, Cheilostomata). II. *Cah. Biol. mar.* **22**, 207–219.
Dendy, A. 1924 *Outlines of evolutionary biology*. New York: Appelton & Co.

Devlin, R. 1975 *Plant physiology*, 3rd ed. New York: D. Van Nostrand Co.
Erwee, M. G. & Goodwin, P. B. 1985 Symplast domains in extrastelar tissues of *Egeria densa* Planch. *Planta* **163**, 9–19.
Fedele, M. 1923 Simmetria ed unità dinamica nelle catene di salpa. *Boll. Soc. nat. Napoli* **36**, 20–32.
Fehilly, C. B., Willadsen, S. M. & Tucker, E. M. 1984 Interspecific chimaerism between sheep and goat. *Nature, Lond.* **307**, 634–636.
Gordon, D. P. 1977 The aging process in bryozoans. In *Biology of bryozoans* (ed. R. M. Woollacott & R. L. Zimmer), pp. 335–376. New York: Academic Press.
Gunning, B. E. S. & Overall, R. L. 1983 Plasmodesmata and cell-to-cell transport in plants. *BioScience* **33**, 260–265.
Hardy, A. C. 1956 *The open sea: its natural history*, part I. *The world of plankton*. London: Collins.
Harper, J. L. 1984 Foreword. In *Perspectives on plant population ecology* (ed. R. Diržo & J. Sarukhán), pp. xv–xviii. Sunderland, Massachusetts: Sinauer Associates.
Harper, J. L. & Bell, A. D. 1979 The population dynamics of growth form in organisms with modular construction. In *Population dynamics* (ed. R. M. Anderson, B. D. Turner & L. R. Taylor), pp. 29–52. Oxford: Blackwell Scientific Publications.
Harvell, C. D. 1984 Predator-induced defense in a marine bryozoan. *Science, Wash.* **224**, 1357–1359.
Heron, A. C. 1972 Population ecology of a colonizing species: the pelagic tunicate *Thalia democratica*. II. Population growth rate. *Oecologia* **10**, 294–312.
Horridge, G. A. 1957 The co-ordination of the protective retraction of coral polyps. *Phil. Trans. R. Soc. Lond.* B **240**, 495–528.
Hubbard, J. A. E. B. 1972 *Diaseris distorta*: An 'acrobatic' coral. *Nature, Lond.* **236**, 457–459.
Hughes, T. P. & Jackson, J. B. C. 1985 Population dynamics and life histories of foliaceous corals. *Ecol. Mongr.* **55**, 141–166.
Jackson, J. B. C. 1977 Competition on marine hard substrata: the adaptive significance of solitary and colonial strategies. *Am. Nat.* **111**, 743–469.
Jackson, J. B. C. 1979 Morphological strategies in sessile animals. *Biology and systematics of colonial organisms* (ed. G. Larwood and B. R. Rosen), pp. 499–555. The Systematics Association special volume no. 11. London: Academic Press.
Josephson, R. K. 1961 Repetitive potentials following brief electrical stimuli in a hydroid. *J. exp. Biol.* **38**, 579–593.
Knight-Jones, E. W. & Moyse, J. 1961 Intraspecific competition in sedentary marine animals. *Symp. Soc. exp. Biol.* **15**, 72–95.
de Kruijf, H. A. M. 1977 Individual polyp behaviour and colonial organization in the hydrocorals *Millepora complanata* (Milleporina) and *Stylaster roseus* (Stylasterina). *Proc. Third int. Coral Reef Symp.* pp. 445–451.
Lidgard, S. 1981 Water flow, feeding and colony form in an encrusting cheilostome. In *Recent and fossil Bryozoa* (ed. G. P. Larwood & C. Nielsen), pp. 135–142. Fredensborg: Olsen and Olsen.
Loewenstein, W. R. 1984 Cell individuality and connectivity, an evolutionary compromise. In *Individuality and determinism* (ed. S. W. Fox), pp. 77–87. New York: Plenum Press.
Lutaud, G. 1977 The bryozoan nervous system. In *Biology of bryozoans* (ed. R. M. Woollacott & R. L. Zimmer), pp. 377–410. New York: Academic Press.
Lutaud, G. 1981 The innervation of the external wall in the carnosan ctenostome *Alcyonidium polyoum* (Hassall). In *Recent and fossil Bryozoa* (ed. G. P. Larwood & C. Nielsen), pp. 143–150. Fredensborg: Olsen and Olsen.
Mackie, G. O. 1963 Siphonophores, bud colonies and superorganisms. In *The lower Metazoa* (ed. E. C. Dougherty), pp. 329–337. University of California Press, Berkeley.
Mackie, G. O. 1964 Analysis of locomotion in a siphonophore colony. *Proc. R. Soc. Lond* B **159**, 366–391.
Mackie, G. O. 1973 Coordinated behavior in hydrozoan colonies. In *Animal colonies: development and function through time* (ed. R. J. Boardman, A. H. Cheetham & W. A. Oliver, Jr), pp. 95–106. Stroudsberg, Pennsylvania: Dowden, Hutchinson & Ross.
Mackie, G. O. 1978 Coordination in physonectid siphonophores. *Mar. Behav. Physiol.* **5**, 325–346.
Mackie, G. O. 1984 Fast pathways and escape behavior in Cnidaria. In *Neural basis of startle behavior.* (ed. R. C. Eaton), pp. 15–42. New York: Plenum.
Mackie, G. O. & Boag, A. 1963 Fishing, feeding and digestion in siphonophores. *Pubbl. Staz. Zool. Napoli* **33**, 178–196.
Mackie, G. O. & Bone, Q. 1977 Locomotion and propagated skin impulses in salps (Tunicata: Thaliacea). *Biol. Bull.* **153**, 180–197.
Mackie, G. O. & Bone, Q. 1978 Luminescence and associated effector activity in *Pyrosoma* (Tunicata: Pyrosomida). *Proc. R. Soc. Lond.* B **202**, 483–495.
Mackie, G. O. & Carré, D. 1983 Coordination in a diphyid siphonophore. *Mar. Behav. Physiol.* **9**, 139–170.
Mackie, G. O., Lawn, I. D. & Pavans de Ceccatty, M. 1983 Studies on hexactinellid sponges. II. Excitability, conduction and coordination of responses in *Rhabdocalyptus dawsoni* (Lambe, 1873). *Phil. Trans. R. Soc. Lond.* B **301**, 401–418.
Mackie, G. O. & Mackie, G. V. 1967 Mesogloeal ultrastructure and reversible opacity in a transparent siphonophore. *Vie et Milieu* **18**, 47–71.
Mackie, G. O. & Singla, C. L. 1983a Coordination of compound ascidians by epithelial conduction in the colonial blood vessels. *Biol. Bull. mar. biol. Lab., Woods Hole* **165**, 209–220.

Mackie, G. O. & Singla, C. L. 1983*b* Studies on hexactinellid sponges. I. Histology of *Rhabdocalyptus dawsoni* (Lambe, 1873). *Phil. Trans. R. Soc. Lond.* B **301**, 365–400.

Morin, J. G. 1974 Coelenterate bioluminescence. In *Coelenterate biology* (ed. L. Muscatine & H. M. Lenhoff), pp. 347–438. New York: Academic Press.

Potts, D. C. 1984 Generation times and the Quarternary evolution of reef building corals. *Palaeobiology* **10**, 48–58.

Rees, J. 1971 Paths and rates of flood distribution in the colonial hydroid *Pennaria*. In *Experimental coelenterate biology* (ed. H. M. Lenhoff, L. Muscatine & L. V. Davis), pp. 119–128. University of Hawaii Press.

Rinkevitch, B. & Loya, Y. 1983 Oriented translocation of energy in grafted reef corals. *Coral Reefs* **1**, 243–247.

Rosen, B. R. 1979 Modules, members and communes: a postscript introduction to social organisms. In *Biology and systematics of colonial organisms* (ed. G. Larwood & B. R. Rosen), pp. xiii–xxxv. The Systematics Association special volume no. 11. London: Academic Press.

Ryland, J. S. 1979 Structural and physiological aspects of coloniality in Bryozoa. In *Biology and systematics of colonial organisms* (ed. G. Larwood & B. R. Rosen), pp. 211–242. The Systematics Association Special volume no. 11. London: Academic Press.

Ryland, J. S., Wigley, R. A. & Muirhead, A. 1984 Ecology and colonial dynamics of some Pacific reaf flat Dideminidae (Ascidiacea). *Zool. J. Linn. Soc.* **80**, 261–282.

Sabbadin, A. 1979 Colonial structure and genetic patterns in ascidians. In *Biology and systematics of colonial organisms* (ed. G. Larwood & B. R. Rosen), pp. 433–444. The Systematics Association Special volume no. 11. London: Academic Press.

Shelton, G. A. B. 1979 Coordination of behaviour in cnidarian colonies. In *Biology and systematics of colonial organisms* (ed. G. Larwood & B. R. Rosen), pp. 141–154. The Systematics Association special volume no. 11. London: Academic Press.

Shelton, G. A. B. 1982 Anthozoa. In *Electrical conduction and behaviour in simple invertebrates* (ed. G. A. B. Shelton), pp. 203–242. Oxford: Clarendon Press.

Sibaoka, T. 1966 Action potentials in plant organs. *Symp. Soc. exp. Biol.* **20**, 49–74.

Silén, L. 1944 The anatomy of *Labiostomella gisleni* (Bryozoa: Protocheilostomata) with special regard to the embryo chambers of the different groups of Bryozoa and to the origin and development of the bryozoan zoarium. *K. Svenska Vetensk Akad, Handl.* **21**, 1–111.

Spencer, A. N. & Schwab, W. E. 1982 Hydrozoa. In *Electrical conduction and behaviour in simple invertebrates* (ed. G. A. B. Shelton), pp. 73–148. Oxford: Clarendon Press.

Spencer, H. 1898 *The principles of biology*. London: Williams & Norgate.

Stebbing, A. R. D. 1973 Competition for space between the epiphytes of *Fucus serratus*. *J. mar. biol. Ass U.K.* **53**, 247–261.

Strathmann, R. R. & Chaffee, C. 1984 Constraints on egg masses. II. Effect of spacing, size and number of eggs on ventilation of masses of embryos in jelly, adherent groups, or thin-walled capsules. *J. exp. mar. Biol. Ecol.* **84**, 85–93.

Telfer, W. H. 1975 Development and physiology of the oocyte-nurse cell syncytium. *Adv. Insect Physiol.* **2**, 223–319.

Thorpe, J. P. 1982 Bryozoa. In *Electrical conduction and behaviour in simple invertebrates*. (ed. G. A. B. Shelton), pp. 393–439. Oxford: Clarendon Press.

Thorpe, J. P., Shelton, G. A. B. & Laverack, J. S. 1975 Colonial nervous control of lophophore retraction in cheilostome Bryozoa. *Science, Wash.* **189**, 80–81.

Vogel, S. 1981 *Life in moving fluids: the physical biology of flow*. Boston, Massachusetts: Willard Grant Press.

Vogt, C. 1853 Recherches sur les animaux inférieures de la Mediterranée. I. Sur les siphonophores de la mer de Nice. *Mém. Inst. nat. Genev.* **1**, 1–164.

Warner, A. E., Guthrie, S. C. & Gilula, N. B. 1984 Antibodies to gap-junctional protein selectively disrupt junctional communication in the early amphibian embryo. *Nature, Lond.* **311**, 127–131.

Werner, B. 1979 Coloniality in the Scyphozoa: Cnidaria. In *Biology and systematics of colonial organisms* (ed. G. Larwood & B. R. Rosen), pp. 81–103. The Systematics Association Special volume no. 11. London: Academic Press.

Wetherbee, R. 1980 Transfer connections: specialized pathways for nutrient translocation in a red alga. *Science, Wash.* **204**, 858–889.

White, J. 1984 Plant metamerism. In *Perspectives on plant population ecology* (ed. R. Dirzo & J. Sarukhán), pp. 15–47. Sunderland, Massachusetts: Sinauer Associates.

Wiebe, P. H., Madin, L. P., Haury, L. R., Harbison, G. R. & Philbin, L. M. 1979 Diel vertical migration by *Salpa aspersa* and its potential for large-scale particulate organic matter transport to the deep-sea. *Mar. Biol.* **53**, 249–255.

# The genetic basis of plant form

By L. D. GOTTLIEB

*Department of Genetics, University of California,
Davis, California 95616, U.S.A.*

Plant architecture is relevant to a number of questions in population biology because it affects the number, size, and fecundity of individuals. Architectural differences in wild plants have frequently been described and are presumed to have a genetic basis because the differences are maintained when the plants are grown in uniform gardens, but little genetic research has been done. Studies in crop plants, however, provide substantial information about how plant form can be genetically manipulated. They show that the architecture of many crops has been successfully modified by making a small number of genetic substitutions that affect shoot length, flowering node, branch presence and orientation, habit, and growth determinacy. The changes occur at the level of metamers (leaf–axillary bud–internode) and become multiplied by iteration into the characteristic architecture of the plant. Metamer growth and iteration are tightly coordinated by genetic factors that operate at the level of the whole plant. Evidence supporting this hypothesis includes single gene control of coordinated changes among successive internodes, genetic control of production of metabolites or signals that move from mature tissues to shoot growing points, and allometries connecting organs arising from the same meristem. Since different plant architectures are associated with differences in fitness, information on the genetic basis of the morphological and physiological characters that cause the architectural differences will elucidate how fitness characters evolve.

## INTRODUCTION

The primary data in plant demography are counts of individual plants over time. In clone-forming plants (as many as two-thirds of the common perennial species in Great Britain according to Abrahamson (1980)) such counts have uncertain meaning because the number of individuals need not equal the number of genotypes. Since correct genotype assignment is essential for the analysis of population turnover, individual fitnesses, and the likelihood of adaptive response to environmental change, plant demographers have had to consider how shape or architecture affects the size, number, and distribution of individuals.

Much attention has been paid to the iterative or metameric structure of the plant body (White 1979, 1984; Harper & Bell 1979; Harper 1981) because it provides the architectural basis for the vegetative fragmentation of clonal plants and establishment of the ramets as separate individuals. The 'rapprochement' of morphology and ecology is significant because it may stimulate more intensive investigations to identify the morphological traits contributing to fitness differentials within populations and to the adaptive differences between geographically separate populations. To facilitate this effort, I review present knowledge of the genetic control of plant architecture, and discuss the usefulness of the concept of plant metamerism from the standpoints of genetics and development.

## Morphological divergence and genetic analysis

Plant development is characterized by an open and plastic pattern of morphogenesis in which growth and organ differentiation are initiated at meristems that occur at the apices of all shoots and roots. Other meristems function in actively growing organs such as intercalary meristems in the internodes and leaf sheaths of many monocotyledons and the marginal meristems of dicotyledon leaf blades. In some plants new meristems can also be generated after predation or injury. Cell divisions in the shoot apical meristem lead to the serial formation of lateral primordia from which leaves, buds, or floral parts develop. Further cell divisions in the apical and subapical meristems contribute to the increase in shoot length.

The repetitive structure of the plant shoot, based on the node–leaf–axillary bud–internode complex (the metamer of White (1984)), permits the plant to adjust its morphology through time according to local environmental conditions. The responses appear to be mediated by locally synthesized growth substances as well as those produced elsewhere in the plant body and need not be related to the response or condition of other meristems on the same individual. The result is often a marked plasticity that can lead to many genotypes having a similar appearance or single genotypes displaying multiple phenotypes as a function of age, size, or microhabitat. Such capability is important for stationary organisms, which must adapt to their environment and respond rapidly to both advantageous and deteriorating conditions.

The central role of meristems in plant development may be a major factor facilitating the evolution of morphological differences which are discrete or discontinuous in expression rather than gradual (Gottlieb 1984). Thus many differences in morphological characters involve presence versus absence of structures, or sharp alternatives in shape, and architectural position or orientation. Other morphological characters differ in continuous fashion, showing changes in dimensions, mass, or numbers, the classical components of agricultural yield. The two categories of morphological change appear to involve different numbers of genetic changes (Gottlieb 1984). Discrete changes are governed by one or two genes. Continuous changes, which often involve end products of growth such as height, leaf length, or seed number, are influenced by numerous genes with slight effects individually, though many probably act only indirectly via general effects on growth and vigour. The categories are not exclusive alternatives but rather represent the extremes of a spectrum of morphological changes. Indeed in complex characters, the differences may include aspects that are discrete and others that are continuous. For example, in *Aquilegia* a single gene governs presence versus absence of petal spurs but the length of the spurs is influenced by numerous genes (Prazmo 1965). Before genetic studies of architectural differences are done, the specific underlying morphological changes must be identified.

Architectural differences in wild plants are frequently described, and are presumed to have a genetic basis since the different phenotypes are maintained when grown in uniform gardens, but only a few genetic analyses have been made. These include studies in *Layia* on orientation of branches and presence versus absence of central stem (Clausen 1951) and prostrate or erect habit in several species (references in Gottlieb 1984). No studies appear to have been reported on clone-forming plants. The paucity of genetic research on wild (non-domesticated) plants reflects the complexity of the analysis, and the fact that most geneticists working with plants study agricultural species in which true-breeding lines are available (or can be constructed) and large progenies readily grown.

Consequently, it is necessary to study the results of research with crops to learn how genes affect plant architecture. Varietal differences within crop species include a large array of different architectural forms. Breeders have determined the genetic basis of these differences and have been able to design plants for particular environments by transferring genes and combining genes in novel combinations. Thus, although cultivated fields are not similar to natural habitats, the results with crops reveal how plant phenotypes can be genetically manipulated and thereby help to inform us about what is possible. Morphological changes in crops will certainly have counterparts in wild plants, and there is no reason to believe that the types of genetic control will be much different.

One of the most critical problems in a genetic analysis is the design of test environments that enhance or magnify the effects of particular genes or gene combinations that might otherwise be obscured by environmental variables. For example, flowering time in pea (*Pisum sativum*) was originally thought to be a quantitative trait affected by an additive polygenic system. However when $F_2$ plants from a cross, later shown to involve the *Sn/sn* allelic pair (see below), were grown under short days and mild temperatures, the segregation was sharp with *Sn* conditioning flowering at a high (late) node and *sn* at a low (early) node (Murfet 1977a). The segregation was blurred if cool temperatures were used or if the photoperiod was increased, and the allelic difference disappeared altogether if the plants were vernalized and exposed to continued cold nights. The incorrect conclusion that flowering was largely polygenic in pea resulted from growing plants under cool temperatures in field conditions. 'The problem is to devise methods and circumstances which will enable us to trek back from the phenotypic (visable) periphery of gene action, where the effects of many genes are often integrated into a single parameter, through the maze of interlocking biochemical wheels in the primary level of gene action where every gene has an individual effect' (Murfet 1977a).

Another profound problem is how to decide which phenotypic character is the correct one for genetic analysis. Plant demographers emphasize life-history traits such as survivorship, age of reproduction, and fecundity, and the plant breeder often measures only yield. Characters of this sort are abstract constructs that add up an uncertain number of genetic inputs and are closely affected by the environmental conditions in which the plants are examined. Some plant breeders are acutely aware of these problems. 'Genes do not exist for yield *per se*. Genetic control is indirect, through control of the physiological components...' (Wallace *et al.* 1972) and, one may add, the morphogenetic components. To make sense of genotypic diversity in wild plant populations it will be necessary to pay close attention to the primary components of morphological development and physiological performance.

One of the best examples of phenotypic analysis known to me is the identification of the primary determinant of shape difference between elongate and spherical fruits of *Capsicum annuum* (red pepper) by Sinnott & Kaiser (1934) and Kaiser (1935). They plotted increase in ovary length against increase in ovary diameter on a double logarithmic plot and found that the allometric line connecting length and diameter changed slope sharply following fertilization in the elongate type but maintained a constant slope in the spherical type (figure 1). The change in slope segregated as a single gene difference. But the mature fruits were continuously variable from elongate to spherical, depending on the duration of their growth (when allometric slope is greater than unity, the fruits become more elongate as growth continues), effectively masking the discrete allometric change (figure 2). Analysis of the mature fruits only, the likely procedure in most quantitative genetic studies, would fail to reveal the simple but significant developmental input to the shape difference.

200  L. D. GOTTLIEB

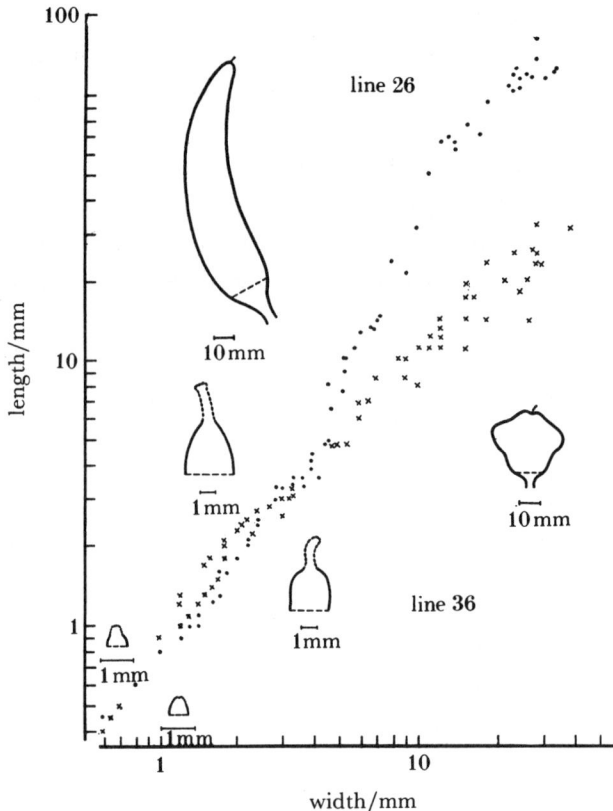

FIGURE 1. Double logarithmic plot of length against width from ovary primordia to mature fruit in two varieties of pepper (*Capsicum annuum*) with elongate and approximately spherical mature fruit shapes. Note change in allometry in the elongate variety, at an ovary width of approximately 5 mm, which is initiated shortly after anthesis (from Sinnott & Kaiser 1934).

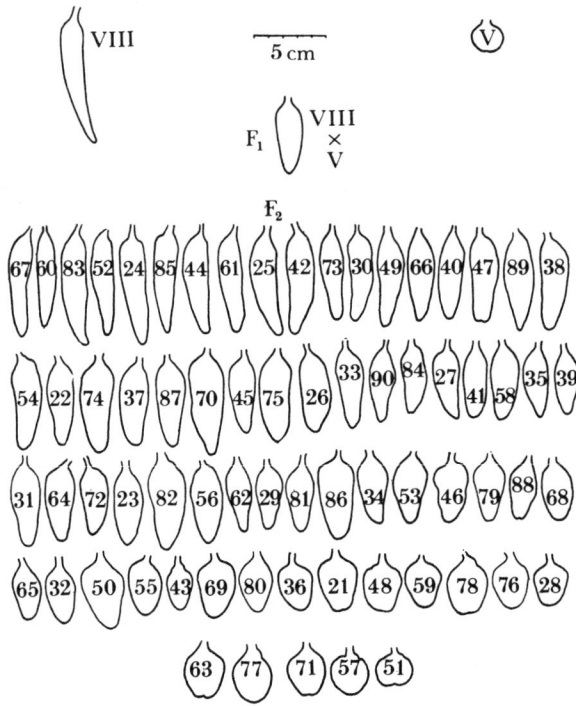

FIGURE 2. Trace drawings of typical mature elongate and spherical fruits from two varieties of pepper (*Capsicum annuum*), the $F_1$ hybrid between them, and $F_2$ progeny. Note absence of discrete segregation in shape (from Kaiser 1935).

## GENETIC MODIFICATION OF ARCHITECTURE IN CROP PLANTS

Remarkable phenotypic changes have been made in the architecture of rice (*Oryza sativa*), tomato (*Lycopersicon esculentum*), pea (*Pisum sativum*), and wheat (*Triticum* spp.) by very simple genetic substitutions involving a few major genes. Subsequent analyses of the interactions of these genes in the diverse genetic backgrounds required for particular environments often led to incorporation of additional gene modifications. In the following case studies, I emphasize the consequences of the initial changes brought about by the genes that had large effects.

### Rice

Probably the most significant genetic modification of plant architecture was the development in the early 1960s of the dwarf varieties of wheat and rice (Coyne 1980). In rice, three- to fourfold increases in yield were achieved by making use of a single recessive gene from Taichung Native 1, a Taiwanese Indica variety, which conferred a number of beneficial effects: a short, stiff, wide, upright leaf; short internodes; and a high number of short panicles (Aquino & Jennings 1966). The dwarf varieties averaged about half the height of standards. Dwarf stature had been previously known in rice, but the phenotype had not been useful because it invariably exhibited complex patterns of inheritance and was subject to marked environmental influence. The great value of the Taichung Native 1 gene was that it had a major effect on stature and was not correlated with undesirable agronomic traits. This made it possible to transfer the gene to already acceptable varieties (Aquino & Jennings 1966).

The increased yield associated with the gene was clearly related to the new architecture. The erect leaves permitted a more uniform distribution of incident radiant energy over the entire leaf canopy, including the lower leaves, which resulted in substantially higher accumulated dry seed mass per unit leaf area per unit time (reviewed by Wallace *et al.* 1972). The short plant stature also conditioned a strongly positive response to nitrogen fertilization, and without the lodging that nitrogen typically caused in tall varieties. In addition, the dwarfs partitioned a larger proportion of total photosynthate to their seed rather than to additional vegetative structures. But the yield advantage of dwarf stature did not simply reflect a novel plant form. The advantage became evident only with appropriate cultural practices and only in pure stand since the dwarf plants were unable to compete with tall plants for light.

### Processing tomato

Most varieties of tomato normally have indeterminate growth and sprawl indefinitely, continuing to flower and mature fruits further and further out on the stem and branches. The decision in California in the 1950s to design machines to harvest tomatoes for the processing industry required major changes in tomato architecture and in the pattern of fruit set and ripening. The primary changes depended on the use of the recessive *sp* gene (Rick & Butler 1956) which confers a compact, orderly, and determinate growth form (Rick 1978), and makes possible the harvesting of an entire field at one time.

The stems of varieties with the *sp* gene develop only one or two leaves between successive inflorescences and terminate in an inflorescence whereas normal stems have three leaves between reproductive structures. Branches of such varieties terminate their growth at approximately equal distances from the central stem and the plant flowers abundantly in short time period. Both normal and *sp* plants produce a similar number of leaves before the first inflorescence and show similar rates of growth.

The *sp* gene is often used in conjunction with the recessive gene *j* which conditions the absence of the joint in the fruit pedicel, at which fruit abscission normally takes place. This causes the fruit to remain tightly attached to the calyx. The inflorescences of *j* also develop leaves in addition to flowers, resulting in plants with a dense proliferation of leaves, branches, attenuated stems, and frequent unfruitfulness (Rick & Savant 1955). However, both problems are relieved by combining *j* with *sp* and when this is done the attachment of the fruits remains sufficiently strong that the vibration associated with mechanical harvesting does not cause the fruit drop which may occur when *sp* is used alone.

Major genes controlling branching habit and fruit position in the related genus *Capsicum* (peppers) have also been identified and may prove useful to developing a mechanical harvesting system (McCammon & Honma 1984).

## Dried pea

The availability of two major recessive genes in pea that change leaf structure and thereby canopy structure has stimulated a number of studies to improve the yield in the dried pea crop (Hedley & Ambrose 1981). The pea leaf of normal plants is pinnately compound and composed of two large foliaceous stipules, one or more pairs of subterminal leaflets, and terminal tendrils. Incorporation of the *af* gene converts the leaflets into tendrils and the *st* gene reduces stipule size. When both genes are present in the homozygous state, the phenotype is said to be 'leafless', and when the *af* mutation is present by itself, the phenotype is designated 'semi-leafless'. Both types have improved standing ability compared with peas with the usual foliage because their greater number of tendrils interweave and provide support for the plants. The crop dries more rapidly and has a reduced risk of disease. Light penetration through the canopy is also improved.

Anatomical studies (Meicenheimer *et al.* 1983) showed that the multiple tendrils of *af* plants develop from numerous secondary primordia that form at the position of the leaflet primordia. The tendrils develop from radial marginal meristems like those of normal tendrils, and both adaxial and marginal meristems are absent. Thus the *af* gene appears to establish tendrils in new positions without causing changes in their appearance. The *st* gene reduces stipule size by reducing the duration of cell division in both the abaxial and adaxial stipule meristems. Knowledge of the ontogenetic basis of mutant development is important because it assigns the time and location at which the morphological changes are initiated.

The leafless plants appear to have efficient net $CO_2$ fixation and are able to translocate sufficient photosynthate to the developing pods (Hedley & Ambrose 1981). At relatively low planting densities (16 plants per square metre) canopies produced by leafless plants have half the total dry mass attained by leafed plants but at densities in excess of 100 plants per square metre the leafless canopies were more productive (Hedley & Ambrose 1981). The leafless phenotypes, however, showed a reduced relative growth rate compared with the leafed plants and as a result produced less biomass per unit area after 60 days, reducing their value as a crop (Pyke & Hedley 1983). In contrast growth of the semi-leafless type (with normal stipules) was similar to that of leafed plants, suggesting to Pyke & Hedley (1983) that it might provide a better crop phenotype than the double mutant. Their proposal has been supported by independent studies carried out in Wisconsin (Wehner & Gritton 1981).

The attempt to define a crop phenotype for dried peas, like the research with dwarf rice and determinate processing tomatoes, has involved substantial analyses of interactions between plants and the environment in which they are grown. The genetic manipulations proved generally straightforward compared with the intensive investigations of the crop ecology that followed, since after the genes were incorporated into a wide range of genetic backgrounds the breeder had to assess how the new genotypes performed in different environments.

Population biologists who undertake genetic analysis of the differences between plants already growing in nature have to tackle the reverse problem. Because the plants they study could generally be considered successful, though perhaps divergent in fitness, their task is to identify the genetic basis of the morphological and physiological traits that confer success. This might be done by examining the relative performance of $F_2$ segregants and identifying the plants that differed greatly in fitness. Correlating the fitness differences with differences in specific characters would permit 'the character states to be "mapped" onto fitness', as suggested by Antonovics (1984).

## Shoot elongation

Other studies in pea concern the effects of gene interaction on shoot elongation and furnish one of the most detailed genetic analyses of plant architecture. In this species, internode length is governed by five major loci and internode number by five additional loci. Most of the phenotypes produced by different combinations of alleles at the ten genes have been examined in a series of elegant studies reported in Murfet (1977a, b), Reid et al. (1983), and Reid & Murfet (1984).

The genes governing internode length are associated with gibberellin (GA) metabolism. The tall *Le*/dwarf *le* alternative was one of the classic traits studied by Mendel. Homozygosity of the recessive *le* allele reduces internode length by 40–60% and causes the stem to have a zigzag appearance. Biochemical studies, making use of gas chromatography and mass spectrometry to identify GAs, established that *le le* plants cannot carry out the conversion of $GA_{20}$ to $GA_1$ in the shoot apical region, presumably because they lack the necessary enzyme (Ingram et al. 1983, 1984). The unavailability of $GA_1$ is the specific cause of reduced internode length, but how this is accomplished is not understood. Of interest is that $GA_{20}$, which is an intermediate metabolite in the elongating shoot, is apparently the active GA in the elongating fruit pod. ($GA_1$ is not present in this organ.)

Alleles at the *La*, *Cry*, *Lm*, and *Na* loci also affect internode length and, in different combinations with *Le* or *le*, produce a series of length phenotypes. The anatomical effects of *La* and *Cry* mimic the effects of applied GA, which is thought to mean that they are also involved in GA metabolism (Reid et al. 1983). The *na* allele eliminates the phenotypic difference between *Le* and *le* and may block an early step in GA biosynthesis (see below). The five genes illustrate both the complexity of genetic control of internode length and the ease with which single gene substitutions can change the phenotype.

A comparable set of genes affecting internode lengths has been identified in maize (*Zea mays*). Mutants at these genes cause dwarfing by reducing internode lengths as much as five- to tenfold, and at least four of them appear to block particular steps in the pathway leading to synthesis of $GA_1$, which Phinney & Spray (1982) believe is the only active GA in maize shoot elongation. One of these mutants ($d_1$) prevents the conversion of $GA_{20}$ to $GA_1$ (Spray et al. 1984). Its resemblance to *le* in pea suggests the two gene loci may be homologous.

Node number also affects shoot length. In pea, five loci govern node number by affecting the time of transition from vegetative to reproductive growth. The loci act to determine whether plants are early developing (flowering at a low node), intermediate, or late developing (flowering at a high node) (Murfet 1977b; Reid & Murfet 1984). Separation of the phenotypes governed by the five loci depended on close attention to conditions of cultivation. Complete diagnosis required growth under both short and long days and particular temperature régimes. *Sn* and *Hr* govern response to photoperiod perhaps by controlling the synthesis of a flowering inhibitor in the shoot. Gene *E* promotes flowering by reducing expression of *Sn* in the cotyledons. The four alleles identified at *Lf*, in homozygous condition, result in a minimum flowering node of 15, 11, 8, or 5, respectively, indicating a remarkable precision of effect, postulated to occur by influencing the sensitivity of the shoot apex to flowering stimuli (Murfet 1975). The *veg* gene at the fifth locus prevents flowering over a wide range of photoperiod and temperature régimes and overrides the effects of the other loci. However, segregation at *Sn* and *Hr* was still identifiable in *veg* plants by attention to pleiotropic effects other than flowering such as the appearance of the apical bud, production of lateral branches, or change in growth rates. The differences among the *Lf* alleles, however, were completely obliterated by *veg* (Reid & Murfet 1984).

It is not unusual for architectural traits to be affected by genes that govern the expression of a number of characters. It is sometimes possible to separate the several effects of these genes by placing them on different genetic backgrounds (Williams 1960). Such experiments probably occur in nature frequently following hybridization and could be studied once major genes are identified.

In contrast to the mutants that block the synthesis of GA in the pea shoot, an apparent lesion in the GA-sensing mechanism or the presence of GA antagonists may be responsible for the semidwarf stature of many high yielding wheats. In the 1950s dwarfing genes from the Japanese variety Norin 10 were incorporated into many wheat cultivars. Initial attempts to determine the number of genes conditioning the dwarf phenotype were not successful because final plant height was influenced by many genes in addition to the dwarfing ones (Law & Gale 1979). The situation was clarified when it was shown that Norin 10 and related strains exhibited a modified response to exogenous GA, which in other wheats causes stem and leaf elongation, but in these strains stimulated tillering without affecting elongation (Gale & Marshall 1975). The semidwarf plants were found to have high endogenous levels of GA. The recognition that they were insensitive to applied GA led to the identification of the responsible genes, now designated *Rht1* and *Rht2*. The result demonstrated that the quantitative difference in plant height was actually a qualitative difference in response to GA (Law & Gale 1979).

*Other genes affecting plant form*

A number of other genes that affect architecture in various crop plants have been identified: determinacy in soybean (*Glycine max*) (Hartung et al. 1981; Bernard 1972); climbing in dry bean (*Phaseolus vulgaris*) (Kretchmer et al. 1979); growth habit in lentil (*Lens culinaris*) (Ladizinsky 1979); determinacy in muskmelon (*Cucumis melo*) (Paris et al. 1984); leaf number and determinacy in cucumber (*Cucumis sativus*) (Miller & George 1979); internode length in watermelon (*Citrullis lanatus*) (Liu & Loy 1972); and branching in sunflower (*Helianthus annuus*) (Hockett & Knowles 1970). For a general review of other genes modifying crop plant architecture see Coyne (1980).

## Discussion

The studies described above demonstrate that the architecture of many crop plants has been successfully modified by changing only a few morphological traits such as length and number of internodes, presence and orientation of branches, habit, and determinacy. Breeders have studied these traits because genes that affected them in large ways were available and they could be placed into a range of different genetic backgrounds to assess how they affected yield. But the characters examined represent only a small proportion of those that influence plant architecture. Little or no information, for example, is available about the genetic basis of branch angle, location of abscission zones, pattern (relative timing) of lateral bud release, relative rates of branch and stem elongation, orientation of parts (nodding, horizontal, erect), or ability to form adventitious roots.

Genetic analyses of the differences in expression of these and other characters will have to be done to understand how changes that operate at the level of particular metamers are multiplied into the architecture of the whole plant. Interactions between the two phenotypic levels were not directly addressed in the crop studies. Nevertheless, they do furnish important evidence. Nearly all of it was consistent with the hypothesis that the growth and differentiation of metamers are tightly coordinated by factors that operate at the whole plant level. Evidence for the lack of metamer independence can be found in the studies of changes in internode length, graft transmissibility, and also from considerations of allometric correlations.

Reid & Murfet (1984) described a cross in pea between a 'cryptodwarf' with genotype $le\ la\ cry^c\ Na\ Lm$ and a tall line with genotype $Le\ La\ cry^c\ Na\ Lm$ which segregated four distinct height classes in the $F_2$ in the 9:3:3:1 ratio expected for two independently assorting loci. The stem internodes of plants in each of the four classes showed different patterns of length increase (figure 3). The significant point is that within each pattern, successive internodes elongated in a regular and correlated fashion until a particular maximum length was attained, and then successive internodes maintained this same length. Thus the effect of a single gene was able to coordinate changes involving many internodes. Similar correlated increases in successive stem

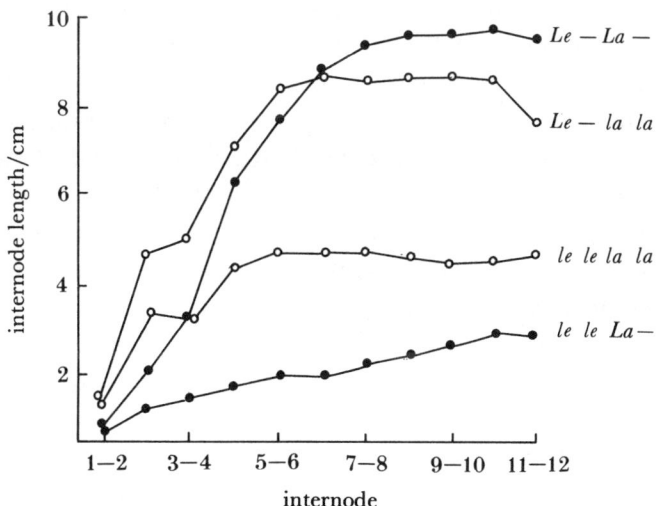

FIGURE 3. Mean internode length against internode number in an $F_2$ progeny in pea (*Pisum sativum*) segregating for $Le/le$ and $La/la$ loci (modified from Reid *et al.* 1983, figure 2).

internode lengths of different genotypes were also documented in *Cucumis melo* (Paris *et al.* 1984) and *Trifolium repens* (Booysen & Laude 1964). But in these latter examples the genetic situation is less certain.

I am aware of only a single example in which the length of a particular internode is controlled by a gene locus independently of other internodes on the same stem. The recessive gene *eui* (elongated uppermost internode) in rice nearly doubles the length of the uppermost internode but has no affect on other internodes (Rutger & Carnahan 1981). Since panicle length is also increased in plants with the *eui* gene, it is possible that the increase in uppermost internode length is a pleiotropic consequence of increased levels of GA produced in the larger panicles. The unusual rice internode is in the same position as the long internode basipetal to the flowering node in *Trifolium repens* which was shown to be subject to GA influence (Booysen & Laude 1964). GA biosynthesis presumably generally uses substrates from different parts of the plant.

A second example of correlative differentiation is the study of graft-transmissibility associated with the *Na* locus in pea (Reid *et al.* 1983; Ingram *et al.* 1983). *Na* was found to be transmissible across grafts but its effect was lost when the recessive allele *na* was substituted. In contrast, transmissible effects were not found for any alleles at the *Le*, *La*, and *Cry* loci. The transmissible substance in *Na* plants, which is produced in largest quantities in mature stem and leaf tissues, was able to promote elongation in dwarf *na* scions. Since the addition of *na* eliminates the phenotypic difference between *Le* and *le* plants (long versus short internodes), and since *Le* is responsible for the specific conversion of $GA_{20}$ to $GA_1$ in the upper shoot, a final step in GA biosynthesis, it is likely that the *na* allele blocks the transmission or synthesis of a signal or a substrate that comes from the mature tissues. Thus *na* appears to govern an early step in GA metabolism. Such evidence also suggests whole-plant correlation.

Consideration of allometric relationships reveals additional examples of developmental correlations. A remarkable allometry is that between leaf number and trunk diameter in trees (Rothacher *et al.* 1954; see also White 1979). Among tree species, for trunks of a given diameter leaf number is inversely related to leaf size (White 1979). A more apparent allometry is that between the size of a plant organ and the size of the meristem from which it develops, a relationship first noted by Sinnott, according to Grafius (1978). Grafius (1978) added the valuable addendum that organs that develop from the same meristem are less subject to genetic manipulation than those arising from different meristems. For example, the areas of successive leaves on a single culm of barley show strong positive correlations, with the coefficients decreasing with distance between leaves (Fowler & Rasmusson 1969). Grafius (1978) pointed out that the area of such leaves could not be changed independently by breeding. Likewise the rate of growth in length and width of the same organ (fruit shape or leaf shape) could not be independently selected. Their shapes might be changed by selecting a different allometric constant between length and width or by changing the duration of growth. In contrast allometries between organs arising on different meristems are more readily changed by genetic manipulations because they are largely physiological (hormonal control, source–sink, inhibitory, etc.) and change with the environment. Grafius (1978) gives the example from a breeding programme in barley in which he was able to uncouple number of tillers per unit area from number of seeds per head. The developmental evidence in these examples suggests that although 'a plant is a population of parts' (Harper 1984), the growth and differentiation of the parts are coordinated to a large extent.

The primary reason why plant population biologists should study the genetic basis of architectural differences is to find out the number of gene changes they represent. This information is necessary to account for differences in fitness that can be assigned to architectural characters. The literature contains many excellent examples of such fitness differences (Dirzo & Sarukhán 1984) but, since the genotypes that exhibit them have not been intercrossed in a genetic design, the number and type of differences among the genotypes are not known. Our ignorance in this area is nearly complete.

The many studies in crop plants indicate that genetic analyses can also be done in wild plants. Since the expression of architectural differences is often discrete, many of them may prove to have simple patterns of inheritance (this review and Gottlieb 1984). The likelihood of detecting genes with large phenotypic effects will be increased by careful attention to the morphological components including their anatomy and development as well as to devising environmental conditions that maximize genotypic differences in response.

I thank Vera S. Ford for her thoughtful comments on the manuscript.

## References

Abrahamson, W. G. 1980 Demography and vegetative reproduction. In *Demography and evolution in plant populations* (ed. O. T. Solbrig), pp. 89–106. Oxford: Blackwell Scientific Publications.

Antonovics, J. 1984 Genetic variation within populations. In *Perspectives on plant population ecology* (ed. R. Dirzo & J. Sarukhán), pp. 229–241. Sunderland, Massachusetts: Sinauer Associates Inc.

Aquino, R. C. & Jennings, P. C. 1966 Inheritance and significance of dwarfism in an Indica rice variety. *Crop Sci.* **6**, 551–554.

Bernard, R. L. 1972 Two genes affecting stem termination in soybeans. *Crop Sci.* **12**, 235–239.

Booysen, P. de V. & Laude, H. M. 1964 Influence of flower initiation and development on internode growth in the Ladino clover stolon. *Crop Sci.* **4**, 520–524.

Clausen, J. 1951 *Stages in the evolution of plant species.* Cornell University Press: Ithaca.

Coyne, D. P. 1980 Modification of plant architecture and crop yield by breeding. *Hort. Sci.* **15**, 244–247.

Dirzo, R. & Sarukhán, J. (eds) 1984 *Perspectives on plant population ecology.* Sunderland, Massachusetts: Sinauer Associates Inc.

Fowler, C. W. & Rasmusson, D. C. 1969 Leaf area relationships and inheritance in barley. *Crop Sci.* **9**, 729–731.

Gale, M. D. & Marshall, G. A. 1975 The nature and genetic control of gibberellin insensitivity in dwarf wheat grains. *Heredity* **35**, 55–65.

Gottlieb, L. D. 1984 Genetics and morphological evolution in plants. *Am. Nat.* **123**, 681–709.

Grafius, J. E. 1978 Multiple characters and correlated response. *Crop Sci.* **18**, 931–934.

Harper, J. L. 1981 The concept of population in modular organisms. In *Theoretical ecology* (ed. R. M. May), 2nd edn, pp. 53–77. Princeton University Press.

Harper, J. L. 1984 Foreword. In *Perspectives on plant population ecology* (ed. R. Dirzo & J. Sarukhán), pp. xv–xviii. Sunderland, Massachusetts: Sinauer Associates Inc.

Harper, J. L. & Bell, A. D. 1979 The population dynamics of growth form in organisms with modular construction. In *Population dynamics* (ed. R. M. Anderson, B. D. Turner & L. R. Taylor), pp. 29–52. Oxford: Blackwell Scientific Publications.

Hartung, R. C., Specht, J. E. & Williams, J. H. 1981 Modification of soybean plant architecture by genes for stem growth habit and maturity. *Crop Sci.* **21**, 51–56.

Hedley, C. L. & Ambrose, M. J. 1981 Designing 'leafless' plants for improving yields of the dried pea crop. *Adv. Agron.* **34**, 225–277.

Hockett, E. A. & Knowles, P. F. 1970 Inheritance of branching in sunflowers, *Helianthus annuus* L. *Crop Sci.* **10**, 432–436.

Ingram, T. J., Reid, J. B., Murfet, I. C., Gaskin, P., Willis, C. L. & MacMillan, J. 1984 Internode length in *Pisum*. The *Le* gene controls the 3β-hydroxylation of gibberellin $A_{20}$ to gibberellin $A_1$. *Planta* **160**, 455–463.

Ingram, T. J., Reid, J. B., Potts, W. C. & Murfet, I. C. 1983 Internode length in *Pisum*. IV. The effect of the *Le* gene on gibberellin metabolism. *Physiol. Plant.* **59**, 607–616.

Kaiser, S. 1935 The factors controlling shape and size in *Capsicum* fruits; a genetic and developmental analysis. *Bull. Torrey Bot. Club* **62**, 433–454.

Kretchmer, P. J., Laing, D. R. & Wallace, D. H. 1979 Inheritance and morphological traits of a phytochrome-induced single gene in bean. *Crop Sci.* **19**, 605–607.

Ladizinsky, G. 1979 The genetics of several morphological traits in the lentil. *J. Hered.* **70**, 135–137.

Law, C. N. & Gale, M. D. 1979 Cytological markers and quantitative variation in wheat. In *Quantitative genetic variation* (ed. J. N. Thompson & J. M. Thoday), pp. 273–293. New York: Academic Press.

Liu, P. B. W. & Loy, J. B. 1972 Inheritance and morphology of two dwarf mutants in watermelon. *J. Am. Soc. Hort. Sci.* **97**, 745–748.

McCammon, K. R. & Honma, S. 1984 Genetics of the 'umbrella' branching habit in *Capsicum annuum* L. *Theor. appl. Genet.* **68**, 541–545.

Meicenheimer, R. D., Muehlbauer, F. J., Hindman, J. L. & Gritton, E. T. 1983 Meristem characteristics of genetically modified pea (*Pisum sativum*) leaf primordia. *Can. J. Bot.* **61**, 3430–3437.

Miller, G. A. & George, W. L. 1979 Inheritance of dwarf and determinate growth habits in cucumber. *J. Am. Soc. Hort. Sci.* **104**, 114–117.

Murfet, I. C. 1975 Flowering in *Pisum*: multiple alleles at the *Lf* locus. *Heredity* **35**, 85–98.

Murfet, I. C. 1977a Environmental interaction and the genetics of flowering. *A. Rev. Pl. Physiol.* **28**, 253–278.

Murfet, I. C. 1977b The physiological genetics of flowering. In *The physiology of the garden pea* (ed. J. F. Sutcliffe & J. S. Pate), pp. 385–430. London: Academic Press.

Paris, H. S., Nerson, H. & Karchi, Z. 1984 Genetics of internode length in melons. *J. Hered.* **75**, 403–406.

Phinney, B. O. & Spray, C. 1982 Chemical genetics and the gibberellin pathway in *Zea mays*. In *Plant growth substances* (ed. P. F. Wareing), pp. 101–110. New York and London: Academic Press.

Prazmo, W. 1965 Cytogenetic studies in the genus *Aquilegia*. III. Inheritance of the traits distinguishing different complexes in the genus *Aquilegia*. *Acta Soc. Bot. Pol.* **34**, 403–437.

Pyke, K. A. & Hedley, C. L. 1983 The effect of foliage phenotype and seed size on the crop growth of *Pisum sativum*. *Euphytica* **32**, 193–203.

Reid, J. B. & Murfet, I. C. 1984 Flowering in *Pisum*: a fifth locus, *Veg. Ann. Bot.* **53**, 369–382.

Reid, J. B., Murfet, I. C. & Potts, W. C. 1983 Internode length in *Pisum*. II. Additional information on the relationship and action of loci *Le*, *La*, *Cry*, *Na* and *Lm*. *J. exp. Bot.* **34**, 349–364.

Rick, C. M. 1978 The tomato. *Scient. Am.* **245**, 77–87.

Rick, C. M. & Butler, L. 1956 Cytogenetics of the tomato. *Adv. Genet.* **8**, 267–382.

Rick, C. M. & Savant, A. C. 1955 Factor interactions affecting the phenotypic expression of the jointless character in tomatoes. *Proc. Am. Soc. Hort. Sci.* **66**, 354–360.

Rothacher, J. S., Blow, F. E. & Potts, S. M. 1954 Estimating the quantity of tree foliage in oak stands in the Tennessee Valley. *J. For.* **52**, 169–173.

Rutger, J. N. & Carnahan, H. L. 1981 A fourth genetic element to facilitate hybrid cereal production – a recessive tall in rice. *Crop Sci.* **21**, 373–376.

Sinnott, E. W. & Kaiser, S. 1934 Two types of genetic control over the development of shape. *Bull. Torrey Bot. Club* **61**, 1–7.

Spray, C., Phinney, B. O., Gaskin, P., Gilmour, S. J. & MacMillan, J. 1984 Internode length in *Zea mays* L. The dwarf-1 mutation controls the 3β-hydroxylation of gibberellin $A_{20}$ to gibberellin $A_1$. *Planta* **106**, 464–468.

Wallace, D. H., Ozbun, J. L. & Munger, H. M. 1972 Physiological genetics of crop yield. *Adv. Agron.* **24**, 97–146.

Wehner, T. C. & Gritton, E. T. 1981 Horticultural evaluation of eight foliage types of peas near isogenic for the genes *af*, *tl*, and *st*. *J. Am. Soc. Hort. Sci.* **106**, 272–278.

White, J. 1979 The plant as a metapopulation. *A. Rev. Ecol. Syst.* **10**, 109–145.

White, J. 1984 Plant metamerism. In *Perspectives on plant population ecology* (ed. R. Dirzo & J. Sarukhán), pp. 15–47. Sunderland, Massachusetts: Sinauer Associates Inc.

Williams, W. 1960 The effect of selection on the manifold expression of the 'suppressed lateral' gene in the tomato. *Heredity* **14**, 285–296.

## The influence of neighbours on the growth of modular organisms with an example from trees

By M. Franco†

*Unit of Plant Population Biology, School of Plant Biology,*
*University College of North Wales, Bangor, Gwynedd LL57 2UW, U.K.*

The growth and form of a modular organism is determined by the rigid rules of iteration (branching) and the differential response of each growing point to the local conditions around it. The degree of response of each individual module is itself dependent on the degree of physiological integration of the whole organism. Morphological continuity is a requisite but not a guarantee of physiological integration. In general, 'phalanx' growth-forms show more physiological integration than 'guerrilla' growth-forms. Trees, as an example of morphologically integrated modular organisms, show a variety of responses to the presence of modules both of the same and of different species. When two modules interact, three extreme responses are possible: (i) both modules stop growing or change their orientation, or both; (ii) one module is inhibited while the other continues its growth; (iii) neither is affected by the presence of the other.

The first case produces a clear separation ('shyness') between neighbouring modules both within and between trees. The second case produces a hierarchy of dominance-suppression. Finally, failure to 'recognize' the presence of a neighbour module may result in physical damage by abrasion of both participants. Under certain circumstances, this can also produce a visual impression of 'shyness'. The importance of this 'recognition' mechanism is discussed for both modular animals and plants.

### Introduction

Many modular organisms are sessile and grow by the iteration of some basic units of construction (modules *sensu* Harper 1981). In contrast to free-living organisms, sessile forms cannot usually escape the effects of competition by moving away from zones of high density of neighbours and establishing themselves in areas of lower density. The complex behavioural patterns exhibited by mobile animals are absent. One finds instead that both 'search' for resources and 'escape' from competitors occur through the adjustment of the morphological arrangement of modules. Two extremes of this morphological display are represented by the so-called 'guerrilla' and 'phalanx' growth forms (Lovett Doust 1981). Modules of 'guerrilla'-type organisms are relatively widely scattered and separated from each other by long stolons, rhizomes or internodes; they are often produced by species that colonize and explore recently open habitats where neighbours are sparse. The 'phalanx'-type form is more common in crowded environments; modules are densely packed and resist the invasion of space by modules of other organisms; they do not tend to explore but exploit the habitat already occupied.

Regardless of their position along this morphological continuum, as the population of modules grows bigger the effects of density are sensed in the immediate surroundings of the

† Present address: Instituto de Biología, Universidad Nacional Autónoma de México, Ciudad Universitaria, Deleg. Coyoacán, 04510 México, D.F., México.

organism. The critical factor for sessile forms is the capture of space as a means of gaining access to more resources.

The response to local density has been studied in both sessile animals and plants. For marine animals, it has generally been assumed that competition occurs when there is limited physical space for attachment of organisms (Paine 1984). Buss (1980) has shown, however, that this competition for space can be mediated by competition for food. Neighbourhood models that predict the effect of local density on the performance of individuals have been developed for plants (Pacala & Silander 1985; Silander & Pacala 1985). From these studies it is clear that the performance of a modular organism is a function of the space that it captures through the multiplication of its constituent parts. An extension of this approach to account for space capture by individual modules belonging to the same and to different genetic individuals (genets) has been presented by Bülow-Olsen et al. (1984).

Since module display and resource capture are intimately related, one might expect individual modules to respond locally to the conditions in their immediate neighbourhood. Those modules that find themselves in favourable conditions should grow and multiply, while those in less propitious environments should produce copies of themselves at a lower rate or have increased risks of mortality. A modular organism with several axes of growth radiating into neighbourhoods containing different amounts of resources should grow at different speeds in different directions.

A growth programme that is responsive to local conditions and is not constrained by fixed, inflexible rules might seem the most efficient iterative method of growth. The adaptive or opportunistic reiteration of growth in angiosperms (Hallé et al. 1978) might then be interpreted as an improvement on the rather inflexible pattern of shoot module display of most gymnosperms. A programme that modifies the rules of growth with time has been called non-stationary by Waller & Steingraeber (1986) (and see Bell in this volume). If non-stationarity is expressed locally, as the conditions in different parts of the modular organism become more dissimilar, this is termed Markovian (Waller & Steingraeber 1986). This kind of growth programme implies a high degree of independence in the behaviour of each individual module.

The mechanisms by which modular organisms interact are still not well documented. A first step in understanding these mechanisms is the identification of resource depletion zones (RDZs) (Harper 1985). The RDZ of a module is the area of space around it where resources have been depleted to such an extent that there are insufficient resources to support the unimpeded growth of additional modules. In the simplest case, resource depletion is represented by situations in which one or a series of modules produce an umbra on modules positioned on the lee side with respect to the direction of resource movement. Examples of this kind are provided by upstream colonies of bryozoans that, by capturing food particles, deplete downstream colonies of essential resources (see, for example, Okamura 1984). The light depletion gradient in plant communities (Monsi & Saeki 1953) is an expression of this semicontinuous diminution of resources along a feeding hierarchy determined by the spatial position of modules. Within a physically interconnected modular organism (for example, a tree) there can be a strong hierarchy, determined not only by the spatial position of modules along the gradient of resources (height of the canopy) but also by the physiologically imposed hierarchy of (meristem) dominance (Maillette 1982 a, b).

## Degree of integration of modules

The degree of integration of a modular organism varies depending on the particular species and the conditions under which an individual develops (Winston 1978, 1979; Jackson 1979; McKinney 1984). Among colonial marine animals and pasture herbs the degree of integration is correlated with their growth form: guerrilla-type forms show less colony integration than phalanx-type forms (Jackson 1979; Schmid 1986). It is also correlated with the degree of specialization of individual modules: the presence of polymorphic zooids with specialized functions in bryozoans require a high degree of interdependence among members of the clone (McKinney 1984).

Even in highly integrated modular organisms, however, (such as siphonophores among animals, and trees among higher plants) the question of whether or not modules react individually to the conditions around them is far from settled. Morphological continuity between modules does not necessarily imply a high degree of physiological interdependence (see, for example, St Pierre & Wright 1972). In perennial forest herbs the degree of integration seems to be correlated with the risks of death of the individual ramet (Ashmun *et al.* 1982). Some experiments with grasses and other herbs indicate that physiological re-integration of physically connected parts can occur after defoliation or shading, at least for a brief period of time. The disturbed parts act in this case as sinks for assimilates (Ashmun *et al.* 1982).

The degree to which individual modules are independent or interdependent in their carbon economy has been reviewed for plants by Watson & Casper (1984). They concluded that, although there may be physical communication through very closed vascular systems, the movement of carbon between leaves is in general restricted by the vascular anatomy to those in neighbouring orthostichies. In experiments with grasses and sedges it has been observed that the flow of assimilates goes preferentially in the direction from parent to offspring meristems, although mature offspring can contribute reciprocally to parent (Marshall & Sagar 1968; Ong & Marshall 1979; Noble & Marshall 1983; Tietema 1980). The assumption that this flow determines the performance of daughter modules has been made in the modelling of branching in trees (Honda *et al.* 1981; Borchert & Honda 1984). Other simulation models do not include this assumption (see, for example, Aono & Kuni 1985) but can simulate branching structures that look like real trees (see also Bell, this symposium). These similarities do not, however, resolve the issue of whether the assumption of physiological interdependence of modules is necessary to mimic the morphology of the organism.

## Growth and form in modular organisms

While growth may be defined as the change in the number of modules comprising an organism, form is a more elusive quantitative concept. To make comparisons on a continuous numerical scale, form can be defined as 'a relationship between real numbers' (Medawar 1945). If for example, we define form as the ratio between length ($l$) and width ($w$), a square will always be equal in form to another square, no matter how different in size they might be. However, a rectangle with $l > w$ will have a different form because this quotient will be greater than one. The reference point can be an external one and, for example, two rectangles with equal $l$ and $w$, but the one lying horizontal and the other standing vertically, can be said to be different in form. The change in form from a rectangle with infinite width and finite length to a rectangle of finite width and infinite length is then a continuous function which can be used as a reference model in the study of forms on the plane.

This approach may be useful for unitary organisms but not in general for modular organisms. Since growth in the latter is potentially unlimited and branching (budding) can produce many different shapes, form cannot easily be defined. In modular organisms the most useful approach has been to identify the form of the individual module and then tackle questions on the nature of the self-similarity of the whole organism (for theories of self-similarity see Günther (1975); MacDonald (1983)). Two examples of this approach are the pipe-model theory of tree form (Shinozaki *et al.* 1964) and a similar approach to the study of growth form in fossil arborescent cheilostome bryozoans by Cheetham *et al.* (1981). These studies aim at the description of overall form as the result of a tightly controlled process that reproduces itself at all levels within the organism.

### Neighbour effects on the growth and form of modular organisms

When modular organisms compete with each other the effect can be expressed as a change in the birth and death of individual modules (Bazzaz & Harper 1977; Whitney 1982; Franco 1985a; Hartnett & Bazzaz 1985; Jones 1985). The net result is a decrease in the growth rate of the population of modules. This, however, does not necessarily modify the self-similarity of the whole organism. So, for example, the linearity between the cross-sectional area of conducting tissues and the amount of leaves supplied, predicted by the pipe model, remains unaltered when trees are grown either isolated or in the presence of neighbours (Franco 1985b).

#### *An approach to the analysis of growth and form of competing modular organisms*

In this paper, I propose that a convenient way of summarizing the changes in growth and form of modular organisms, as determined by interference with neighbours, may be provided by focusing on the symmetry of the whole organism. The method used is the analysis of the distribution of modules around the central axis, the point of initial attachment to the substrate of the larvae or propagule, by using circular statistics (Mardia 1972; Batschelet 1981).

In the ideal situation, the polar coordinates of each module are recorded (figure 1). When modules are too numerous or when they are grouped on axes radiating horizontally from the central, vertical axis, the size ($w$, mass, diameter, length, etc.) of each secondary axis is measured and its orientation ($\theta$) with respect to a reference point (north) is recorded. The rectangular coordinates of each secondary axis inserted in a reference circle of radius equal to one are calculated and weighted by their respective size. The rectangular coordinates of the mean angle are a weighted mean of the rectangular coordinates of each axis and are calculated according to the formulae:

$$x = \sum_i w_i \cos\theta_i / \sum_i w_i; \quad y = \sum_i w_i \sin\theta_i / \sum_i w_i.$$

The degree of asymmetry of the organism is estimated by the parameter $r$. This parameter is calculated as the Euclidian distance:

$$r = (x^2 + y^2)^{\frac{1}{2}}.$$

In physical terms, the parameter $r$ represents the length of the standardized vector joining the geometrical centre of the unit circle with the centre of mass of the circular distribution of axes. When the circular distribution of the axes is perfectly uniform the value of $r$ equals zero (figure 1a). If these axes are bigger or tend to reorient themselves in a particular direction, the centre of mass will be eccentric to the geometric centre of the unit circle. This will be detected as an increase in the value of $r$ (figure 1b). A value of $r$ equal to one means that all the axes grow in the same direction (figure 1c). In practice, the value of the vector $r$ will always lie between zero and one. The statistical significance of $r$ as a measure of concentration depends on the sample size (number of modules or, for grouped data, number of axes of the organism). With the exception of figure 3, a value of $r$ significantly different from zero ($p < 0.10$) is represented in this study by a continuous arrow of length $r$. A dotted arrow means that $r$ is not significantly different from zero.

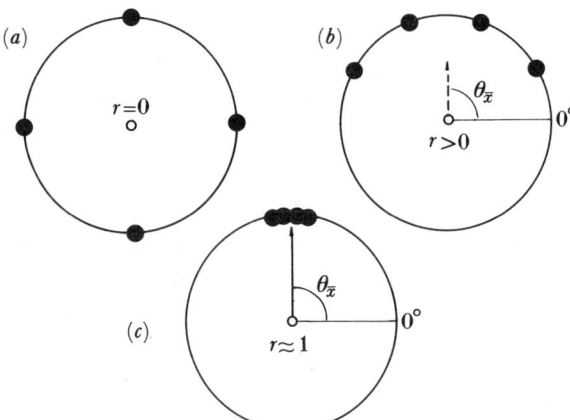

FIGURE 1. Mean angle ($\theta_{\bar{x}}$) and angular concentration ($r$) of three groups of four hypothetical points centred in a circle of radius 1.

The angular orientation of this vector ($\theta_x$) can then be calculated (Batschelet 1981) as:

$$\begin{aligned}
&\arctan(y/x) \quad \text{if} \quad x > 0, \\
&180° + \arctan(y/x) \quad \text{if} \quad x < 0, \\
&90° \quad \text{if} \quad x = 0 \quad \text{and} \quad y > 0, \\
&270° \quad \text{if} \quad x = 0 \quad \text{and} \quad y < 0, \\
&\text{undetermined} \quad \text{if} \quad x = 0 \quad \text{and} \quad y = 0,
\end{aligned}$$

$r$ is then an index of the degree of asymmetry of the organism and hence a measure of the relative intensity of competition on different sides of it. The method is similar to the vector analysis of Watson & Cook (1982) but is simpler and more amenable to statistical analysis and interpretation.

### Neighbour effects on the symmetry of branching

The degree of asymmetry of competing modular organisms was investigated by using the annual plant, *Kochia scoparia* var. *culta* (Chenopodiaceae), as an experimental model of a tree, and three conifer species, *Picea sitchensis*, *Pinus contorta*, and *Larix kaempferi*.

Plants of *Kochia* were grown in a greenhouse in pots 23 cm in diameter and 18.0 cm in depth, filled with John Innes no. 2 compost. There were six pots and nine plants in the design. Three plants, approximately one-month-old, were each transplanted to the centre of separate pots. The other six plants were grown in pairs in three similar pots. These paired plants were placed opposite to each other along the diameter of the pot and separated by a distance of 8 cm (4 cm from the centre of the pot). By growing single and paired plants it was intended to compare the asymmetry created by the presence of a neighbour with the radial symmetric growth of isolated plants. After transplanting, the plants were allowed to grow from 18 May to 7 August 1984. To prevent plants growing in the direction of a particular, unidirectional light source, the pots were periodically rotated. They were also periodically randomized on the bench in the glasshouse. On 7 August each branch was individually separated and weighed. Because the 2/5 phyllotaxis of *Kochia* determines rather precisely the position of each branch around the central stem, it was decided to divide the crown into eight equal sectors of 45° and the position of each branch was assigned to the sector containing its main bulk. The resulting data were grouped in eight mean angular directions. For grouped data the parameter $r$ has to be corrected and for eight groups this correction factor is $\times 1.0262$ (Batschelet 1981). To relate the response in the reorientation of crown growth of paired plants a canonical correlation (Johnson & Wehrly 1977) was performed on the weighted sines and cosines of the angular orientation of the branches of each pair of plants.

In a second study, the growth response of neighbouring conifer trees was studied in Gwydyr Forest, North Wales. The intersection of three monospecific stands was chosen to investigate the response of individual branches to the presence of neighbour trees both of the same and of different species. The trees are located in the Ty'n-y-Mynydd block (national grid reference SH 768594). They had been planted in 1959 and the plot was first thinned in the winter of 1983. The data were collected in the summer of 1983, before thinning. When the data were collected the canopy was closed and in places impenetrable. This suggested that interference between neighbour trees had taken place for several years.

To investigate the degree of branch asymmetry of these trees as a cumulative process throughout their life, the size variables used were the diameter and cross-sectional area, 10 cm from the insertion of the branch on the stem, and the total length of the network of each branch, that is the length of the main branch plus the length of second and higher order branches. Since leaves are produced uniformly along the branches in *Picea sitchensis* and *Pinus contorta*, network path length is directly proportional to the total number of leaves produced during the life of the branch (assuming no loss of dead shoots). For *Larix kaempferi* this is true for long shoots, but not necessarily for the short shoots. On the other hand, the cross-sectional area of a branch is assumed to be proportional to the amount of leaves present on it (that is, the prediction of the pipe-model theory (Shinozaki *et al.* 1964; Waring *et al.* 1982)). Since both branch diameter and network path length increase with the age of the branch, all three variables yield similar results. Therefore only the results using branch cross-sectional area are presented.

The orientation of each branch was measured by using a specially built protractor which could be opened and positioned around each tree. Attached to it was a compass which permitted the placing of the instrument in the same direction in all trees. Since both spruce and larch produce fairly linear stiff branches, only one measure of direction was necessary. In contrast, pine branches can become curved and reorient themselves. In this case, two measures of direction were recorded: (i) at the branch insertion; and (ii) at the branch tip, both with respect to the line formed by the centre of the tree and the magnetic north.

The position (height) of each branch along the stem and whether the branch as a whole was dead or alive were also recorded. For spruce and pine, branches were grouped in tiers. In the case of Japanese larch groups were formed every 50 cm along the stem.

The results are presented for (i) the whole tree; (ii) live and dead branches separated; (iii) individual tiers or groups of branches along the stem. In this last case it was not possible to perform a separate analysis for live and dead branches because the sample size was too small.

## Results

### Kochia scoparia var. culta

Plants of *Kochia* growing singly in pots in the absence of neighbours develop a symmetric crown whose angular concentration ($r$) is not significantly different from zero (figure 2; table 1). Consequently, their mean angle of direction (with respect to an arbitrarily defined north (0°) at the beginning of the experiment) does not have any meaning. These isolated plants, however, seem to develop slightly more asymmetric crowns as their total size increases (table 1). This means that small differences in branch growth rate are accentuated with time.

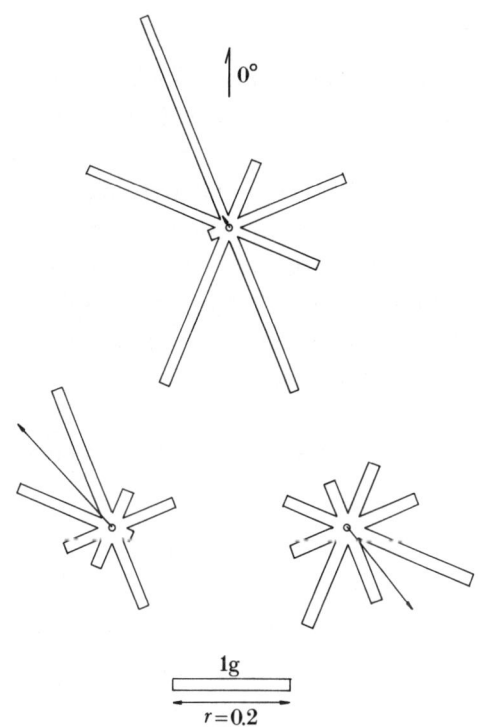

Figure 2. The symmetry of growth of the crown of one isolated and two competing plants of *Kochia scoparia* var. *culta*. The vector arrow projecting from the centre of the distribution of leaf mass in grams represents the mean angle; its length is equal to the angular concentration ($r$) of leaf mass in the crown. For scale, a vector equal to 0.2 and a bar equal to 1 g are shown.

When *Kochia* plants are grown in pairs they grow in opposite directions and away from each other (figure 2; table 1). This direction of growth is not necessarily on the expected 270–90° plane formed by the centres of the two plants and seems to be related to the direction of their spiral phyllotaxis. The degree of asymmetry of these paired plants is accentuated by their total size, bigger pairs being on the whole more asymmetric (table 1). When one of the plants dominates over the other, the smaller plant has the higher degree of asymmetry (table 1).

TABLE 1. TOTAL LEAF MASS, ANGULAR CONCENTRATION $(r)$, MEAN ANGLE $(\theta_{\bar{x}})$ AND DOMINANT CANONICAL CORRELATION $(R_c)$ FOR CONTROL AND PAIRED PLANTS OF *KOCHIA SCOPARIA* VAR. *CULTA* GROWN IN A HEATED GLASSHOUSE AT PEN-Y-FFRIDD FIELD STATION, BANGOR, NORTH WALES, IN THE SUMMER OF 1984

| control plants | | | paired plants | | | |
|---|---|---|---|---|---|---|
| leaf mass / g | $r$ | $\theta_{\bar{x}}$ / deg | leaf mass / g | $r$ | $\theta_{\bar{x}}$ / deg | $R_c$ |
| 8.786 | 0.019 | 325 | 4.615 | 0.048 | 351 | 0.836*** |
| | | | 2.893 | 0.094 | 143 | |
| 9.357 | 0.043 | 49 | 4.759 | 0.230 | 317 | 0.718*** |
| | | | 5.469 | 0.176 | 140 | |
| 12.892 | 0.141 | 20 | 4.935 | 0.156 | 354 | 0.914*** |
| | | | 6.969 | 0.115 | 211 | |

***, $p < 0.001$.

It is to be expected that neighbouring plants will grow into each other's resource depletion zones at a time determined by their size and their distance apart. Since these are continuous variables, the symmetry of branching may vary continuously from a uniform circular distribution when the plants are small or very isolated from each other, to a maximal asymmetry when the plants are large and close together. For this reason and to test the hypothesis that growth is reoriented towards the open side of the crowns, a canonical correlation between the weighted sines and cosines (Johnson & Wehrly 1977) of the branch angles for each pair of plants was performed. In all three cases, the canonical correlation was highly significant (table 1) indicating that individual plants do not integrate and average their growth in all directions but tend to grow in the direction where neighbours interfere least. This is despite the fact that the ordering of branches is only an approximation to the closeness of two 'equivalent' branches in the two plants.

*Conifer species*

Although the complexity of the interactions among several species in the natural environment might make an interpretation of dominance difficult, it proved possible to detect general patterns of crown response to neighbours in the three species studied. These responses are illustrated with three individual trees, one of each species.

Figure 3 shows the distribution of some of the trees studied at the intersection of three monospecific stands. The plot is close to an access road and the edge of the forest canopy is represented by the dotted line. To the left of this dotted line the forest spreads for more than 50 m. Only the trees in close proximity to those analysed in this study are shown. Their stems, with diameters drawn to scale, are represented by circles with letters standing for pine (P), spruce (S), and larch (L). The plot containing lodgepole pines is found to the upper left of the dashed line. The Sitka spruce stand is to its lower right. Japanese larch is restricted to the edge of the forest, along the road. The trees discussed in this and the next two figures have subscript 1.

Lodgepole pine seems to be the species that suffers most from the presence of either of the other two species. In particular, the individual at the centre of figure 3 shows clear signs of

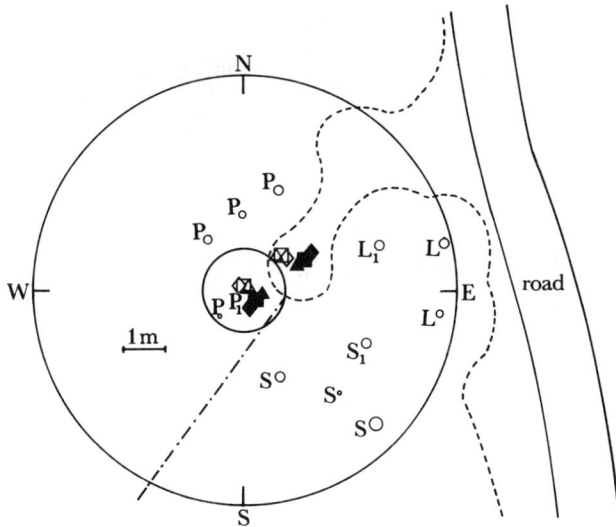

FIGURE 3. The growth response, as estimated by the angular concentration of branches ($r$), of a suppressed lodgepole pine ($P_1$) and the distribution of closest neighbour trees: (L) Japanese larch, (S) Sitka spruce, (P) lodgepole pine. The outer, bigger circle is an arbitrarily drawn circle, centred on $P_1$, of radius 1. The inner circle has radius 0.2, which, given the amount of live (71) and dead (48) branches, is an approximate significance level ($0.05 < p < 0.10$) for values of $r$ being greater than this value by chance. The triangles, squares, and diamonds represent the results of the circular analysis, that is, the values of $r$, inscribed within the unit circle. Their orientation with respect to the line formed by the centre of the circle and the magnetic north is the mean angle. The two groups of symbols inside the inner circle are the results for the angles of the branches measured at their insertion on the stem. The two groups outside the inner circle represent the results for the branch angles measured at the tips of the branches. ($\triangle$, $\blacktriangle$) Live branches; ($\diamond$, $\blacklozenge$) dead branches; ($\square$, $\blacksquare$) total number of branches; ($\triangle$, $\diamond$, $\square$) $r$ calculated by disregarding cross-sectional areas of branches; ($\blacktriangle$, $\blacklozenge$, $\blacksquare$) $r$ calculated by incorporating cross-sectional areas of branches.

suppression. This tree is surrounded by members of all the three species. Its stem diameter is only 7.1 cm compared with the diameters of the trees of the other two species which are on the average over 20 cm. All the individual pines around the experimental plot have diameters under 20 cm. Since the angular concentration of live and dead branches is almost identical, a look at the whole set of branches is sufficient to describe the direction of growth. Similarly, whether their sizes (branch cross-sectional area) are taken into account or not, the branches seem to grow and orient themselves in the same direction. The only difference in the results is between the orientation of the insertions of the branches and the orientation of their tips. For the former, no difference from a uniform circular distribution is detected (figure 3, symbols inside the inner, approximate significance level circle). This means that branches are produced in all directions. However, when the direction of the tips of the branches is used in the calculation of the rectangular coordinates of the mean angle, a significant direction of growth is observed (figure 3, symbols outside the inner circle). This direction of growth is so well established that it can be detected with or without weighting each individual branch by its cross-sectional area. As one would expect from a very shade-intolerant species, the branches of this tree have grown towards the only opening in the forest, the gap formed by the death of a (pine) tree 1.5 m northeast of it. This tree was the only one removed in this part of the forest and, counting the number of rings in the remaining stump, this removal occurred probably five years before this study was done.

Sitka spruce is the most shade-tolerant species of the three. This is reflected in the analysis

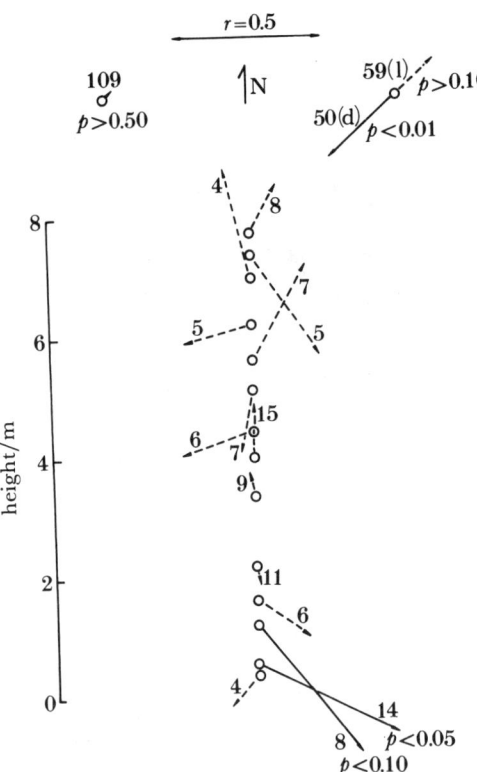

FIGURE 4. The response to neighbours of a dominant Sitka spruce ($S_1$ in figure 3). Mean angle (orientation of arrows with respect to the magnetic north) and angular concentration ($r$ is the length of the arrows) of branches are shown for: (i) all the branches in the tree (upper left; $n = 109$); (ii) live ($n = 59$) and dead ($n = 50$) branches separated (upper right); (iii) each tier of branches along the stem ($n$ shown for each tier). For visual aid in the interpretation, continuous lines are different from zero at a significance level of 0.10. For comparison a vector whose length is equal to 0.5 is also shown.

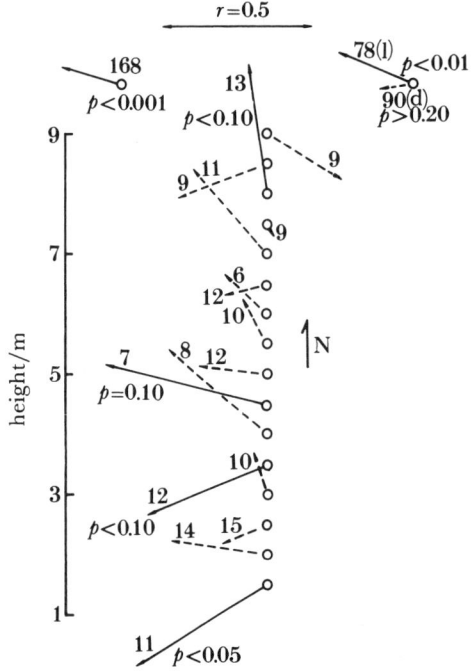

FIGURE 5. The response to neighbours of a codominant Japanese larch ($L_1$ in figure 3). Mean angle and angular concentration ($r$) of branches are shown for: (i) all the branches in the tree (upper left; $n = 168$); (ii) live ($n = 78$) and dead ($n = 90$) branches separated (upper right); (iii) each tier of branches along the stem ($n$ shown for each tier). For visual aid in the interpretation, continuous lines are different from zero at a significance level of 0.10. For comparison a vector whose length is equal to 0.5 is also shown.

of the circular distribution of branches as a lack of preference in the direction of growth, except for the two tiers of branches near the bottom of the tree (figure 4). These lower branches, growing in the direction of the edge of the forest, have slowly gained access, throughout the years, to the east side of the plot where old branches of larch have started to die. None of the other tiers in the tree shows any particular, preferred direction of growth. When all the branches are considered together it is obvious that the tree grows and exerts its dominance in all directions (figure 4). It has a relatively thick trunk and branches profusely, being one of the dominant trees in this group. When live and dead branches are considered separately, however, dead branches seem to be grouped in the direction of the thickest part of the forest, that formed by the Sitka spruce canopy. Before the thinning of trees in the winter of 1983–4, this part of the forest was impenetrable, and only by breaking the old, dead branches could access be gained to climb and measure the trees.

The third example shows the results for a Japanese larch tree (figure 5). In this case the whole portion of the tree in close contact with neighbour trees (whole tree, live branches, and individual tiers) shows a preferential direction of growth towards the gap in the forest. Interestingly, this tree does not have a mean direction of growth pointing towards the north,

TABLE 2. (a) NUMBER OF LIVE AND DEAD SHORT SHOOTS ON NINE MIDDLE-CANOPY BRANCHES OF JAPANESE LARCH ORIENTED TOWARDS DIFFERENT NEIGHBOURHOODS

|  | neighbourhood | | |
| --- | --- | --- | --- |
|  | open space | larch | spruce |
|  | 289/146 | 66/92 | 324/98 |
|  | (1.979) | (0.725) | (3.306) |
| live shoots/dead shoots | 633/305 | 195/102 | 490/237 |
|  | (2.075) | (1.912) | (2.068) |
|  | 1005/431 | 292/152 | 546/316 |
|  | (2.332) | (1.921) | (1.728) |

Notice how the ratio live to dead shoots (in parentheses) increases with the size of the branch for the open space and larch neighbourhoods but decreases in the spruce neighbourhood; shoot mortality increases by direct physical damage of long branches in the latter.

TABLE 2. (b) TEST OF INDEPENDENCE† OF THE THREE FACTORS INVOLVED: NEIGHBOURHOOD ($N$), BRANCH SIZE ($S$), AND SHOOT CONDITION ($C$)

| interaction | degrees of freedom | residual deviance | | |
| --- | --- | --- | --- | --- |
| $N \times C \times S$ | 4 | 57.69*** | | |
| $N \times S$ | 4 | live | dead | |
|  |  | 76.43*** | 21.42*** | |
| $C \times S$ | 2 | open space | larch | spruce |
|  |  | 2.76 n.s. | 30.73*** | 24.22*** |
| $N \times C$ | 2 | small | medium | large |
|  |  | 61.58*** | 0.37 n.s. | 11.26** |

n.s., Non-significant at $p = 0.05$; **, $p < 0.01$; ***, $p < 0.001$.

Figures are residual deviance of the maximal model and as such they represent a measure of interaction. The residual deviance is approximated by a $\chi^2$ distribution with the degrees of freedom shown. One three-way model and eight two-way models were fitted. The latter were done on the corresponding two-way contingency tables for each factor level of the third factor.

† Log–linear model (Sokal & Rohlf 1981).

where interference is and has been presumably lower. The reason for this has to be looked for elsewhere: Japanese larch appears to continue its growth in the presence of the other two species but strongly reduces it in the presence of members of its own species (table 2). Branches of lodgepole pine stop or redirect their growth away from branches of Japanese larch and these in turn continue their growth. When branches of Sitka spruce and Japanese larch meet, physical damage by abrasion is produced causing the death of the leading shoots. New shoots arise from lateral buds and these in turn meet the same fate. The continuous production of new shoots in these 'stubborn' branches is accompanied by the laying down of new xylem. Consequently, their cross-sectional area also increases continuously. The tree has accumulated branch cross-sectional area in all directions except where new buds are not being produced, that is, where there are neighbours of the same species. One has to infer that this individual tree is not directing its growth towards the opening in the forest but that growth is severely inhibited by conspecific trees present on its east side.

## Discussion

It seems clear that trees are capable of responding locally to the interference imposed by neighbours both of the same and of different species. This response manifests itself both as a change in the birth and death rates of individual modules (Jones 1985) and an increased degree of asymmetry of the whole tree. The response seems however to be different depending on the neighbour species. The order of dominance observed in this study was Sitka spruce > Japanese larch > lodgepole pine. These results suggest that the artificial community formed by these three species in this plantation more closely resembles a hierarchy than a network of competitive interactions (for a definition of hierarchy and network kind of communities see Buss (1979) and Buss & Jackson (1979)). In theory, other factors being equal, Sitka spruce would eventually replace the other two species. It is known among foresters that the order of shade tolerance of these three species is spruce > larch > pine. This is reflected in the results.

It is tempting to make an analogy between the hierarchy observed and the continuum guerrilla–phalanx (Lovett Doust 1981). However, at least in conifers, the basic morphological patterns of module display are similar and differences are mainly expressed as different numbers of leaves or of short shoot modules and in frequency of branching by long shoots.

Looking at the interactions among the three species at a finer scale, the rank of dominance is not immediately obvious, at least for the dominant and intermediate species. This is because, in contrast to *Pinus contorta* which can change the orientation of its branches, the branches of *Larix kaempferi* and *Picea sitchensis* tend to grow straight and collide with each other before their bud birth rate is reduced and a reorientation of growth can take place (M. Franco & J. L. Harper, unpublished results). This means that Sitka spruce attains dominance after several years of enduring the constant clashing and death of meristems in the middle canopy, only because its higher rate of stem elongation permits it eventually to overtop the larch trees.

The branches of Sitka spruce are not very responsive to the presence either of members of its own or of different species. Although a reduction of growth occurs after several years of constant interference with neighbour trees, this reduction does not affect the degree of symmetry of the crown. This happens because by the time growth is diminished these branches are in parts of the crown that are heavily shaded anyway, if not by other trees, by the upper part of its own crown.

Japanese larch, with a relatively sparse crown, does not seem to be very responsive to the presence of the other two species. The shoots of Japanese larch are, however, very responsive to the presence of shoots of its own species. Although Sitka spruce produces a heavier shadow than Japanese larch, the elongation of axes of the latter is not suppressed by the presence of the former during the first years of growth, that is, in the upper parts of the canopy. Modules of these two species can then crash into each other causing physical damage by abrasion, with subsequent death of meristems. It is not until these meristems have died that the branch increases the frequency of long shoots on its lateral, secondary axes and spreads perpendicular to the direction of the neighbour tree.

Both Sitka spruce and Japanese larch retain dormant buds in the lower parts of their crowns but, since larch is less able to tolerate the shading imposed by both its own and neighbour's branches, they are less likely to produce long shoots and hence significant amounts of foliage later in life.

The differential response of branches growing in different neighbourhoods may provide some clues about the mechanisms that produce 'crown shyness' in trees (Ng 1980). This phenomenon occurs when a clear boundary gap is visible between the crowns of two trees, or between the branches of a single tree, producing a kind of loose jigsaw puzzle. The explanation advanced by Jacobs (1955) is that, by rubbing against each other in the wind, abrasion and death of branch tips is responsible for the spacing between crowns and branches (the 'phytosadism' of Harper 1985). Putz *et al.* (1984) found a significant correlation between the amplitude of crown- and branch-swaying in the wind and gap size in still air conditions in mangrove forests of *Avicennia germinans* in Costa Rica. Ng (1980) argued, however, that abrasion alone was not sufficient to account for the regularity of crown shyness observed in tropical rain forests of Malaysia. He suggested that, by detecting the changes in the light conditions around them, branch tips stop growing before any physical damage is produced. Since the existence of crown shyness is based on a subjective visual impression, deciding when this phenomenon occurs may be difficult in some circumstances. One might expect a whole range in the degree of 'shyness' of different species of trees. The results of the present study are interesting in this context because both kinds of crown interaction (abrasion and inhibition of growth) occur in different pairwise combinations of species. Three extreme outcomes are possible: (i) inhibition of the growth of both plants; (ii) inhibition of only one of them; and (iii) lack of inhibition of both members. Clearly, only the first of them will invariably produce crown shyness. The second case will normally be seen as the kind of hierarchy of shade-tolerant and shade-intolerant species. The last one will produce crown shyness only in cases where wind-swaying plays an important role in shoot mortality; otherwise the branches will tend to intermingle to some extent. The observation that crown shyness is common between members of the same species but not between members of different species may be due to a symmetrical response (either (i) or (iii) above) of the interacting trees or branches. This is, however, only a particular instance of the continuum of crown responses.

A similar phenomenon to crown shyness occurs in modular marine animals. Francis (1973 a, b; 1976) reports the agonistic behaviour of the sea anemone *Anthopleura elegantissima*. In this species clones can be produced by longitudinal fission of the individual anemones. When two clones of the same genotype meet, fusion between them is possible. When clones of two different genotypes meet, however, a strongly aggressive behaviour is elicited, involving the penetration by nematocysts of the tissues in close contact. This tissue becomes necrotic and the damaged anemone may move or simply lean away from the aggressor. The result is something that looks very much like interclone 'shyness'. On a broader scale, Lang (1973) has reviewed the agonistic interactions between different species of corals. As in trees, the variety of interaction is not limited to 'shyness'. The presence of a 'pecking order' or hierarchy of species seems to be common.

The fact that each of the three conifer species studied showed differential response towards the other two raises the question of whether modules of a particular plant species are able to detect and respond differently to the presence of modules of the same and of different species. The term 'recognition' has been used to describe many different phenomena in both plant and animal studies (Heslop-Harrison 1978). In studies of animals it has mainly been used to describe the immunological identification by animal tissues of alien cells. This identification can involve two extremes in the amount of information and complexity required. On the one hand, the simple identification of self as opposed to non-self and, on the other, the specific recognition of, for example, a particular antigen. The simplest recognition mechanism, that of self and non-self, is common in colonial (modular) organisms (Burnet 1971). While modular animals are able to recognize the presence of genetically similar somatic tissues (Buss *et al.* 1985), in plants recognition seems to be restricted to the prevention of self-fertilization and, in horticultural practice, the restriction of the range of possible interspecific grafting.

One common characteristic of these recognition mechanisms is that they require the coupling of allosteric sites at the cell membrane level. The presence of a cell wall is obviously a strong physical impediment to the development of this capacity in plants. Very commonly, recognition at a distance occurs through the production and detection of chemical substances (Grant & Mackie 1974). This chemical recognition has been shown to be important in the detection and overgrowth of arborescent gorgonians by milleporid hydrocorals (Wahle 1980) providing a mechanism for the identification of space occupied by a competitively inferior species.

The kind of recognition to which we are alluding here is of a slightly different nature. It presumably involves changes in the physical (light) environment that could be detected by the photoreceptors of plants (although they also could probably act at the membrane level (Raven 1983)). Holmes (1983) has shown that plants can distinguish between the shade produced by other plants and the shade produced by inanimate objects. This distinction is possible through phytochrome detecting the increase in the ratio of red to far-red light underneath a live, green canopy.

Deregibus *et al.* (1985) have shown that tillering in *Paspalum dilatatum* and *Sporobolus indicus*, two common grass species of the Argentine Pampas, can be enhanced by artificially increasing the amount of red light underneath their closed canopies. Their results are reinforced by experiments with *Trifolium repens* by Solangaarachchi & Harper (1986), who grew plants of this species under three different amounts of photosynthetically active radiation: 100, 50 and 30% of incident light. For each of the two reduced levels of light, shade was produced by two different objects. In one case, pieces of black polythene floating in a water tank provided the canopy above the experimental plants. In the other, green leaves of *T. repens* floating in an identical water tank were used. Their results, for all the measures of plant performance recorded, are consistent with the hypothesis that at similar levels of photosynthetically active radiation a green canopy inhibits plant growth more than an inert canopy.

Holmes' (1983) argument is based on the importance to a plant of distinguishing between the shade cast by a neighbour plant and that produced by rocks or soil. In a highly competitive environment, the importance of this recognition mechanism may rest more on the advantage of detecting when a taller neighbour has died so that stem extension rate can be rapidly modified, and the chance of access to the upper levels of the canopy improved. One could also hypothesize that the blue-light-absorbing photoreceptor (Briggs & Iino 1983; Holmes 1983) in conjunction with phytochrome may provide the plant with a more accurate picture of its surroundings.

Although different species have different degrees of response to changes in phytochrome photoequilibrium (Holmes 1983), the possibility of specific recognition of and response to broadly different kinds of shade cast by different species or groups of species with consequent modification of both extension rate of internodes and of module birth rate remains to be explored in greater detail.

I am indebted to Professor J. L. Harper for his critical advice and continuous encouragement during the course of this research. I greatly acknowledge Dr J. White for his thorough and constructive reading of the manuscript and some valuable references. In addition, Dr T. Jayasingam and Dr M. Watson made important improvements on the original manuscript. I thank Mr C. Whitaker for statistical advice and Dr B. Okamura and Dr R. Oswald for some useful references. Dr M. Solangaarachchi and Professor J. L. Harper allowed me to quote their not yet published results. Mr C. Ellis and the staff at Pen-y-Ffridd Experimental Station provided the facilities to conduct the experiments on *Kochia*. Mr W. Taylor of the Forestry Commission granted permission to work in Gwydyr Forest. I acknowledge the field assistance provided by Mr D. Gale and Dr G. Mattlack. I owe thanks to Professor F. Hallé and Professor G. O. Mackie who pointed out the similarity of this work to the phenomenon of 'shyness' in trees and modular animals. This work was supported by the Universidad Nacional Autónoma de México and the British Council.

## References

Aono, M. & Kunii, T. L. 1984 Botanical tree image generation. *Computer Graph. Appl.* **4**, 10–34.

Ashmun, J. W., Thomas, R. J. & Pitelka, L. F. 1982 Translocation of photoassimilates between sister ramets in two rhizomatous forest herbs. *Ann. Bot.* **49**, 403–415.

Batschelet, E. 1981 *Circular statistics in biology*. London: Academic Press.

Bazzaz, F. A. & Harper, J. L. 1977 Demographic analysis of the growth of *Linum usitatissimum*. *New Phytol.* **78**, 193–208.

Borchert, R. & Honda, H. 1984 Control of development in the bifurcating branch system of *Tabebuia rosea*. A computer simulation. *Bot. Gaz.* **145**, 184–195.

Briggs, W. R. & Iino, M. 1983 Blue-light absorbing photoreceptors in plants. *Phil. Trans. R. Soc. Lond.* B **303**, 347–359.

Bülow-Olsen, A., Sackville-Hamilton, N. R. & Hutchings, M. J. 1984 A study of growth form in genets of *Trifolium repens* L. as affected by intra- and interplant contacts. *Oecologia, Berl.* **61**, 383–387.

Burnet, F. M. 1971 'Self-recognition' in colonial marine forms and flowering plants in relation to the evolution of immunity. *Nature, Lond.* **232**, 230–235.

Buss, L. W. 1979 Habitat selection, directional growth and spatial refuges: why colonial animals have more hiding places. In *Biology and systematics of colonial organisms* (ed. G. Larwood & B. R. Rosen), pp. 459–497. Systematics Association special volume no. 11. London: Academic Press.

Buss, L. W. 1980 Bryozoan overgrowth interactions – the interdependence of competition for space and food. *Nature, Lond.* **281**, 475–477.

Buss, L. W. & Jackson, J. B. C. 1979 Competitive networks: nontransitive competitive relationships in cryptic coral reef environments. *Am. Nat.* **113**, 223–234.

Buss, L. W., Moore, J. L. & Green, D. R. 1985 Autoreactivity and self-tolerance in an invertebrate. *Nature, Lond.* **313**, 400–402.

Cheetham, A. H., Hayek, L. C. & Thomsen, E. 1981 Growth models in fossil arborescent cheilostome bryoans. *Paleobiology* **7**, 68–86.

Deregibus, V. A., Sanchez, R. A., Casal, J. J. & Trlica, M. J. 1985 Tillering responses to enrichment of red light beneath the canopy in a humid natural grassland. *J. appl. Ecol.* **22**, 199–206.

Francis, L. 1973a Clone specific segregation in the sea anemone *Anthopleura elegantissima*. *Biol. Bull.* **144**, 64–72.

Francis, L. 1973b Intraspecific aggression and its effect on the distribution of *Anthopleura elegantissima* and some related anemones. *Biol. Bull.* **144**, 73–92.

Francis, L. 1976 Social organization within clones of the sea anemone *Anthopleura elegantissima*. *Biol. Bull.* **150**, 361–376.

Franco, M. 1985a A modular approach to tree production. In *Studies on plant demography: a Festschrift for John L. Harper* (ed. J. White), pp. 257–272. London: Academic Press.

Franco, M. 1985b The architecture and dynamics of tree growth. Ph.D. Thesis, University of Wales.

Grant, P. & Mackie, A. (eds) 1974 *Chemoreception in marine organisms*. London: Academic Press.

Günther, B. 1975 Dimensional analysis and theory of biological similarity. *Physiol. Rev.* **55**, 659–699.

Hallé, F., Oldeman, R. A. A. & Tomlinson, P. B. 1978 *Tropical trees and forests: an architectural analysis*. Berlin: Springer-Verlag.

Harper, J. L. 1981 The concept of population in modular organisms. In *Theoretical ecology: principles and applications* (ed. R. M. May), pp. 53–77. Oxford: Blackwell Scientific Publications.

Harper, J. L. 1985 Modules, branches, and the capture of resources. In *Population biology and evolution of clonal organisms* (ed. J. B. C. Jackson, L. W. Buss & R. Cook). New Haven: Yale University Press.

Hartnett, D. C. & Bazzaz, F. A. 1985 The integration of neighbourhood effects by clonal genets in *Solidago canadensis*. *J. Ecol.* **73**, 415–427.

Heslop-Harrison, J. 1978 *Cellular recognition systems in plants*. London: Edward Arnold.

Holmes, M. G. 1983 Perception of shade. *Phil. Trans. R. Soc. Lond.* B **303**, 503–521.

Honda, H., Tomlinson, P. B. & Fisher, J. B. 1981 Computer simulation of branch interaction and regulation by unequal flow rates in botanical trees. *Am. J. Bot.* **68**, 569–585.

Jackson, J. B. C. 1979 Morphological strategies of sessile animals. In *Biology and systematics of colonial organisms* (ed. G. Larwood & B. R. Rosen), pp. 499–555. Systematics Association special volume no. 11. London: Academic Press.

Jacobs, M. R. 1955 *Growth habits of the eucalypts*. Commonwealth of Australia: Forestry and Timber Bureau.

Johnson, R. A. & Wehrly, T. 1977 Measures and models for angular correlation and angular-linear correlation. *J. R. Stat. Soc.* B **39**, 222–229.

Jones, M. 1985 Modular demography and form in silver birch. In: *Studies on plant demography: a Festschrift for John L. Harper* (ed. J. White), pp. 223–237. London: Academic Press.

Lang, J. 1973 Interspecific aggression by scleractinian corals. 2. Why the race is not only to the swift. *Bull. mar. Sci.* **23**, 260–279.

Lovett Doust, L. 1981 Population dynamics and local specialization in a clonal perennial (*Ranunculus repens*). I. The dynamics of ramets in contrasting habitats. *J. Ecol.* **69**, 743–755.

MacDonald, N. 1983 *Trees and networks in biological models*. Chichester: John Wiley & Sons.

Maillette, L. 1982a Structural dynamics of silver birch. I. The fates of buds. *J. appl. Ecol.* **19**, 203–218.

Maillette, L. 1982b Structural dynamics of silver birch. II. A matrix model of the bud population. *J. appl. Ecol.* **19**, 219–238.

Mardia, K. V. 1972 *Statistics of directional data*. London: Academic Press.

Marshall, C. M. & Sagar, G. R. 1968 The distribution of assimilates in *Lolium multiflorum* Lam. following differential defoliation. *Ann. Bot.* **32**, 715–719.

McKinney, F. K. 1984 Feeding currents of gymnolaemate bryozoans: better organization with higher colonial integration. *Bull. mar. Sci.* **34**, 315–319.

Medawar, P. B. 1945 Size, shape and age. In *Essays on growth and form presented to D'Arcy Wentworth Thompson* (ed. W. E. le Gros Clark & P. B. Medawar), pp. 157–187. Oxford: University Press.

Monsi, M. & Saeki, T. 1953 Über den Lichtfaktor in den Pflanzengesellschaften und seine Bedeutung für die Stoffproduktion. *Jap. J. Bot.* **14**, 22–52.

Ng, F. S. P. 1980 Shyness in trees. *Nat. Malays.* **2**, 34–37.

Noble, J. C. & Marshall, C. 1983 The population biology of plants with clonal growth. II. The nutrient strategy and modular physiology of *Carex arenaria*. *J. Ecol.* **71**, 865–877.

Okamura, B. 1984 The effects of ambient flow velocity, colony size, and upstream colonies on the feeding success of bryozoa. I. *Bugula stolonifera* Ryland, an arborescent species. *J. exp. mar. Biol. Ecol.* **83**, 179–193.

Ong, C. K. & Marshall, C. 1979 The growth and survival of severely-shaded tillers in *Lolium perenne* L. *Ann. Bot.* **43**, 147–155.

Pacala, S. W. & Silander, J. A. 1985 Neighborhood models of plant population dynamics. I. Single-species models of annuals. *Am. Nat.* **125**, 385–411.

Paine, R. T. 1984 Ecological determinism in the competition for space. *Ecology* **65**, 1339–1348.

Putz, F. E., Parker, G. G. & Archibald, R. M. 1984 Mechanical abrasion and intercrown spacing. *Am. Midl. Nat.* **112**, 24–28.

Raven, J. A. 1983 Do plant photoreceptors act at the membrane level? *Phil. Trans. R. Soc. Lond.* B **303**, 403–417.
Schmid, B. 1986 Spatial dynamics and integration within clones of grassland perennials with different growth form. *Proc. R. Soc. Lond.* B **228**, 173–186.
Shinozaki, K., Yoda, K., Hozumi, K. & Kira, T. 1964 A quantitative analysis of plant form – the pipe model theory. I. Basic analysis. *Jap. J. Ecol.* **14**, 97–105.
Silander, J. A. & Pacala, S. W. 1985 Neighbourhood predictors of plant performance. *Oecologia, Berl.* **66**, 256–263.
Sokal, R. R. & Rohlf, F. J. 1981 *Biometry: the principles and practice of statistics in biological research* (2nd edn). San Francisco: W. H. Freeman.
Solangaarachchi, S. M. 1985 The nature and control of branching pattern in *Trifolium repens*. Ph.D. thesis, University of Wales.
St Pierre, J. C. & Wright, M. J. 1972 Distribution of $^{14}$C photosynthates in timothy (*Phleum pratense* L.) during vegetative growth. *Crop Sci.* **12**, 191–194.
Tietema, T. 1980 Ecophysiology of the sand sedge, *Carex arenaria* L. II. The distribution of $^{14}$C assimilates. *Acta Bot. Neerl.* **29**, 165–178.
Wahle, C. M. 1980 Detection, pursuit, and overgrowth of tropical gorgonians by milleporid hydrocorals: Perseus and Medusa revisited. *Science, Wash.* **209**, 689–691.
Waller, D. M. & Steingraeber, D. A. 1986 Branching and modular growth: theoretical models and empirical patterns. In *Population biology and evolution of clonal organisms* (ed. J. B. C. Jackson, L. W. Buss & R. Cook), pp. 225–258. New Haven: Yale University Press.
Waring, R. H., Schroeder, P. E. & Oren, R. 1982 Application of the pipe model theory to predict canopy leaf area. *Can. J. For. Res.* **12**, 556–560.
Watson, M. A. & Casper, B. B. 1984 Morphogenetic constraints on patterns of carbon distribution in plants. *Ann. Rev. Ecol. Syst.* **15**, 233–258.
Watson, M. A. & Cook, C. S. 1982 The development of spatial pattern in clones of an aquatic plant, *Eichhornia crassipes* Solms. *Am. J. Bot.* **69**, 248–253.
Whitney, G. G. 1982 A demographic analysis of the leaves of open and shade grown *Pinus strobus* L. and *Tsuga canadensis* (L.) Carr. *New Phytol.* **90**, 447–453.
Winston, J. E. 1978 Polypide morphology and feeding behavior in marine ectoprocts. *Bull. mar. Sci.* **28**, 1–31.
Winston, J. E. 1979 Current-related morphology and behaviour in some Pacific-coast bryozoans. In *Advances in bryology* (ed. G. P. Larwood & M. B. Abbott), pp. 247–268. London: Academic Press.

# Defensive strategies of modular organisms

By P. E. J. Dyrynda

*Marine Research Group, School of Biological Sciences, University College of Swansea, Singleton Park, Swansea, SA2 8PP, U.K.*

[Plates 1 and 2]

Convergences concomitant with the occurrence of modular growth among systematically remote plant and invertebrate taxa not only reflect similar optimal ways of exploiting resources such as space, but also common defensive requirements among such organisms.

This paper analyses the kinds of unfavourable interspecific interactions, principally predation, epibiosis, and endobiosis, which are found among the major aquatic invertebrate groups that may be considered to be modular (Porifera, Bryozoa, and some of the Coelenterata and Tunicata). Most of the organisms are also non-locomotory, and in extreme cases, virtually immotile. The defence mechanisms of organisms exhibiting the opposing traits of (i) modular and unitary organization, and (ii) motility and immotility, are compared and contrasted. There is a more widespread occurrence of defence (i) by means of consolidated and unconsolidated skeletal reinforcement, and (ii) by actively and passively dispensed secondary substances, in less motile than in more motile organisms. These defensive modes represent alternatives to 'fight' and 'flight' responses seen within the more motile invertebrates. Lack of motility is of greater significance in correlating defensive modes than is modularity.

The balance between physical and chemical mechanisms used in defence can vary, even among closely related taxa. A more particular pattern of significance is the more widespread occurrence of defence by the use of passively dispensed chemical substances within modular, rather than unitary non-locomotory invertebrate groups. This may be a response to the increased risks of pathogenic infection which modular biota face through their susceptibility to frequent large scale wounding and partial mortality.

## 1. Introduction

### (a) Aims

Striking convergences of form and perhaps function have accompanied the acquisition of modular growth by systematically remote non-locomotory taxa (figures 1–5 and 7–12, plates 1 and 2). These convergences may primarily reflect superior space-competitive abilities (Jackson 1977; Harper 1985) (figures 1–3), but some are more easily related to optimal exploitation of resources like light and food (Ryland & Warner, this symposium), or defence against the biological risks of predation, epibiosis or endobiosis. Remarkably similar or even identical secondary metabolites with defensive potential have been found among taxa as different as vascular plants and sessile invertebrates.

This paper aims (i) to identify (but not comprehensively review) the range of potential or actual defensive mechanisms found in non-locomotory modular invertebrates in response to

TABLE 1. SUGGESTED RANGES OF MECHANISMS OF DEFENCE AMONG SOME MAJOR GROUPS OF MODULAR AND UNITARY AQUATIC INVERTEBRATES CATEGORIZED ACCORDING TO WHETHER (GENERALLY) NON-LOCOMOTORY, LOCOMOTORY BUT SLOW MOVING, OR LOCOMOTORY BUT AGILE

(This scheme is neither definitive nor exhaustive in coverage; only reflecting evidence, views and suggestions as appear in the text. Some exceptions to the scheme are referred to in footnotes. Defence may not be the sole, or even the major function listed mechanisms. For each specific group, a mechanism is only referred to if it is believed to have the potential to serve in defence.)

| group | modular cloner | unitary cloner | unitary non-cloner | defence by locomotion | defence by flexion | defence by reinforcement | | | defence by secondary substance | | |
|---|---|---|---|---|---|---|---|---|---|---|---|
| | | | | | | consolidated skeleton | unconsolidated skeleton | density of spines | passive metabolites | venoms | other |
| non-locomotory | | | | | | | | | | | |
| Porifera (sponges) | + | – | – | – | PE | PEI | PI | PE | PEI | – | PEI 8 |
| Coelenterata | | | | | | | | | | | |
| Alcyonacea (soft corals) | + | – | – | – | PE | PI | PI | – | PEI | PE 5 | PE 5 |
| Gorgonacea (sea fans) | + | – | – | – | PE | PI | PI | – | PEI | PE 5 | PE 5 |
| Madreporaria (hard corals) | + | 1 | – | – | PE | PE | – | PE | – | PE 5 | PEI 5, 8 |
| Actiniaria (anemones) | – | + | – | E 2 | PE | – | – | – | – | PE 5 | PEI 5, 8 |
| Polychaeta | | | | | | | | | | | |
| Sedentaria (in part) (tube worms) | – | 3 | + | – | PE | PI | – | – | – | – | – |
| Crustacea | | | | | | | | | | | |
| Cirripedia (barnacles) | – | – | + | – | PE | PI | PI | – | – | – | – |
| Mollusca | | | | | | | | | | | |
| Bivalvia (in part) (non-locomotory spp.) | – | – | + | – | PI | PI | – | – | – | – | – |
| Bryozoa (sea mats) | + | – | – | E 2 | P | PI | – | PE 6 | PEI | – | PE 6 |

TABLE 1 (cont.)

| | | | | | | | | |
|---|---|---|---|---|---|---|---|---|
| Tunicata | | | | | | | | |
| Ascidiacea: group A (modular taxa) | + | — | — | I | PI | — | — | — |
| Ascidiacea: group B (unitary taxa) | — | + | — | I | PI | PEI | — | — |
| Locomotory, slow moving | | | | | | | | |
| Coelenterata | | | | | | | | |
| Siphonophora (for example, *Physalia*) | + | — | P 4 | — | — | — | P 5 | P 5 |
| Crustacea Decapoda–Reptantia (crabs, lobsters, etc.) | — | + | — | PE | PI | — | — | — |
| Mollusca | | | | | | | | |
| Prosobranchia (whelks, etc.) | — | + | — | PE | PI | — | — | I 8 |
| Nudibranchia (sea slugs) | — | + | — | PE | — | PI | P 5 | PEI 5, 8, 9 |
| Echinodermata (starfish, urchins sea cucumbers) | — | + | — | PE | PEI | PEI | P 7, 8 | PE |
| Tunicata Thaliacea (in part) (modular spp.) | + | — | — | P 4 | — | — | — | — |
| Locomotory, fast moving | | | | | | | | |
| Polychaeta Errantia (locomotory worms) | — | + | — | PE | — | — | — | — |
| Crustacea Decapoda–Natantia (prawns, etc.) | — | + | — | PE | PI | — | — | — |
| Mollusca Cephalopoda (octopus, squids) | — | + | — | PE | — | — | P | — |

P, defence against predation; E, epibiosis; I, endobiosis.
1, A very few scleractinians, for example *Fungia*, are unitary cloners like some anemones; 2, some anemones and bryozoans are capable of relatively very slow locomotion; 3, a very few tubicolous polychaetes, for example *Filograna*, are unitary cloners; 4, pelagic and modular; 5, nematocysts; 6, avicularia; 7, pedicellariae; 8, surface mucus; 9, acid glands.

biological risks; (ii) to identify common trends that may exist among these; and (iii) to consider whether they are a function of modularity or of some other trait. Problems arise in the analysis of these traits because those most likely to be correlated with defensive requirements are not mutually exclusive. This is true for the main categories discussed here, namely, modular versus unitary, and locomotory versus non-locomotory habits.

### (b) Modular organization

The concept of modularity, when less specifically defined, can accommodate a broad range of plants and invertebrates with similar growth patterns, although there is the view that it should be more restricted in definition (Boardman *et al.* 1973, Harper & Bell 1979; Larwood & Rosen 1979; Rosen 1979; Chapman & Stebbing 1980; Harper 1985). For the purposes of this paper, modular species are recognized more broadly as macrobiota growing by the addition of repeated blocks, units, or modules from growing points or fronts, that remain interlinked. The resulting continuous or intact clones, hereafter referred to as *individuals*, have theoretically indeterminate growth, size and longevity (figure 4). Organisms regarded as modular include macroalgae and vascular plants, and among the aquatic Invertebrata, all Porifera and Bryozoa and many of the Coelenterata and Tunicata (table 1).

### (c) Degree of motility

Modular organisms are generally non-locomotory (unable to move from place to place under their own power) (table 1), and usually sessile, that is, they grow anchored to a substratum

---

#### DESCRIPTION OF PLATE 1

FIGURE 1. Strong tidal flow and intermediate conditions of illumination facilitate the development of a mixed assemblage of modular forms. A sponge, *Halichondria bowerbanki*, competes for space with various erect macroalgae. (The Fleet Lagoon, southern England, chart datum C.D. −4 m.)

FIGURE 2. Thin, sheet-encrusting modular forms: *Cryptosula pallasiana* (Cheilostomata, Bryozoa) (Cp) and a coralline algal species (Rhodophyceae) (Ca) compete for space with unitary forms, the barnacle *Balanus balanus* (Bb) and the tubicolous polychaete *Pomatoceros* sp. (Ps). Space occupancy by such sessile biota is kept well below 100% by heavy grazing pressure from the echinoid *Echinus esculentus* (compare figure 6). The above species prevail by virtue of appropriate anti-predatory defences: all are characterized by heavy surface skeletal reinforcement (calcification), and the modular forms also by their ability to survive large-scale partial mortality. (Loch Fyne, Scotland, C.D. −12 m.)

FIGURE 3. Modular-cloning contrasted with unitary non-cloning. A recently established, originally monospecific cover of the barnacle *Semibalanus balanoides* (Cirripedia, Crustacea) is now being overgrown by a colony of the sheet-encrusting modular ascidian *Trididemnum tenerum* (Ascidiacea, Tunicata). A proportion of the barnacles survive the epibiosis in that whereas their non-vital calcified exoskeleton is overgrown, the orifice through which feeding and other vital exchange takes place is physically kept clear (arrow) by the movement of opercular plates or the cirral beat. (Kepple Pier, Firth of Clyde, Scotland, C.D. −3 m.)

FIGURE 4. *Flustra foliacea* (Cheilostomata, Bryozoa) grows to a combined, sheet-encrusting and frondose-erect gross morphology. By virtue of indeterminate growth, both elements are added to in annual increments (arrowed). Individual fronds have been known to survive for 12 years (Stebbing 1971), and the clone in its entirety has the potential to survive very much longer. (Swanage Pier, southern England, C.D. −4 m.)

FIGURE 5. Detail of the surface of a thin, sheet-encrusting sponge *Microciona atrasanguinea* showing the barrage of pointed spines consisting of siliceous spicules (unconsolidated skeletal elements) and believed to provide defences against predation and epibiosis. (Scanning electron micrograph, sample from Swanage Pier, Dorset.)

FIGURE 6. *Echinus esculentus* (Echinodermata) grazing sessile invertebrates colonizing a vertical rock face (compare figure 2). The tube-feet, the means of locomotion of this slow-moving invertebrate, are clearly visible, as are the barrage of calcareous spines, which supplement the consolidated exoskeletal reinforcement as antipredatory defences. (Loch Fyne, Scotland, C.D. −12 m.)

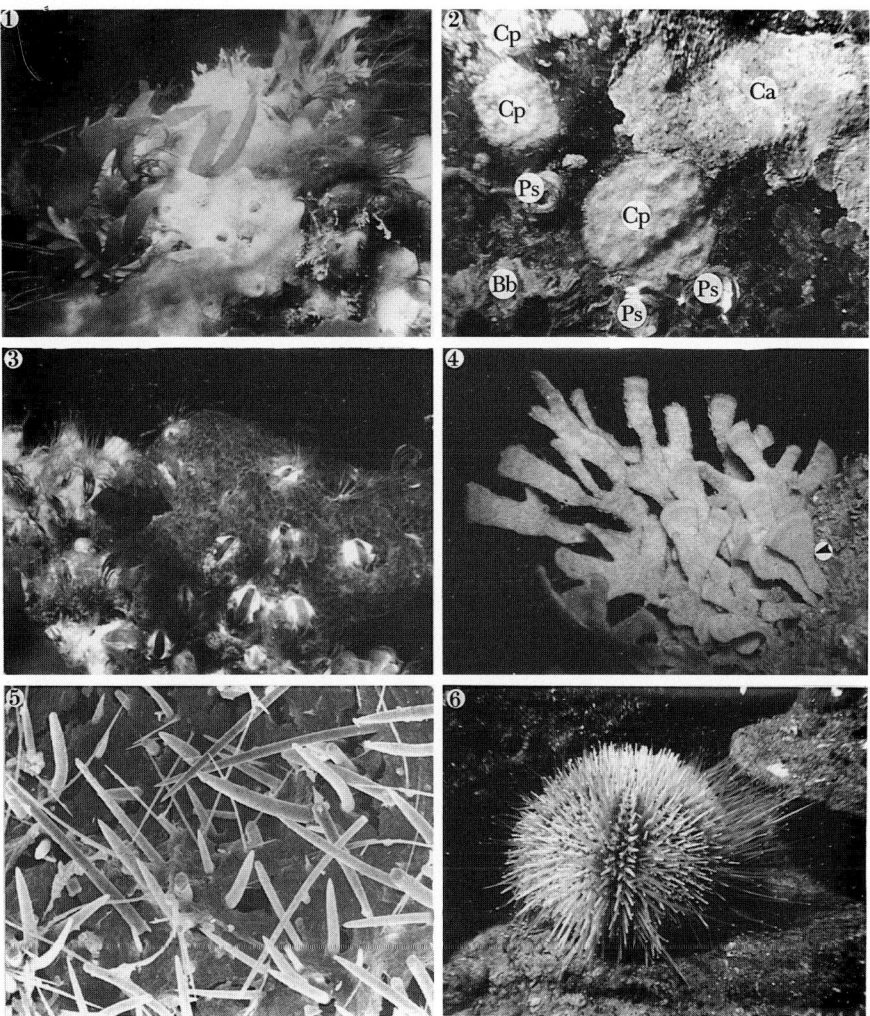

FIGURES 1–6. For description see opposite.

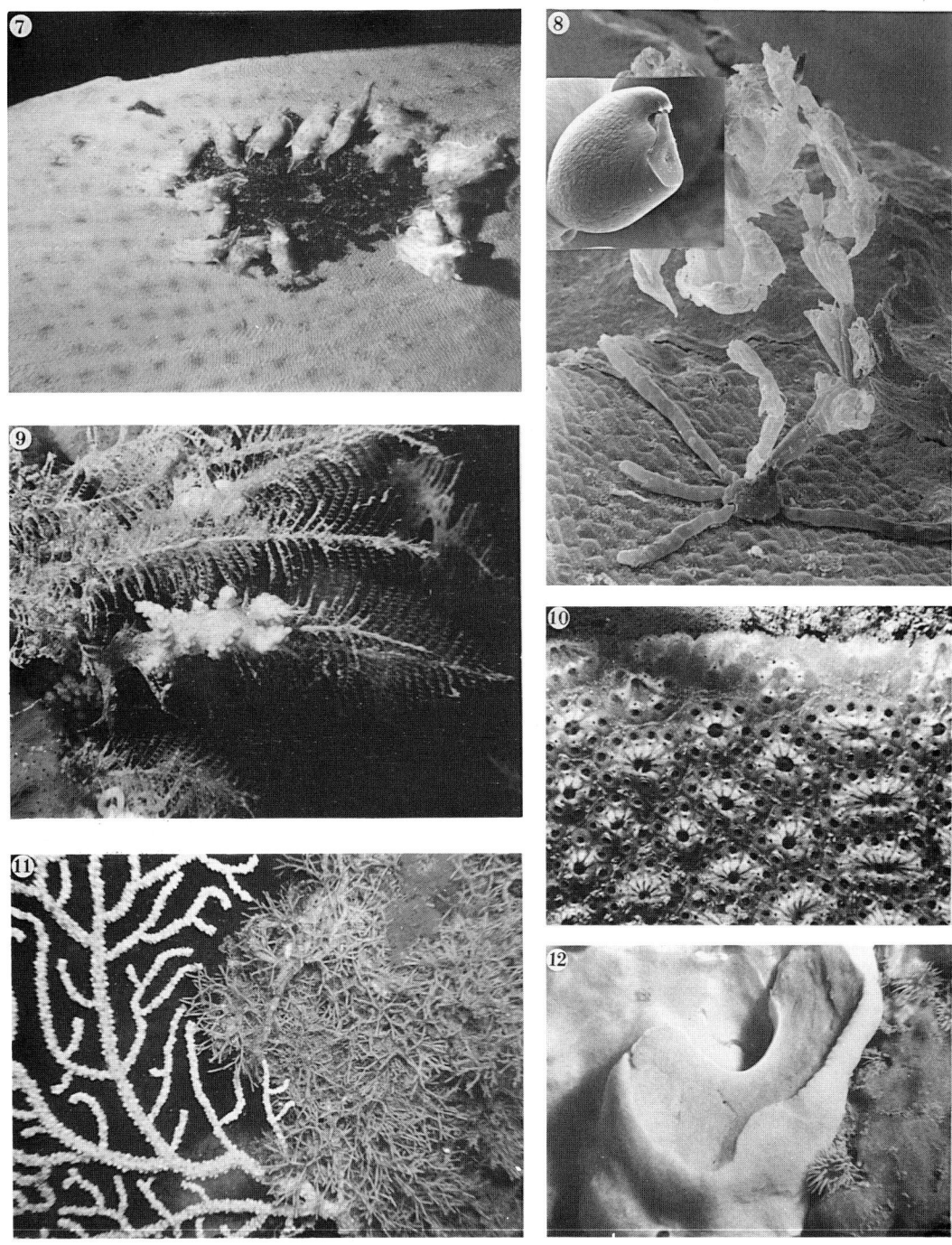

FIGURES 7–12. For description see opposite.

consisting of rock, sediment, soil, etc. However, some that are non-locomotory and benthic, are non-sessile, lying unattached upon the substratum, whereas other non-locomotory forms are planktonic, drifting in air or water currents. Some modular invertebrates like the Siphonophora (Coelenterata) and Thaliacea (Tunicata) have secondarily re-acquired locomotion. Both of these taxa are pelagic but the bryozoans *Cristatella* and *Selenaria* and the ascidian *Diplosoma virens* are all benthic (Mundy 1980; Cook & Chimonides 1978; Ryland *et al.* 1984). Although incapable of true locomotion, some sessile and encrusting poriferans, scleractinians, bryozoans and ascidians appear to 'creep' by directional growth and die-back. Whereas many non-locomotory modular species are motile, in that they can effect appreciable gross or local body flexions, in common with the majority of plants, others are immotile. It is also true that not all non-locomotory invertebrates are modular. The anemones (Actiniaria, Coelenterata) (barely locomotory) are unitary and can clone, wheras most other unitary, non-locomotory forms do not clone, for example, most tube worms (Sedentaria, Polychaeta), barnacles (Cirripedia, Crustacea), and some bivalves (Lamellibranchia, Mollusca).

Among locomotory, unitary forms, the degree of locomotion is highly variable, for example, many reptant decapod crustaceans, prosobranch and opisthobranch molluscs, and also echinoderms, are relatively slow-moving; whereas errant polychaetes, natant decapod crustaceans, and cephalod molluscs tend to be much more agile (table 1).

*(d) Categories of defence*

Interactions unfavourable to a *defender*, that is, predation, epibiosis, and endobiosis (often occurring in combination), are kinds of sequestration, that is, the *offender* (any micro-

---

DESCRIPTION OF PLATE 2

FIGURE 7. An extensive colony of the thin, sheet-encrusting bryozoan *Membranipora membranacea* (growing as an epibiont on *Laminaria digitata*) is being predated by a group of nudibranch molluscs *Polycera quadrilineata*. Zooids (modules) beyond the immediate area of injury are unaffected, so that despite loss by partial mortality, the colony of this fast-growing species survives. The nudibranch is marked with clear orange and black spots on a white background, which is probably aposematic (see text). (Oxwich Bay, South Wales, c.d. −3 m.)

FIGURE 8. Early colony of *Epistomia bursaria* (Cheilostomata, Bryozoa) growing on a rhodophyte. The first zooid (ancestrula) is clearly visible. From this, two erect shoots have arisen (one much smaller than the other). Feeding and reproductive activity are confined to shoots (Dyrynda & King 1982), each of which consists of alternating pairs of autozooids (feeding) and avicularia (zooids on shoots have partly collapsed during preparation; intact giant avicularium shown in inset). Avicularia probably function as do those of *Bugula* (see text). Four encrusting (and heavily calcified) runners have also been produced by the ancestrula. These will subsequently give rise to further runners and shoots. (Scanning electron micrographs prepared from specimens collected at Swanage Pier, southern England, c.d. −4 m).

FIGURE 9. A species of the nudibranch *Doto* climbs a shoot of the modular hydroid *Kirchenpaueria pinnata* on which feeding zooids are confined to side branches (Swanage Pier, southern England, c.d. −3 m.)

FIGURE 10. Detail of the surface of the sheet-encrusting modular ascidian *Botryllus schlosseri* showing the replication of zooids (in constellations), resulting in such a high-density of vital orifices that any degree of epibiosis would constitute interference. *Botryllus* is known to deter direct and proximal settlement by the propagules of competitors (see text). (Swanage Pier, southern England, c.d. −3 m.)

FIGURE 11. Partial mortality of this sea fan *Eunicella verrucosa* (Gorgonacea, Coelenterata) has exposed its gorgonin core (right, which persists long after polyp mortality). Whereas the living gorgonian keeps itself totally clear of sessile macroepibionts, the exposed endoskeleton supports a heavy epibiotic cover initiated by larval settlement and dominated by erect bryozoans. (Gulland Rock, southwest Britain, c.d. −35 m.)

FIGURE 12. Sponge *Suberites domuncula* with an extensive laceration (cause unknown). This was healed completely within seven days. (Swanage Pier, southern England, c.d. −2 m.)

or macrospecies) attempts to capture a primary resource, whether nutritional, or habitational space, already possessed by the defending modular macrobiont. This is in contrast with competition, in which the resource in question is presumed not to be possessed by either interacting individual.

Defensive arrays deployed against particular risks consist of 'physical' or 'chemical' mechanisms, or both. They might operate externally and before contact; or *at the surfaces* upon contact with an offender; or *internally*, if an offender penetrates the defender. Although defence is usually *intrinsic* (that is, the defender itself has its own defensive mechanisms), it is also sometimes *extrinsic* if it has a symbiont which acts as a defensive agent. Symbionts serve in the defence of many large immotile modular organisms.

A defence mechanism, as recognized here, is any mechanism that is capable of providing defence against biological risks, even though this may not be the sole, or even the major function of the mechanism in question. The main risks discussed below are predation, epibiosis and endobiosis. Mechanisms of defence are as categorized in table 1. In the text below, reinforcement and secondary substances are concentrated upon.

## 2. Reinforcement

### (a) Categories of reinforcement and their occurrence

'Panclonal' skeletal systems, internal or external, constructed or produced by the intermodular interaction of 'hard' acellular elements characterize all modular groups, as they do numerous unitary invertebrates. Such skeletal systems provide support or defence, or both. In *consolidated* reinforcement, skeletal elements are continuous or fused between modules (for example, the internal spongin fibre networks of some demosponges, or the siliceous frameworks of some hexactinellid sponges). Scleractinian corals produce rigid consolidated, external calcareous bases secreted by the overlying polyps, into which the polyps can usually withdraw. Modular hydroids (Coelenterata) and bryozoans possess external chitinous skeletal reinforcement, supplemented in many bryozoans by calcification. Except for athecate hydroids, inactive or threatened polyps (or polypides) withdraw into the protection of the exoskeleton. Such responses are analogous to the 'flight' escape reactions of locomotory invertebrates. Modular ascidians have mucopolysaccharide exoskeletons, ranging from delicate to heavy in constitution. The 'panclonal' (or 'colony-wide') closure of vital orifices (that is, siphons), to seal the organism within the protection of the exoskeleton, characterizes some ascidians, and constitutes an even more specialized kind of physical 'flight' than retraction alone.

Unconsolidated skeletal reinforcement is provided by spicules (granules or sclerites), not fused or continuous between modules (though sometimes linked by connective tissues). Calcareous or siliceous spicules characterize most poriferans. Alcyonacean soft corals contain calcareous spicules: individuals that are soft when dilated with water while feeding, may become rigid upon contraction as spicules 'interlock' to form a framework into which the polyps withdraw.

Combined reinforcement occurs in demosponges with both spongin networks and calcareous spicules. Some gorgonacean sea fans combine a tough endoskeletal consolidated core (gorgonin) with calcareous spicules in the enclosing polyp sheet. In many didemnid ascidians, the mucopolysaccharide test is supplemented by calcareous spicules.

Surface spines are common among non-locomotory modular invertebrates (for example, pointed spicules project through the epidermis in the form of spines in some sponges (figure 5)).

Consolidated skeletal elements also include spines as in many scleractinian corals and in bryozoans. Ascidians do not produce spines.

Coelenterate nematocysts might also be included in this reinforcement category, if we regard them as defensive, surface structures. They are chitinous and, when appropriately triggered, 'explosively' discharge an often penetrant and injurious filament, in many cases containing a toxic agent (see §3$d$). Avicularia are specialized non-feeding zooids characteristic of the cheilostomate bryozoans. Some support pincer-like structures which may function defensively (Cook, 1979; Winston 1984).

Many non-locomotory and slow-moving unitary invertebrates possess consolidated skeletons, usually external and calcified, (for example, tube worms, barnacles, many bivalve molluscs, decapod crustaceans, prosobranch molluscs, and echinoids (figure 6)). Unitary, sessile ascidians have mucopolysaccharide tests similar to their modular equivalents. Unconsolidated skeletal reinforcement is uncommon among non-locomotory forms, except for some barnacles; but spicules do occur within the slow-moving dorid nudibranch molluscs (which sequester their spicules from their sponge prey (Todd 1981)), and in most asteroid and holothurian echinoderms. Among locomotory groups, spines (at density) are more characteristic of slower-moving groups, for example, some reptant decapod crustaceans, and echinoids in general (figure 6). Nematocysts are ubiquitous and characteristic within all coelenterates, including unitary forms like anemones (Actiniaria). They also occur in aeolid nudibranchs, sequestered from their coelenterate prey (Todd 1981). Echinoderm pedicellariae are remarkably similar to some defensive bryozoan avicularia in form and perhaps function. They possess gripping pincers, and some are venomous (Russell 1984).

External reinforcement, if present, among more motile forms including the modular thaliaceans and some siphonophores, tends to be less substantial than in sessile forms. This is even more true for more agile locomotory organisms like errant polychaetes and natant decapods.

### (b) *Skeletal reinforcement as a defence against predation*

Many modular, non-locomotory species exhibit intermodular functional delegation, more vulnerable vital systems (for example, feeding or reproduction) being confined to less vulnerable parts of the whole clonal individual (for example, centrally in sheet-encrusters, or upon the shoots of erect species). Attack may be prevented by barrages of spines. Thus small gastropod molluscs are unable to climb the spiny surfaces of the sponge *Microciona atrasanguinea* (personal observation)(figure 5). Such defences are usually passive in that they exist irrespective of threat, but some are active, for example, the sheet-encrusting bryozoan *Membranipora membranacea* produces peripheral spines in response to grazing pressure by nudibranch molluscs (Harvell 1984) (figure 7).

Jackson (1979) considers 'escape' from predation to be a major advantage of the erect gross morphology adopted by many colonial invertebrates. Some thecate hydroids and cheilostomate bryozoans (for example, *Dynamena* and *Bugula* or *Epistomia*, respectively) have combined runner-and-erect gross morphologies (figure 8)). The runners, highly vulnerable to substratum-bound predators, are protected by heavy skeletal reinforcement. Feeding and reproductive activity is usually confined to erect 'shoots', the proximal sections of which can be difficult to climb, often being smooth and very slender. Nevertheless, there may be specifically coadapted predators like nudibranch molluscs of the genus *Doto*, which are specialized in their ability to climb such features, (figure 9). Nematocysts and avicularia have the potential to prevent a surface-bound predator from reaching its feeding site. *Trivia* (Prosobranchia, Mollusca) sharply withdraws its foot when it comes into contact with *Alcyonium digitatum* (Alcyonacea, Coelenterata) (personal observation). Nematocysts may entrap or toxify the predator. N. Ravenscroft (unpublished) observed that the avicularia that line the shoots of *Bugula* (Cheilostomata, Bryozoa) can immobilize various climbing predators like crustaceans, mites and pycnogonids. The avicularia can maintain their grip for days, even causing predators to shed appendages or die (other examples are cited by Winston 1984) (figure 8).

If the predators can actually reach their attempted feeding site, skeletal features may prevent them from commencing to feed. The mass withdrawal of vulnerable polyps or polypides to within the skeletal framework, or the closure of orifices, may safeguard vital systems. Sharp spines may wound the feeding predator. Skeletal reinforcement, consolidated or unconsolidated, may render the surface impenetrable. For example, the exoskeletal morphology of scleractinian corals protects polyps from chaetodontid and other coral-grazing fish. Polyps raised on protuberances are more vulnerable than when recessed within foveolate skeletal structures. Best & Winston (1984) found exoskeletal strength of sheet-encrusting cheilostomate bryozoans, and hence vulnerability to penetration by grazing molluscs, to be highly variable.

Internal skeletal reinforcement, consolidated or otherwise, may curtail feeding and so limit potential damage: internal 'vital' structures may be impenetrable, and angular spicules may irritate the gut of the predator. Modular species are at an advantage in this respect in their ability to survive a large degree of partial mortality such that internal mechanisms may constitute a major line of antipredatory defence. Although metameric annelids and asteroid echinoderms, for example, can appreciably enhance their survivorship by regeneration, this potential does not approach that of modular biota. The prospect of such organisms 'sacrificing' a proportion of themselves as part of a defensive strategy would carry a high risk of outright mortality.

### (c) *Skeletal reinforcement as a defence against epibiosis*

The physical (and chemical) nature of a substratum is a primary factor influencing larval settlement. The extreme delicacy of the surfaces of some species (for example, the modular ascidian *Diplosoma listerianum*) render them unsuitable for epibiosis. Spines can also prevent colonization, both by larval settlement and lateral overgrowth. The spines of *Microciona atrasanguinea* ensnare larvae of *Bugula* (Cheilostomata, Bryozoa) (personal observation) (figure 5). As an 'active' equivalent to this mode of defence, the sheet-encrusting bryozoan *Electra pilosa* produces a peripheral barrage of spines when 'threatened' by lateral overgrowth of adjacent biota (Stebbing 1973). In addition to advantages in terms of anti-predation, the adoption of an erect gross morphology can also be considered an 'escape' from epibiosis initiated by lateral growth (Jackson 1979), particularly when proximal sections are slender, since sheet-encrusters tend to grow around rather than up such substrates (figures 1, 8, 9). On the other hand, runner networks of genera such as *Sertularia*, *Bugula* and *Epistomia* are not harmed by being overgrown since they are heavily skeletalized, and are not sites of vital exchange (feeding, etc.).

In consolidating its surface by skeletal reinforcement, an organism usually becomes more susceptible to epibiotic colonization. It presumably gains, however, from better self-support, or better resistance to other kinds of attack like predation, or both. Symbiotic epibionts, however, may prevent further epibiosis of their host by having defences of their own.

## (d) Skeletal reinforcement as a defence against endobiosis

Surface skeletal reinforcement, particularly when consolidated, may be important in deterring the entry of both potential micro- and macroendobionts (Ratcliffe 1986).

## 3. SECONDARY SUBSTANCES

### (a) General

Defensive secondary substances may be categorized into those that are essentially physical in their action and effect (for example, colour, adhesion or lubrication) and those that are chemical (for example, signal, noxious, or toxic). Both can be subdivided into those passively dispensed, that is, irrespective of threat; and those actively so, that is, in response to a stimulus associated with a specific threat.

### (b) Secondary substances that act physically

A number of poriferan taxa (for example, *Myxilla*) produce substantial quantities of mucus, which accumulates throughout the individual, and is released in quantity from sites of injury. Scleractinian corals and zoanthids are also major producers of mucus. This is secreted from epithelia and flows as a film over the coral surface. Sponge and coelenterate mucus can have lubricant or adhesive physical effects but may also carry, for example, noxious or toxic substances. Epithelial mucus is not known to be produced by Bryozoa or Ascidiacea. It is doubtful, however, if mucus serves as a defence against macropredators, except by hindering the approach of substratum-bound climbers (for example, gastropod molluscs or pycnogonids) by adhesive or lubricating effects. Noxious or toxic substances carried within mucus may be antipredatory.

Surface mucus is more likely to be a defence against epibiosis, especially inhibiting larval settlement. Larvae of *Bugula* (Cheilostomata, Bryozoa) settling on *Myxilla* become incapacitated when their ciliary mechanisms become clogged by mucus (personal observation). This action may also affect larvae which might settle on scleractinian corals. Surface mucus is also likely to provide a major defence against micro-endobiosis (Ratcliffe 1986). Antonius (1981) believes this is so for scleractinian corals.

Of the unitary groups, surface mucus is particularly common among anemones (Actiniaria, Coelenterata) and gastropod molluscs. It is also produced by echinoderms (Russell 1984) and other groups.

Pigments can provide defensive colour or pattern, carotenoids being widespread among modular sessile groups, as they are among invertebrates in general. Strong, unpatterned pigmentation is characteristic of many modular species, patterning being less common. Aposematic coloration is common among some modular groups. Potential predators may associate the bright colours of many toxic species with their unsuitability as prey; though not all toxic species are distinctive in this way, *Flustra foliacea*, for example, (figure 4) being a dull plain brown.

Anemones are one unitary immotile group in which aposematism is significant (as a forewarner of nematocyst protection). The same is true for nudibranch molluscs with noxious or toxic chemical protection (Todd 1981) (figures 7, 9) and many echinoderms with skeletal reinforcement or toxicity (figure 6).

*(c) Secondary substances that act chemically: passively dispensed substances*

Many modular, non-locomotory invertebrates are known to accumulate secondary substances of relatively low molecular mass that are passively dispensed, that is, whether present within the organism, on the surface, or actually released from the organism, their occurrence exists irrespective of specific threats (contrast with venoms, discussed in (*d*)). Many of these substances have the potential to exert noxious or toxic effects. Evidence is growing that their occurrence in marine species is on a scale comparable to that known for plants, not only marine macroalgae but also vascular plants (Whittaker & Feeny 1971; Rice 1974; Rosenthal & Janzen 1979). Alkaloids and terpenoids, for example, are particularly prevalent in both (Halstead 1978; Scheuer 1978, 1983; Hashimoto 1979; Rosenthal & Janzen 1979; Russell 1984), and there are striking examples of the very same substances being common to higher plants and modular invertebrates (for example, the monoterpenoids citral and geraniol occur both in vascular plants and the cheilostomate bryozoan *Flustra foliacea* (Christopherson & Carle 1978)).

Invertebrates, like plants, can accumulate closely related metabolites that are species-specific either individually or in combination (for example, the terpenoid chemistry of sponge groups is sufficiently characteristic for their use in chemotaxonomy (Bergquist & Wells 1983)). These kinds of substances can accumulate sufficiently to constitute a significant proportion of an animal's biomass (for example, the alcyonacean coelenterate *Lobophytum crassospiculatum* contains up to 5 % dry mass of diterpenoids (Coll *et al.* 1985). The occurrence of such metabolites at concentration is patchy among the non-locomotory modular invertebrates. In part at least, this reflects a research bias, but there are well-screened groups for which absence is probably genuine. Passive secondary substances are nearly universal among poriferans (Minale 1978; Bergquist & Wells 1983), whereas among coelenterates they are abundant in alcyonaceans and gorgonaceans and probably scarce among scleractinians (Hashimoto 1979). They are not reported for the pelagic Siphonophora. The substances concerned are quite different from the higher molecular mass polypeptides that occur within venomous nematocysts (Russell 1984). The few bryozoans that have been investigated are rich sources of secondary substances (Christopherson & Carle, 1978; Carle *et al.* 1982; Wulff *et al.* 1982). Perhaps least is known about modular ascidians. Although vanadium and sulphuric acid have been claimed as chemical defensive agents that are widespread among both modular and unitary representatives of this group (Stoecker 1978, 1980); as yet, evidence for the presence of organic equivalents is limited.

Although such potentially defensive metabolites also occur in unitary invertebrates, those that are typical of modular biota, like alkaloids and terpenoids, are not well represented, except in organisms that sequester them via the food chain. Among the major non-locomotory, or nearly non-locomotory unitary groups, anemones (Actiniaria, Coelenterata), tubicolous polychaetes, barnacles (Cirripedia, Crustacea) and bivalve molluscs are not particularly known for such substances, and only unitary ascidians are claimed to be (Stoecker 1978, 1980). Among locomotory unitary forms, slow-moving groups are mostly characterized by defensive metabolites, for example, some opisthobranch molluscs sequester them from their diet which is often based on modular organisms (for example, *Phyllidia* acquires sesquiterpenoids from the sponge, *Hymeniacidon* (Todd 1981)). Many echinoderms produce saponins with antipredatory actions (Russell 1984). Defensive metabolites are less common among faster-moving invertebrate groups.

Defensive potential is dependent on potency and levels of dose: a function of concentration and duration of exposure. At very low doses, a barely perceptible substance may have a signalling role (compare with aposematism in (b)). At intermediate doses, it may have noxious roles (that is, unpleasant but causing no damage), and at higher doses, it may be sublethally or lethally toxic. Concentrations are likely to be greatest internally, less at the surface of the organism, and, if release takes place, very much less in the surrounding water column. Although within some species metabolites may be generally distributed, in others, there is appreciable localization. Gradients of antimicrobial activity occur along fronds of *Flustra foliacea* (Al-Ogily & Knight-Jones 1977) (figure 4), and levels of palytoxin (one of the most potent marine toxins known) within colonies of the zoanthid coelenterate *Palythoa* are greatest in female zooids, and particularly in their developing eggs (Hashimoto 1979). Although one would expect internal and surface metabolites to be released at the sites of injury it does not follow that they are released from intact uninjured individuals. This phenomenon, however, has been demonstrated for a sponge (J. Thompson, in Russell 1984), and also by *in situ* experimentation on the alcyonarian soft coral *Sarcophyton*, which is known to release detectable levels of a monoterpenoid (Coll *et al.* 1982a).

Whereas noxious or toxic doses of a substance may well be generated within or on the surface of an organism, it is difficult to believe that the same levels of potency could be achieved by external release into the surrounding water column other than within the confines of a tide-pool. It is certainly difficult to envisage this in the kinds of current-scoured localities where modular invertebrates proliferate (figure 1).

(i) *Passive chemical defence against macropredators*

Noxious doses of secondary substances, whether present externally, on the surface, or within internal tissues, may provide antipredatory defence by evoking aversive responses from potential predators. Predatory invertebrates often select their prey from a distance by chemoreception, a fact well established by food preference tests (for example, Chadwick & Thorpe 1981). Coll *et al.* (1985) have shown that purified soft coral metabolites (impregnated into pellets) cause fish to reject at the 'tasting' or 'mouthing' stage.

In view of their more prolonged and intimate contact with the prey, climbing, substrate-bound predators are more likely to be toxified by noxious or toxic doses of surface metabolites than are non-climbers. The action of surface metabolites (before feeding) or of internal metabolites (after the onset of feeding) may be external in their effects on predators (via gills, sensory organs, etc.) or internal (via the gut). Ichthyotoxic substances are common within sponges, alcyonacean soft corals, and occur in bryozoans (Green 1977; Bakus, 1981; Coll *et al.* 1982b; Dyrynda 1985). More chronic antipredatory effects like reduced fecundity, and indeed carcinogenesis, may select against predator populations feeding on chemically protected prey, as is well recognized for plants (Keeler & Tu 1983).

Passively dispensed secondary metabolites serve in the antipredatory defence of several unitary invertebrate groups, all locomotory, including nudibranch molluscs and echinoderms (Todd 1981; Hashimoto 1979; Russell 1984).

(ii) *Passive chemical defence against epibiosis*

Since invertebrate larvae widely use chemoreception to select their settlement surfaces, signal doses of metabolites, whether on the surface or released from an organism constituting a potential settlement site, may inhibit settlement of motile propagules. Whereas most studies have been concerned with positive chemotaxis, Grosberg (1981) has demonstrated the converse: larvae of many species do not settle on surfaces more heavily colonized by the sheet-encrusting ascidian *Botryllus schlosseri* (figure 10), possibly because *Botryllus* releases inhibitors. In this and in other cases, settlement may be prevented by the initiation of an avoidance response or by sublethal toxification, for example, incapacitation of ciliary mechanisms leading to disorientation.

Such metabolites would be most important for species at risk from epibionts, that is, those with reinforced surfaces. Burkholder (1973) suggested that the strong antimicrobial activity shown by gorgonacean coelenterates may reflect larvicidal defence. Although gorgonaceans are rigid, they are usually notably devoid of epibionts (figure 11). The same was proposed for the bryozoan *Flustra foliacea* (Al-Ogily & Knight-Jones 1977), for which larvotoxicity was subsequently demonstrated (Dyrynda 1985) (figure 4). Bandurrage & Fenical (1985) have separated and identified specific larvicidal factors (saponins) from the alcyonacean *Muricea fruticosa*.

Similar defences against epibiosis by larval settlement occur among unitary, non-locomotory invertebrates. Larvotoxic agents occur on the surface of the unitary ascidian *Ascidia nigra* (suggested by Stoecker (1978) to be sulphuric acid or vanadium, as in modular species). In contrast, tubicolous polychaetes and cirripede crustaceans, also at risk, are probably free of such chemical defences. Motile invertebrates that are incapable of maintaining skeletalized surfaces free of epibionts by flexion, scraping, or other physical means like spines and pedicellariae, are also vulnerable. Periodic moulting solves this for Crustacea and echinoderms may use passively dispensed metabolites in conjunction with pedicellariae and spines (Russell 1984).

Secondary substances may also inhibit colonization by lateral overgrowth. Some modular species release, or present at their surface, metabolites suppressing the progress of such colonizers, for example, the growing fronts of some thinly encrusting sponges are preceded by bands of necrosis within their opponents (Bryan 1973; Jackson & Buss 1975; Ayling 1983). Assays have shown that some tropical sponges and ascidians contain agents that are toxic to competing bryozoans (Jackson & Buss 1975). Alcyonacean coelenterates administer doses of toxic secondary metabolites to neighbouring scleractinian corals, sufficient to retard respiratory and growth rates, or even to kill them (Coll *et al.* 1985).

Defences against epibiosis initiated by lateral overgrowth are unnecessary for locomotory invertebrates, since even the slowest movers can move away faster than any rate of lateral overgrowth by potential epibionts.

(iii) *Passive substances as defences against endobiosis*

Toxic secondary metabolites are involved in defence against endobiosis throughout the Invertebrata (Ratcliffe 1986). However, most modular species differ from unitary ones in their greater susceptibility to pathogenic microbial invasion, associated with the tendency for partial mortality often following large-scale wounding (figure 12). Despite the disruption of surface defences, many species are able to maintain themselves free of microepibionts and endobionts (Burkholder 1973), and it may be no coincidence that antimicrobial activity is the most widely documented characteristic known for purified secondary substances (see, for example, Burkholder 1973; Amade *et al.* 1982).

*(d) Actively dispensed substances: venoms*

In contrast to most passively dispensed substances, actively dispensed venoms are typically of relatively high molecular mass, for example, polypeptides. Although these substances are often primarily concerned with feeding (Halstead 1978), venomous systems can also be defensive in some organisms. Many coelenterate nematocysts are venomous, containing often highly toxic polypeptides (Halstead 1978; Hashimoto 1979; Russell 1984). For many species, their sole function is prey capture, their defensive role being secondary or insignificant, although those not located near the oral disc of feeding polyps are more likely to be defensive. Nematocysts may be the main line of anti-predatory defence for siphonophores (Halstead 1978; Hashimoto 1979; Russell 1984). A crude venomous system is encountered in 'aggressive' scleractinian corals which toxify their neighbours that are competing for space, with digestive enzymes released from extruded mesenterial filaments (Lang 1973). Venomous systems remain little known for the other modular groups.

Nematocysts provide the same sort of protection for anemones, as they do for modular coelenterates. Among slow-moving locomotory groups, aeolid nudibranchs are also venomous; some of them sequestering undischarged nematocysts from their coelenterate prey whereas others secrete acidic venoms (Todd 1981). Among the prosobranchs, species of *Conus* are highly venomous (Halstead 1978; Hashimoto 1979; Russell 1984). The spines and pedicellariae of some echinoderms are also known to produce venoms (Russell 1984). The use of venomous systems for defence may be less common among the more agile locomotory invertebrate groups. However, many cephalopods produce venomous saliva which may be used for antipredation in addition to their major function of prey capture (Russell 1984).

## 4. Discussion

Common themes of defence are shared by sessile invertebrates that combine immotility with modularity. Some defensive mechanisms, notably the use of passively dispensed secondary substances of relatively low molecular mass, are strikingly similar to those of other modular groups, that is macroalgae and vascular plants. Rather than being a function of modularity, these common themes are a function of being immotile. A trend can be identified of the increased application of defensive options *not* dependent on motility (movement by flexion or locomotion) as the latter declines.

Highly motile invertebrates like errant polychaetes, natant decapod crustaceans and cephalopod molluscs are more able to 'fight' potential predators by body flexions alone, or are quick enough to escape by locomotion as a 'flight' option. They can also easily escape epibiosis initiated by lateral overgrowth by locomotion. They can deter larval settlement by body flexions or by abrasion between their body surfaces and their surroundings as they move.

Slow-moving, locomotory biota (that is, some reptant decapod crustaceans, prosobranch and opisthobranch molluscs and echinoderms as unitary examples; and the modular thaliacean tunicates and some siphonophoran coelenterates) are less able to use motion in 'fight and flight'. Their alternative means of 'fighting' predators include increased skeletal reinforcement, particularly at their surface (for example, reptant decapods, prosobranch molluscs and echinoderms), and spines are a common elaboration of this within reptant decapods and echinoderms. Passively dispensed noxious or toxic secondary substances of relatively low molecular mass are more common (for example, within nudibranch molluscs and echinoderms) than in more agile organisms. The same is true for actively dispensed venoms of relatively high molecular mass (for example, among prosobranch and nudibranch molluscs, echinoderms, siphonophoran coelenterates). Such invertebrates are more at risk from epibiosis by larval settlement than are fast-moving ones, although their motility is still sufficient to avoid colonization by lateral overgrowth. Particularly at risk from epibiosis are species with skeletal surface reinforcement, but among crustaceans, moulting counteracts epibiosis, while molluscs may rely on abrasion, and rigid-surfaced echinoderms may be protected by a combination of spines, pedicellariae and passively dispensed larvicidal agents (many rigid-surfaced species can sustain a degree of epibiosis with little adverse effect). Cryptic camouflage as an alternative to 'flight' by locomotion is a common antipredatory defence among the less locomotory invertebrate groups in general. The siphonophores, pelagic coelenterates that are both modular and locomotory, may rely on nematocysts as their major line of defence against predators (although their tentacular nematocysts are primarily for feeding).

Sessile groups, including those that are modular, exhibit an extreme condition with respect to motility: by definition all are non-locomotory, and in many cases, movements are limited to internal flexions such that alternative mechanisms prevail. Skeletal reinforcement is widespread among tubicolous polychaetes, cirripede Crustacea, Porifera, Coelenterata, Bryozoa, and both unitary and modular Ascidiacea. In many cases, the degree of reinforcement clearly far exceeds requirements for support. Spines are common among all non-locomotory modular groups except ascidians, but are less common among unitary ones. Cheilostomate bryozoans possess avicularia, which in some cases have defensive functions analogous to those of echinoderm pedicellariae. Venomous nematocysts, exclusive to the Coelenterata, constitute a major defence within that group. Passively dispensed metabolites which may be noxious or toxic to predators (among other roles) are common within the Porifera, Coelenterata, Bryozoa, and probably the Ascidiacea, but are not so within tubicolous polychaetes and barnacles. In conjunction with the above, the bright pigmentation of many species may serve in aposematism. More generally, different members of modular invertebrate groups show an emphasis on different defences. For example, whereas consolidated skeletal reinforcement, mucus, and venomous nematocysts constitute major aspects of the defensive array of scleractinian 'hard' corals, within the alcyonaceans the defensive array consists of unconsolidated reinforcement and, especially, passively dispensed metabolites, nematocysts being of reduced significance.

'Flight' options that require movement do exist among non-locomotory taxa in the form of a mass withdrawal of vital feeding structures to within the skeletal framework, or the closure of vital orifices. These are seen within representatives from all unitary and modular non-locomotory groups. Cryptic camouflage occurs in some groups, but others have distinct patterning and pigmentation, possibly aposematic, for example, signalling toxicity to potential predators.

Non-locomotory invertebrates are highly vulnerable to epibiosis, whether initiated by larval settlement or lateral overgrowth. Some tolerate a cover of epibionts, keeping only 'vital' orifices and surfaces clear; others remain totally clear (for example, by using 'fight' options like body flexions). Although surface reinforcement tends to increase risks of epibiosis, its supplementation by spines may help to deter it. Venomous systems are unlikely to be of significance in deterring settlement, but passively dispensed substances are effective within representatives from all major modular groups, together with unitary ascidians, but not tubicolous polychaetes or barnacles.

In defence against endobiosis, surface and internal barriers (skeletally reinforced or otherwise), antimicrobial, passively dispensed metabolites, and cell-mediated defences, are all common among motile invertebrates, and they probably also feature within all of the immotile groups. Passive chemical defences, however, are particularly widespread among modular as opposed to unitary non-locomotory groups (and unitary groups more generally). This may reflect the increased risks of microbial invasion associated with the proportionally massive injuries and partial mortality that occur commonly within the lifespan of species from all four major modular groups, particularly those with encrusting forms (Hughes & Jackson 1980, 1985; Jackson & Coates, this symposium).

By virtue of their immotility and often large size, many modular species support a diversity of associated biota, many of which provide extrinsic, symbiotic defence (for example, a heavy cover of sessile epibionts may provide camouflage or other antipredatory or anti-epibiotic mechanisms, whereas locomotory epibionts may provide antipredatory defence for their non-locomotory hosts). Microendobionts may synthesize the defensive metabolites accumulated by their hosts.

The ability to survive a high degree of partial mortality is itself a further form of defence for modular species (figures 2, 7, 11 and 12). The replication of vital systems within modules across a continuous (or intact) clone, spreads the risk of fatal damage occurring. Should some vital systems within the clone be irrevocably damaged, then only the particular modules dependent on them would die. Whether the interference is predatory, epibiotic, endobiotic, or indeed abiotic, the intact clone will survive as long as it contains the minimal modular configuration necessary for this (this configuration may be species-specific). A few unitary groups are also capable of partial mortality, for example, annelids and asteroid echinoderms, but their potential to survive is orders of magnitude below that of modular groups.

I am grateful to Professors John Harper and John Ryland, Dr Brian Rosen and Miss P. Cook for advice and encouragement. I thank colleagues at Swansea including Professor N. Ratcliffe, Dr P. J. Hayward, Dr P. E. King, Dr A. Nelson-Smith, Miss N. Yonow; and also participants at the discussion meeting for constructive comment and discussion. I also thank Mr Niel Ravenscroft for permitting inclusion of his unpublished observations. Dr M. Fordy prepared the scanning electron micrographs.

### REFERENCES

Al-Ogily, S. M. & Knight-Jones, E. W. 1977 Anti-fouling role of antibiotics produced by marine algae and bryozoans. *Nature, Lond.* **265**, 728–729.

Amade, P., Pesnado, D. & Chevolet, L. 1982 Antimicrobial activities of marine sponges from French Polynesia and Brittany. *Mar. Biol.* **70**, 223–228.

Antonius, A. 1981 Coral reef pathology: a review. *Proc. 4th int. Coral Reef Symp.* **2**, 3–6.

Ayling, A. L. 1983 Growth and regeneration rates in thinly encrusting Demospongiae from temperate waters. *Biol. Bull. mar. biol. Lab., Woods Hole* **165**, 343–352.

Backus, G. J. 1981 Chemical defense mechanisms on the Great Barrier Reef, Australia. *Science, Wash.* **211**, 497–499.

Bandurraga, M. M. & Fenical, W. 1985 Isolation of muricins. *Tetrahedron* **41**, 1057–1065.

Bergquist, P. R. & Wells, R. J. 1983 *Marine natural products 4* (ed. P. J. Scheuer), pp. 1–50. New York: Academic Press.

Best, B. A. & Winston, J. E. 1984 Skeletal strength of encrusting cheilostome bryozoans. *Biol. Bull. mar. biol. Lab., Woods Hole* **167**, 390–409.

Boardman, R., Cheetham, A. H. & Oliver, W. A. (eds) 1973 *Animal colonies: development and function through time.* Stroudsburg: Dowden, Hutchinson & Ross.

Bryan, P. G. 1973 Growth rate, toxicity, and distribution of the encrusting sponge *Terpios sp.* (Haldromerida: Suberitidae) in Guam, Mariana Islands. *Micronesica* **9**, 237–242.

Burkholder, P. R. 1973 The ecology of marine antibiotics and coral reefs. In *Biology and geology of coral reefs, vol. 2 Biology 1* (ed. O. A. Jones & R. Endean), pp. 117–182. London and New York: Academic Press.

Carle, J. S., Thybo, H. & Christopherson, C. 1982 Dogger bank itch (3). Isolation, structure determination and synthesis of a hapten. *Contact Dermatitis* **8**, 43–47.

Chadwick, S. R. & Thorpe, J. P. 1981 An investigation of some aspects of bryozoan predation by dorid nudibranchs (Mollusca: Opisthobranchia). In *Recent and fossil Bryozoa* (ed. G. P. Larwood & C. Nielsen), pp. 51–58. Fredensborg: Olsen & Olsen.

Chapman, G. & Stebbing, A. R. D. 1980 The modular habit – a recurring strategy. In *Developmental and cellular biology of coelenterates* (ed. P. Tardent & R. Tardent), pp. 157–162. Amsterdam: Elsevier/North-Holland Biomedical Press.

Christopherson, C. & Carle, J. S. 1978 Chemical signals from a marine bryozoan. *Naturwissenschaften*, **65**, 440–441.

Coll, J. C., Bowden, B. F. & Tapiolas, D. M. 1982a In situ isolation of allelochemicals released from soft corals (Coelenterata, Octocorallia). A totally submersible sampling apparatus. *J. exp. mar. Biol. Ecol.* **60**, 293–299.

Coll, J. C., Bowden, B., Tapiolas, D. M., Willis, R. H., Djura, P., Streamer, M. & Trott, L. 1985 Studies on Australian soft corals, XXXV. *Tetrahedron* **41**, 1085–1092.

Coll, J. C., La Barre, S., Sammarco, P. W., Williams, W. T. & Backus, G. J. 1982b Chemical defences in soft corals (Coelenterata: Octocorallia) of the Great Barrier Reef: A study of comparative toxicities. *Mar. Ecol. Prog. Ser.* **8**, 271–278.

Cook, P. L. 1979 Some problems in interpretation of heteromorphy and colony integration in Bryozoa. In *Biology and systematics of colonial organisms* (ed. G. Larwood & B. R. Rosen), pp. 193–210. London: Academic Press.

Cook, P. L. & Chimonides, P. J. 1978 Observations on living colonies of *Selenaria* (Bryozoa, Cheilostomata) 1. *Cah. Biol. mar.* **19**, 147–58.

Dyrynda, P. E. J. 1985 Chemical defences and the structure of subtidal epibenthic communities. In *Proc. 19th Eur. mar. Biol. Symp.* (ed. P. E. Gibbs), pp. 411–424. Cambridge University Press.

Dyrynda, P. E. J. & King, P. E. 1982 Sexual reproduction in *Epistomia bursaria* (Bryozoa: Cheilostomata), an endozooidal brooder without polypide recycling. *J. Zool. Lond.* **198**, 337–352.

Green, G. 1977 Ecology of toxicity in marine sponges. *Mar. Biol.* **40**, 207–215.

Grosberg, R. K. 1981 Competitive ability influences habitat choice in marine invertebrates. *Nature, Lond.* **290**, 700–702.

Halstead, B. W. 1978 *Poisonous and venomous animals of the world 1*. Washington D.C.: U.S. Govt Printing Office

Harper, J. L. 1985 Modules, branches and the capture of resources. In *Population biology and evolution of clonal organisms* (ed. J. B. C. Jackson, L. W. Buss & R. E. Cook), pp. 1–33. New Haven: Yale University Press.

Harper, J. L. & Bell, A. D. 1979 The population dynamics of growth form in organisms with modular construction. In *Population dynamics* (ed. R. M. Anderson, B. D. Turner & L. R. Taylor), pp. 211–242. Oxford: Blackwell Scientific Publications.

Harvell, C. D. 1984 Predator-induced defense in a marine bryozoan. *Science, Wash.* **244**, 1357–1359.

Hashimoto, Y. 1979 *Marine toxins and other bioactive marine metabolites*. Tokyo: Japan Scientific Societies Press.

Hughes, T. P. & Jackson, J. B. C. 1980 Do corals lie about their age? Some demographic consequences of partial mortality, fission and fusion. *Science, Wash.* **209**, 713–715.

Hughes, T. P. & Jackson, J. B. C. 1985 Population dynamics and life histories of foliaceous corals. *Ecol. Monogr.* **55**, 141–166.

Jackson, J. B. C. 1977 Competition on marine hard substrata: the adaptive significance of solitary and colonial strategies. *Am. Nat.* **111**, 743–767.

Jackson, J. B. C. 1979 Morphological strategies of sessile annimals. In *Biology and systematics of colonial organisms* (ed. G. Larwood & B. R. Rosen), pp. 499–555. London: Academic Press.

Jackson, J. B. C. & Buss, L. 1975 Allelopathy and spatial competition amongst coral reef invertebrates. *Proc. natn. Acad. Sci. U.S.A.* **72**, 5160–5163.

Keeler, R. F. & Tu, A. T. 1983 *Handbook of natural toxins*. New York: Marcel Dekker Inc.

Lang, J. C. 1973 Interspecific aggression by scleractininan corals. 2. Why the race is not only to the swift. *Bull. mar. Sci.* **23**, 260–279.

Larwood, G. & Rosen, B. R. (ed.) 1979 *Biology and systematics of colonial organisms*. London: Academic Press.

Minale, L. 1978 Terpenoids from marine sponges. In *Marine natural products*, vol. 1 (ed. P. J. Scheuer), pp. 175–240. New York: Academic Press.

Mundy, S. P. 1980 A key to British and European freshwater bryozoans. *Freshwat. biol. Ass. scient. Publ.* **41**, 1–31.

Ratcliffe, N. A. 1986 The antimicrobial systems in invertebrates – An overview. In *Natural antimicrobial systems* (ed. R. G. Board, A. K. Charnley, R. M. Cooper, G. W. Gould & M. E. Rhodes Roberts). Bath University Press. (In the press.)

Rice, E. L. 1974 *Allelopathy*. New York: Academic Press.

Rhoades, D. F. 1985 Offensive–defensive interactions between herbivores and plants: their relevance in herbivore population dynamics and ecological theory. *Am. Nat.* **125**, 205–238.

Rosen, B. R. 1979 Modules, members and communes: a postscript introduction to social organisms. In *Biology and systematics of colonial organisms* (ed. G. Larwood & B. R. Rosen), pp. xiii–xxxv. London: Academic Press.

Rosenthal, G. A. & Janzen, D. H. 1979 *Herbivores – their interaction with plant secondary metabolites.* New York: Academic Press.

Russell, F. E. 1984 Marine toxins and venomous and poisonous marine plants and animals (invertebrates). *Adv. mar. Biol.* **21**, 59–217.

Ryland, J. S., Wigley, R. A. & Muirhead, A. 1984 Ecology and colonial dynamics of some Pacific reef flat Didemnidae (Ascidiacea). *Zool. J. Linn. Soc.* **80**, 261–282.

Scheuer, P. J. 1978 *Marine natural products I and II.* New York: Academic Press.

Scheuer, P. J. 1983 *Marine natural products V.* New York: Academic Press.

Stebbing, A. R. D. 1971 The epizoic fauna of *Flustra foliacea*. *J. mar. biol. Ass. U.K.* **51**, 283–300.

Stebbing, A. R. D. 1973 Observations on colony overgrowth and spatial competition. In *Living and fossil Bryozoa* (ed. G. P. Larwood), pp. 173–183. London: Academic Press.

Stoecker, D. 1978 Resistance of a tunicate to fouling. *Biol. Bull. mar. biol. Lab., Woods Hole* **155**, 615–626.

Stoecker, D. 1980 Relationships between chemical defense and ecology in benthic ascidians. *Mar. Ecol. Prog. Ser.* **3**, 257–265.

Todd, C. D. 1981 The ecology of nudibranch molluscs. *Oceanogr. Mar. Biol. Ann. Rev.* **19**, 141–234.

Tursch, B., Braekman, J. C., Daloze, D. & Kaisin, M. 1978 Terpenoids from coelenterates. In *Marine natural products II* (ed. P. J. Scheuer), pp. 247–296. New York: Academic Press.

Winston, J. E. 1984 Why bryozoans have avicularia – a review of the evidence. *Am. Mus. novit.* **2789**, 1–26.

Whittaker, R. H. & Feeny, P. P. 1971 Allelochemics: chemical interactions between species. *Science, Wash.* **171**, 757–770.

Wulff, P., Carle, J. S. & Christopherson, C. 1982 Marine alkaloids 6. The first naturally occuring bromo-substituted quinoline from *Flustra foliacea*. *Comp. biochem. Physiol.* **71**B, 525–526.

# Retrospect on modular organisms

By G. C. Williams

*Department of Ecology and Evolution, State University of New York, Stony Brook, New York 11794, U.S.A.*

Modularity is a new term, but some associated ideas have been with us for a long time. Animals like bryozoans, traditionally recognized as colonial, would all be considered modular at this meeting. It has also been realized for a long time that reproduction can be asexual and produce individuals that differ only from developmental asynchrony or environmental effects. The concept of metamerism is likewise a venerable one, and involves the repetitive units of construction (Hardwick) or products of iterative growth (Mackie) characteristic of modularity.

Despite the antiquity of many of the concepts, I think it clear that there is something new about the idea of a modular organism. It is an idea that can give a benthic ecologist working on sea anemones a feeling that in some ways he may have more in common with a plant ecologist studying herbaceous perennials than with another benthic ecologist mainly interested in molluscs. It would appear that some participants at the meeting find cloning an unnecessary criterion, or metamerism an insufficient criterion of modularity.

### What do we really mean by a modular organism?

It would be premature and obstructive for me to attempt a list of formal criteria for modularity and to insist that it be strictly applied. Everyone at the meeting would agree that strawberry plants are modular and that mammals and roundworms are not. There might not be much unanimity of opinion on a conceptual basis for such inclusion and exclusion. I will briefly discuss this possibly contentious issue mainly because the organizers neglected (or wisely avoided?) its inclusion in the formal schedule.

Its vegetative cloning process makes the strawberry so clearly modular. Most herbaceous annuals would be considered unitary (non-modular) organisms. They grow and mature where they germinate, bloom, produce genetically diverse seeds, and then die. A strawberry plant can do all these things and more. It can live beyond its first blooming, and it can send out one or more horizontal stolons, stems specialized for asexual reproduction. A stolon can continue its growth indefinitely, producing new plants at about half-metre intervals. Each new plant may also be able to send out additional stolons, so that the vegetative spread is a branching process, like that modelled in figure 2 of Bell (this symposium). It is this iteration of complex multicellular units, each capable of physiological and ecological independence by a branching growth process, that makes the strawberry modular.

In applying the modularity concept to other organisms, questions arise as to what in the strawberry life history is necessary or sufficient. Many of the organisms discussed at this meeting seem quite dissimilar to strawberries in many of the features listed. Is a physically demonstrable branching necessary? Even in the strawberry the stolons degenerate in a few months and no longer serve to connect members of the same clone. No one would insist that there be permanent physical branches, but can the branching be purely conceptual, as in a pedigree diagram for a motile organism that reproduces by fission or budding? Can the buds be internally produced parthenogenetic eggs of maternal genotype? I would opt for including ameiotic parthenogenetic lineages among the modular, but think that at least a conceptual branching process is required. A merely linear repetition of parts like the segments of an earthworm would not suffice. So I would accept all of the models of modularity presented by Hallé except the Chamberlain model, which I would reject as merely metameric, not modular.

If an ameiotic parthenogenetic lineage is accepted as modular, what about a hybridogenetic fish like some forms of *Poeciliopsis* (Vrijenhoek 1984; p. 401)? A haploid set of chromosomes forms a genetic module that is replicated indefinitely, even through each member of the all-female hemiclone carries and expresses paternal genes that are disposed of at meiosis. Each individual is genetically unique, and littermates resemble each other only as much as full sisters normally do, but there is no recombination, and each individual is just as similar genetically to a remote ancestor or descendant as to a sister.

Modules can show different kinds and degrees of independence. The early rudiment of a strawberry plant on a stolon depends on its parent for water and nutrients. Only gradually does it take over the job of providing for itself. Each plant may also form another kind of reproductively specialized branch, one that ends in a flower. A flower depends for its upkeep on the plant that made it, but flowers are structurally similar units. This example shows that the criteria of modularity may depend on the interests of the observer. Even without vegetative cloning, a plant with more than one flower might be modular to a morphologist, or to an ecologist interested mainly in the functioning of flowers. To an ecologist interested in demography or trophic structures, only a cloning species would be modular.

It may prove useful to restrict the term modular organism to those that reproduce asexually, but to recognize that there are modular aspects of development and morphology in most unitary organisms. Apple trees are normally unitary, but apple blossoms are structural modules produced abundantly by each mature tree. Hair follicles and sweat glands on a mammal are structurally similar parts with some degree of functional independence, and they are produced by a cell pedigree that can be represented as a branching process. The gametogenesis of any organism depends on a branching pedigree of meristem or germ-line cells that produce functionally similar and independent gametes.

### Causes, consequences and macroevolution of modularity

Modularity has been regarded as a solution to problems of size and scaling. Up to a point, material invested in the corolla of an apple blossom is more than compensated by expected reproductive payoff. Beyond that point the odds are less favourable, and fitness is better served by making flowers more numerous, rather than larger. Likewise the nutrients captured by a hydra are best invested in growth, up to a point, beyond which investment in a bud is more profitable. A clone of a hundred may have the same ratio of surface to volume as a single hydra. A hydra of a hundred times the mass of a normal one would have only about one fifth as much surface per unit mass. This means only one fifth of the gut surface for absorbing nutrients, and one fifth of external skin for respiratory and excretory exchanges. I doubt that such a hydra could meet its maintenance requirements in even the best of environments.

This kind of thinking can be overdone, and it is important to realize that there are many ways of coping with increasing size, in both ontogeny and phylogeny. If one assumes that increasing maintenance requirements for an increasing mass of tissue will eventually consume all available resources, with none left for growth or reproduction, it is easy to show that it is adaptive for growth to stop short of the maximum attainable size. Sebens (1982) thus modelled the optimum size at maturity for a sea anemone. Such modelling ought to be widely applicable, for instance, to the growth and maturation of a fish like the guppy, as it grows from about 5 mm to several centimetres. If an increasing maintenance burden causes guppies to curtail growth and start reproducing at a few centimetres what about a bluefin tuna? It starts its independent existence considerably smaller than a newborn guppy, and may end it at about a hundred thousand times the mass of an adult guppy. Are we to suppose that a milligram of larval and a milligram of adult liver tissue must have the same maintenance cost?

Ryland and Warner seem to accept Sebens' line of reasoning, although they review a diversity of measurements of uptake efficiencies and maintenance costs in relation to size in a variety of organisms. Most indeed support the idea that potential for resource capture may fail to keep pace with metabolic costs as an organism grows, but the data are so variable as to be compatible with a variety of conclusions. Also some deficiencies pervade the cited studies. Very few investigate even as much as a ten-fold mass difference between the largest and smallest specimens. Some unitary organisms (indeed, some modules, such as aspen trees) grow many orders of magnitude during development, often without obvious allometric changes. A 1 cm tuna and a 1 m tuna have much the same shape and life style.

Another problem is the inclusion of both mature and juvenile specimens in the same study. If a guppy's growth rate is less between 3 and 4 cm than it was between 2 and 3 cm, I would be inclined to attribute it to a shifting of resources from growth to preparations for reproduction, not an adjustment to a changing surface:volume ratio.

If one forgets about maintenance costs and assumes that an organism grows at a rate simply determined by surface area and does not change shape as it grows, its rate of length increase will be constant with time, and its mass proportional to the cube of time. The approximate accuracy of this simple formulation is indicated by the lack of any strong dependence of growth rate on absolute size in fishes. Both juvenile sharks 2 m long and larval fishes a few millimetres long commonly grow about 0.3 mm per day (Taylor & Williams 1984).

My reservations relate only to a facile use of surface:volume considerations in explaining module size, not from any doubt on the importance of scaling effects in evolution and ontogeny. I see no alternative to the use of size-optimization models in explaining the similarity of flowers on a tree, or peas in a pod, or zooids in a benthic colony. My inclination would be to explain size uniformities on the basis of ecological specialization, for special prey species of a sea anemone, or preferred pollen vectors for a flowering plant.

Whatever the reason, the continued growth of a module, or even a phalanx of modules, will eventually produce a size at which further investment in growth will not be the optimum use of resources. A phalanx might escape from this bind by abandoning the phalanx for the guerrilla mode of proliferation. Another escape might be investment in widely dispersed propagules, perhaps sexually produced, that can start new phalanges elsewhere. Still another is to indulge in a higher level of modularity and build a phalanx of phalanges as shown by various corals (Rosen). Mackie and others at the meeting spoke of levels of modularity, with the most basic level that of the cell or even the organelle.

Still another solution to the problem, of maintaining optimal module or phalanx size, is module specialization or even the abandonment of modularity. It is widely accepted that multicellularity in organisms more than a few millimetres in size depends on cell specialization. It is equally understandable that masses of multicellular modules may function better if modules in different regions specialize in various ways for the collective interest of the colony. A siphonophore is made up of what, to a morphologist, are modular elements. Yet in many aspects of its biology, as discussed in Mackie's chapter, a siphonophore is very much a unitary organism: its different parts shown extreme specialization, and a subordination to the interests of the entire unit; information is rapidly transmitted between widely separated regions; and most of the growth is from metameric rather than branching addition of parts. A parallel regression of modularity is shown by salps and by Palaeozoic graptolites, as shown by Bates & Kirk (1986).

Some degree of modularity is the general rule in the plant kingdom, is widespread in many animal phyla, and must have appeared independently in many lines of descent. Yet closely detailed convergences may be noted between phylogenetically remote groups, as several participants pointed out. Trinci and Cutter, for example, listed some remarkable similarities between prokaryote and eukaryote soil inhabitants. Only a small proportion of the topographically possible systems of modular proliferation, as explored in Bell's simulations, have actually been found in living organisms. No detailed explanations for such limits on evolved forms of modularity were advanced at the meeting, although there seemed a general agreement that many mathematically possible systems would be functionally maladaptive for any organism.

The absence of broad ranges of character complexes from the earth's biota shows that the evolutionary process is constrained. A major constraint is natural selection, which prevents the production of organisms of less than some minimum level of viability. To say that only some of the conceivable patterns of modularity would be viable is to invoke natural selection. The idea of evolutionary constraint is sometimes used in more than this sense. Holder (1983) proposed that the machinery of vertebrate development is incapable of producing more than a limited range of modification in tetrapod limb and digit patterns. I suggest that this may be analogous to a comparative anatomist of wheels concluding that manufacturing processes are unable to make wheels of other than nearly perfect circularity.

The logically possible forms of modularity may be compared to the forms of swimming animals. Here also there are some often cited examples of convergence, for instance on streamlining, and the repeated use of only three kinds of thrust: jet propulsion, paddling, and longitudinal undulation. Within each of these modes of movement only a small number of possibilities are repeatedly utilized. Sharks and crocodiles swim by horizontal undulations, whales and leeches by vertical ones. Nothing swims by undulations in any other plane. I suspect that we all feel that we understand these examples of evolutionary constraint. When we know as much about the functional and phylogenetic aspects of modularity as we do about swimming, the limitations may have equally obvious explanations.

I wonder whether it may be that today's forbidding constraint may turn out to be tomorrow's adaptive breakthrough. Biologists in the Ordovician no doubt would have proclaimed that only a rudimentary dry-land biota would ever be possible, in the Silurian that aerial flight could never be evolved, and so on. Rather than say that certain evolutionary developments are impossible, I would prefer to use Bell's phraseology, that there are '...forms yet to be realized'. Maybe the lands will someday be clothed in forests of the cross-braced trees that Hardwick says can't happen.

## ECOLOGY AND MICROEVOLUTION OF MODULAR ORGANISMS

Ecological differences between unitary and modular organisms were emphasized by most of the speakers at the meeting. Many are listed in the first paragraph by Jackson and Coates. The differences really are important and their neglect surprising. There have been great advances in demographic theory, but almost none are applicable to the many species for which we need to count both genets and modules. A module-relevant demography based on size rather than age is only beginning to be developed (Hughes 1984; Taylor & Williams 1984). Such generalizations as those advanced by Jackson and Coates on range of larval dispersal in relation to presence or absence of cloning could only have been proposed by someone convinced that the presence or absence of cloning is an important difference.

Many studies of community ecology assume that sessile organisms disperse only in such early stages as larvae or seeds and thereafter can make no habitat selection. Thus they neglect the fact that some of the sessile forms may be modular and users of the guerrilla strategy. The distribution of some of the plants in a weedy field or invertebrates in a benthic community may be poorly related to the distribution of the seedlings or larval settlers from which they developed, perhaps years (or centuries) before. Jackson and Coates mention a stoloniferous invertebrate that may spread as much as 72 cm a year, and this may be much less than the rate of spread of some terrestrial plants (Harper 1985). This sort of clonal growth, followed by a disappearance of modules from regions not permanently suitable, must have a major effect on the ecological distribution of guerilla strategists.

Franco's observations, that trees can greatly modify their growth patterns in relation to immediate environmental conditions, suggest the additional possibility of an actively biased clonal proliferation in an environmental gradient. The avoidance reactions of soil hyphae discussed by Trinci and Cutter would lead to an active habitat selection, and a similar phenomenon was recently reported for a higher plant. Salzman (1985) showed that rhizomes of *Ambrosia psilostachya* in saline soils grow preferentially towards regions of lower salinity. This means deprivation and perhaps abandonment of members of the clone living in higher salinities. This sort of chemotropic growth of a clone differs from chemotactic habitat selection by a motile animal mainly in the time scale of events.

Modularity can have a major effect on the genetic structure of a population, a matter touched on briefly by Jackson and Coates but otherwise neglected at the meeting. The blossoms on an apple tree are reproductive modules, not much different in principle from the gonozooids on a hydrozoan colony. The great number of these modules on one tree must make it improbable that a bee visiting a flower has just come from another tree. Large apple trees must have greater difficulty than small ones in getting their pollen dispersed and their ovules fertilized by pollen nuclei from other trees. Likewise, a strawberry clone that succeeds in the dense occupation of a large area must get mainly within-clone pollen transfer. A clonally spreading plant must either be self-compatible and little affected by inbreeding depression, or have special mechanisms for increasing the likelihood of outcrossing. Some examples and general discussion are provided by Handel (1985).

Cloning raises the likelihood that associated conspecifics are genetically identical, and such association may be sufficiently predictive of genetic similarity to result in kin-selected altruism and cooperation. The colony integration discussed by Ryland and Warner and the more extreme subordination of originally independent zooids discussed by Mackie are attributable to this factor. Long-term physical communication or contact may not be needed. A reliable sensory connection may suffice, as is shown by the sterile castes of parthenogenetic aphids (Aoki 1977).

The theoretical development of population genetics, like that of demography, was undertaken for unitary organisms. The mutation rates one reads about in the textbooks would inflict serious genetic loads on clones in much less time than they commonly persist. There is some evidence for this effect in parthenogenetic fishes (Vrijenhoek 1984) and it should be looked for in other forms in which a clonal life cycle passes through a single-cell stage. Where cloning is vegetative, mutations can perhaps be weeded out by the diplontic selection discussed by Hardwick. I hope it will not be too long before diplontic selection is discussed in the textbooks.

In the population genetics developed for unitary organisms, an age cohort is expected to be replaced within a few times the period of development from zygote to maturity, to remain close to Hardy–Weinberg equilibrium, and to show only minor changes in gene frequency. It is abundantly clear, from the few genetic studies that have been carried out, that modular organisms need not conform to such expectations. Clones of rapidly developing organisms with genotypic life spans measured in millenia were mentioned by a number of speakers. Gross departures from Hardy–Weinberg ratios have been observed in such diverse groups as sea anemones (Ayre 1984), *Daphnia* (Lynch 1983), and clover (Burdon 1980). It is also clear, as pointed out by others at the meeting, that variation in reproductive success among individual genotypes of modular organisms can be orders of magnitude greater than would be expected of unitary forms. I join Harper (1985) in wondering why '...so much of the study of ecology and evolution has been based on the behaviour of unitary organisms'.

## References

Aoki, S. 1977 A new species of *Colophina* (Homoptera, Aphidoidea) with soldiers. *Kontyu* **45**, 333–337.
Ayre, D. J. 1984 The effects of sexual and asexual reproduction on geographic variation in the sea anemone *Actinia tenebrosa*. *Oecologia* **62**, 222–229.
Bates, D. E. B. & Kirk, N. H. 1986 Graptolites, a fossil case history of evolution from sessile, colonial animals to automobile superindividuals. *Proc. R. Soc. Lond.* B **228**, 207–224.
Burdon, J. J. 1980 Intra-specific diversity in a natural population of *Trifolium repens*. *J. Ecol.* **68**, 717–736.
Handel, S. N. 1985 The intrusion of clonal growth patterns on plant breeding systems. *Am. Natst.* **125**, 367–84.
Harper, J. L. 1985 Modules, branches, and the capture of resources. In *Population biology and evolution of clonal organisms* (ed. J. B. C. Jackson, L. W. Buss & R. Cook), pp. 1–33. New Haven: Yale University Press.
Holder, N. 1983 Developmental constraints and the evolution of vertebrate digit patterns. *J. theoret. Biol.* **104**, 451–471.
Lynch, M. 1983 Ecological genetics of *Daphnia pulex*. *Evolution* **37**, 358–374.
Salzman, A. G. 1985 Habitat selection in a clonal plant. *Science, Wash.* **228**, 603–604.
Sebens, K. P. 1982 Limits to indeterminate growth: An optimal size model applied to passive suspension feeders. *Ecology* **63**, 209–222.
Taylor, P. D. & Williams, G. C. 1984 Demographic parameters at evolutionary equilibrium. *Can. J. Zool.* **62**, 2264–2271.
Vrijenhoek, R. J. 1984 The evolution of clonal diversity in *Poeciliopsis*. In *Evolutionary genetics of fishes* (ed. B. J. Turner), pp. 399–429. New York and London: Plenum.